V International Workshop on Mathematical Models and their Applications 2016

IOP Conference Series: Materials Science and Engineering
Volume 173

Krasnoyarsk, Russia
7 - 9 November 2016

ISBN: 978-1-5108-3676-1
ISSN: 1757-8981

Printed from e-media with permission by:

Curran Associates, Inc.
57 Morehouse Lane
Red Hook, NY 12571

Some format issues inherent in the e-media version may also appear in this print version.

Printed by Curran Associates, Inc. (2017)

For permission requests, please contact the Institute of Physics
at the address below.

Institute of Physics
Dirac House, Temple Back
Bristol BS1 6BE UK

Phone: 44 1 17 929 7481
Fax: 44 1 17 920 0979

techtracking@iop.org

Additional copies of this publication are available from:

Curran Associates, Inc.
57 Morehouse Lane
Red Hook, NY 12571 USA
Phone: 845-758-0400
Fax: 845-758-2633
Email: curran@proceedings.com
Web: www.proceedings.com

Table of contents

Volume 173

V International Workshop on Mathematical Models and their Applications 2016
7–9 November 2016, Krasnoyarsk, Russia

Accepted papers received: 17 January 2017
Published online: 17 February 2017

Preface

Papers

IOP Conf. Series: Materials Science and Engineering **173** (2017) 011001 doi:10.1088/1757-899X/173/1/011001

V International Workshop on Mathematical Models and their Applications 2016 (IWMMA)

November 7-9, 2016

Krasnoyarsk, Russian Federation

V International Workshop on Mathematical Models and their Applications 2016 (IWMMA) provided an International Forum for the presentation of original results in mathematical modeling for software and hardware applications in various fields. It stimulated lively discussions among researchers as well as industrialists.

The workshop covered a broad set of research topics including:

- Mathematical models and their applications
- Mathematical modeling techniques
- Optimization techniques, incl. multi-criterion optimization and decision making support
- Data mining and knowledge discovery
- Computational intelligence and its applications
- Bio-inspired and swarm intelligence
- Text/Web/Data mining
- Human-Computer Interaction
- Natural language processing
- Applications in engineering, natural sciences, social sciences, computer science, etc.

All the papers peer reviewed and accepted by the Organizing Committee were published in the Conference Proceedings.

IOP Conf. Series: Materials Science and Engineering **173** (2017) 011002 doi:10.1088/1757-899X/173/1/011002

Peer review statement

All papers published in this volume of *Journal of Physics: Conference Series* have been peer reviewed through processes administered by the proceedings Editors. Reviews were conducted by expert referees to the professional and scientific standards expected of a proceedings journal published by IOP Publishing.

IOP Conf. Series: Materials Science and Engineering **173** (2017) 012001 doi:10.1088/1757-899X/173/1/012001

Investigation into the efficiency of different bionic algorithm combinations for a COBRA meta-heuristic

Sh Akhmedova and E Semenkin

Reshetnev Siberian State Aerospace University
31, Krasnoyarsky Rabochy Av., Krasnoyarsk, 660037, Russian Federation

E-mail: shahnaz@inbox.ru

Abstract. Previously, a meta-heuristic approach, called Co-Operation of Biology-Related Algorithms or COBRA, for solving real-parameter optimization problems was introduced and described. COBRA's basic idea consists of a cooperative work of five well-known bionic algorithms such as Particle Swarm Optimization, the Wolf Pack Search, the Firefly Algorithm, the Cuckoo Search Algorithm and the Bat Algorithm, which were chosen due to the similarity of their schemes. The performance of this meta-heuristic was evaluated on a set of test functions and its workability was demonstrated. Thus it was established that the idea of the algorithms' cooperative work is useful. However, it is unclear which bionic algorithms should be included in this cooperation and how many of them. Therefore, the five above-listed algorithms and additionally the Fish School Search algorithm were used for the development of five different modifications of COBRA by varying the number of component-algorithms. These modifications were tested on the same set of functions and the best of them was found. Ways of further improving the COBRA algorithm are then discussed.

1. Introduction

Co-Operation of Biology-Related Algorithms or COBRA is a meta-heuristic approach developed in 2013 for solving unconstrained real-parameter optimization problems [1]. Its basic idea consists in the cooperative work of different nature-inspired algorithms. In the original version of COBRA, five well-known bionic heuristics were used as component-algorithms: the Particle Swarm Optimization Algorithm (PSO) [2], the Wolf Pack Search Algorithm (WPS) [3], the Firefly Algorithm (FFA) [4], the Cuckoo Search Algorithm (CSA) [5] and the Bat Algorithm (BA) [6]. All of the mentioned heuristics are similar bionic optimization methods originally developed for continuous variable space. These algorithms mimic the collective behaviour of the corresponding animal groups which allows the global optima of real-valued functions to be found. For example, PSO was inspired by fish and bird swarm intelligence and BA was inspired by the echolocation behaviour of bats.

However, there are still various algorithms which can be used as components for COBRA. In this study additionally the Fish School Search algorithm (FSS) [7] was include. The investigation into the efficiency of these six heuristics was conducted on the set of test functions (the Rosenbrock function, the Griewank function, the Rastrigin functions and so on [8]). Their performance demonstrated that all of them are sufficiently effective for solving optimization problems, and their workability was established. Based on the experimental results, five different modifications of the meta-heuristic approach COBRA were developed and their efficiency was examined on 28 test problems from the CEC'2013 competition [9].

Thus, in this paper firstly the original version of COBRA is described, and then a brief description of the FSS algorithm and comparison of the bionic algorithms are presented. In the next section, experimental results obtained using the proposed modification are discussed and finally some conclusions are given.

2. Co-Operation of Biology Related Algorithms (COBRA)

The meta-heuristic approach Co-Operation of Biology Related Algorithms or COBRA [1] was developed on the basis of five optimization methods, namely Particle Swarm Optimization (PSO) [2], the Wolf Pack Search (WPS) [3], the Firefly Algorithm (FFA) [4], the Cuckoo Search Algorithm (CSA) [5] and the Bat Algorithm (FFA) [6] (hereinafter referred to as "component-algorithms"). The main reason for the development of a new meta-heuristic was the inability to say which of the above-listed algorithms is the best one or which algorithm should be used for solving any given optimization problem [1]. Thus, the idea was to use the cooperation of these algorithms instead of any attempts to understand which one is the best for the current problem in hand.

The originally proposed approach consists in the generating of five populations (one population for each bionic algorithm) which are then executed in parallel, cooperating with each other. The algorithm COBRA is a self-tuning meta-heuristic so there is no need to choose the population size for each component-algorithm. The number of individuals in the population of each algorithm can increase or decrease depending on the fitness value: if the overall fitness value has not improved during a given number of iterations, then the size of each population increases, and vice versa, if the fitness value has constantly improved during a given number of iterations, then the size of each population decreases.

There is also one more rule for adjusting the population size, whereby each population can "grow" by accepting individuals removed from another population. The population "grows" only if its average fitness value is better than the average fitness value of all other populations. Therefore a "winner algorithm" (an algorithm whose population has the best average fitness value) can be determined on each iteration. The described competition among component-algorithms allows the largest resource (population size) to be presented to the bionic algorithm most appropriated on the current generation.

With the aim of bringing up to date information on the best achievements to all component-algorithms and preventing their preliminary convergence to their own local optimum all populations communicate with each other. "Communication" was determined in the following way: populations exchange individuals in such a way that a part of the worst individuals of each population is replaced by the best individuals of other populations. Thus, the group performance of all algorithms can be improved.

The performance of the algorithm COBRA was evaluated on a set of benchmark problems with 5, 10 and 30 variables taken from [1] and the experiments showed that COBRA works successfully and is reliable on different benchmarks. Besides, the meta-heuristic COBRA was compared with its component-algorithms, and simulations and comparison showed that COBRA is superior to these bionic algorithms when the dimension grows and complicated problems are solved.

3. Modifications of the meta-heuristic approach COBRA

As previously noted, the original version of the COBRA approach can be described as the cooperative work of five bionic algorithms, but in this study the FSS algorithm was also used. Therefore, in this section a brief description of the FSS algorithm is presented and the idea of the proposed modifications is explained.

3.1. Fish School Search algorithm (FSS)

Fish School Search or FSS is a swarm intelligence algorithm recently developed by Carmelo J. A. Bastos Filho and Fernando Buarque de Lima Neto, based on the social behaviour of schools of fish

IOP Conf. Series: Materials Science and Engineering **173** (2017) 012001 doi:10.1088/1757-899X/173/1/012001

[7]. FSS was first designed to simulate the process of a school of fish seeking food as an optimization problem.

The following general definitions are used for the FSS algorithm:

- the "aquarium" is the hyperdimensional search space;
- a "fish" is a candidate solution for the optimization; they each have a weight and a position;
- "food" is a metaphor for guiding fish to "good" regions of the aquarium;
- "swimming" is an operator composed of three movements that are vectorially composed before the actual move;
- "feeding" is the operator that updates the fish weight according to the quality of motion after swimming.

By living in swarms, the fish improve the survivability of the whole group due to mutual protection against enemies. Also, the fish perform collective tasks in order to find locations with lots of food. In a similar way to how real fish swim in the aquarium in order to find food, the artificial fish search (swim) the search space (aquarium) for the best solutions (locations with most food).

The FSS algorithm is simple in concept and easy to implement. The basic structure of the original FSS has the following steps:

1. initialize a population of fish with random positions on D dimensions in the problem space;
2. for each fish in the swarm, update its position by applying the individual movement operator and evaluate the fitness function;
3. for each fish in the swarm, apply the feeding operator and update the fish weight;
4. for each fish in the swarm, apply the collective-instinctive movement operator and update the fish position;
5. calculate the barycentre;
6. for each fish in the swarm, apply the collective-volitive movement operator, update the fish position and evaluate the fitness function;
7. decrease the individual and volitive steps linearly;
8. continue from step 2 until a stop condition is met (usually a sufficiently good fitness value or a maximum number of iterations).

A more detailed description of the mentioned movement and feeding operators with corresponding formulas can be found in [10]. It is worth noting that the FSS algorithm is fast and is computationally inexpensive. This fact was a reason why the FSS heuristic was included as a component-algorithm of COBRA.

3.2. Efficiency examination of the bionic algorithms and proposed modifications

The performance of COBRA's original components was evaluated on six well-known test functions, namely the Rosenbrock function, the Sphere function, the Ackley function, the Griewank function, the Hyper-Ellipsoidal function and the Rastrigin function [8] with 2, 3 and 4 variables, and was presented in [1]. Therefore, the same functions were also used for the testing of the FSS algorithm and then all the mentioned bionic algorithms were compared using the obtained results.

The minimal error, achieved within a given number of function evaluations, and the minimal number of function evaluations for obtaining the given error level were used for comparison. The observed results are given below in Table 1 where each cell contains the name of the algorithm that demonstrated the best value for the performance evaluation criteria for a given function and optimization space dimension. For each optimization problem and dimension the number of program runs was equal to 100 so the results were averaged by that number. The maximum number of function evaluations was equal to 10000; but if the obtained result differed from the known optimum by less than 0.001, calculations were stopped.

The analysis of results shows how many times each bionic algorithm demonstrates the best result for a given problem. For example, it was established that the FSS algorithm works more slowly than most of the examined heuristics but frequently it found the best minimal error.

Table 1. Comparison of the bionic algorithm performance.

Function	D	Minimal number of function evaluations	Minimal error
Sphere function	2	CSA	CSA
	3	BA	BA
	4	BA	BA
Griewank's function	2	CSA	FSS
	3	WPS	BA
	4	CSA	BA
Ackley's function	2	FFA	PSO
	3	WPS	CSA
	4	CSA	CSA
Hyper-Ellipsoidal function	2	CSA	FSS
	3	FFA	FSS
	4	CSA	BA
Rosenbrock's function	2	BA	BA
	3	FFA	FSS
	4	CSA	FSS
Rastrigin's function	2	BA	CSA
	3	CSA	FFA
	4	WPS	FFA

Thus, the idea was proposed to vary the number of component-algorithms. More specifically, the following five new versions of the meta-heuristic approach COBRA were developed:

1. COBRA-2 (cooperative work of the CSA and the BA algorithms);
2. COBRA-3 (cooperative work of the CSA, the BA and the FFA algorithms);
3. COBRA-4 (cooperative work of the CSA, the BA, the FFA and the WPS algorithms);
4. COBRA-5 (cooperative work of the CSA, the BA, the FFA, the WPS and the FSS algorithms);
5. COBRA-6 (cooperative work of all six bionic algorithms).

Moreover, inclusion of certain algorithms was based on their successfulness during the experiments described above.

4. Experimental results

The modifications of the meta-heuristic approach COBRA proposed in the previous section were validated by using test functions with analytical or known solutions and then the obtained results were compared. The examination into the efficiency of the algorithms was conducted on the set of 28 benchmark functions that were submitted for the CEC 2013 Special Session on Real-Parameter Optimization [9].

All these functions are minimization problems; also they all are shifted and scalable. The same search ranges are defined for all test functions: the search range for each variable was [-100, 100]. So, the algorithm was tested by using 5 unimodal functions, 15 basic multimodal functions and 8 composition functions. The error values achieved after maximal function evaluations over 51 runs were sorted from the smallest (best) to the largest (worst) and then the best, worst, mean, median and standard variance values of the function error values over 51 runs were presented.

The performance of the algorithms was evaluated on functions with 5, 10 and 30 dimensions. In Table 2 the number of "wins" for each modification is presented, where "win" denotes the best obtained mean value of function error for a given number of variables.

As examples, the results for test optimization problems with 5 dimensions obtained by COBRA-6 and COBRA-4 are summarized in Table 3 and Table 4 respectively. In these Tables the standard variance is denoted as "STD".

Table 2. Comparison of the proposed modifications.

D	COBRA-2	COBRA-3	COBRA-4	COBRA-5	COBRA-6
5	2	6	8	2	*10*
10	0	8	*9*	3	8
30	1	*9*	6	7	5
	3	*23*	*23*	12	*23*

Table 3. Results for dimension $D = 5$ obtained by COBRA-6.

	Best value	Worst value	Mean value	STD
1	2.60E-07	2.90E-06	5.63E-07	4.66E-07
2	2.26E-04	7.72E-04	5.57E-04	1.86E-04
3	3.89E-03	8.52E-02	3.45E-02	2.26E-02
4	9.27E-06	4.72E-05	1.62E-05	6.61E-06
5	1.78E-06	1.05E-04	2.55E-05	2.30E-05
6	3.33E-02	7.82E-01	3.24E-01	2.20E-01
7	9.31E-04	4.31E-03	1.67E-03	1.03E-03
8	4.37E-03	2.00E+01	1.34E+01	9.35E+00
9	1.67E+00	1.67E+00	1.67E+00	1.27E-04
10	9.37E-02	1.05E-01	9.40E-02	1.51E-03
11	9.46E-05	1.78E+00	1.44E-01	3.20E-01
12	9.95E-01	9.95E-01	9.95E-01	6.66E-08
13	3.69E+00	8.69E+00	4.05E+00	1.06E+00
14	3.49E-06	5.77E-02	1.16E-03	8.00E-03
15	2.35E+01	2.67E+02	3.26E+01	4.48E+01
16	6.88E-02	6.91E-02	6.88E-02	4.01E-05
17	5.00E+00	5.00E+00	5.00E+00	5.78E-06
18	5.00E+00	5.01E+00	5.00E+00	1.02E-03
19	9.89E-03	4.94E-02	1.40E-02	1.18E-02
20	3.89E-02	3.27E-01	1.13E-01	6.54E-02
21	2.89E-03	1.33E-02	4.79E-03	1.62E-03
22	2.27E-04	5.09E-04	3.14E-04	1.18E-04
23	2.17E-04	4.96E-04	3.03E-04	1.16E-04
24	3.85E+01	3.85E+01	3.85E+01	1.62E-05
25	1.51E-03	5.28E-02	2.60E-03	7.10E-03
26	1.86E-01	8.14E+00	5.02E+00	3.88E+00
27	1.86E-01	8.14E+00	5.02E+00	3.88E+00
28	5.56E-03	1.32E-02	5.84E-03	1.07E-03

Table 4. Results for dimension $D = 5$ obtained by COBRA-4.

	Best value	Worst value	Mean value	STD
1	2.03E-08	2.29E-06	1.51E-07	3.17E-07
2	1.26E-03	8.06E-03	3.43E-03	2.32E-03
3	7.14E-03	2.49E-01	2.75E-02	4.17E-02
4	9.31E-07	8.39E-06	1.69E-06	1.41E-06
5	8.21E-07	7.16E-05	1.70E-05	1.72E-05
6	1.42E-04	1.50E-04	1.44E-04	2.52E-06

7	7.44E-04	2.22E-03	8.66E-04	2.91E-04
8	2.15E-03	2.81E+00	1.41E+00	1.19E+00
9	7.55E-02	8.04E-02	7.87E-02	2.13E-03
10	9.61E-02	2.50E-01	1.03E-01	3.04E-02
11	2.53E-05	2.62E+00	2.61E-01	6.49E-01
12	1.22E-04	1.06E+01	2.67E-01	1.51E+00
13	3.02E+00	3.02E+00	3.02E+00	1.06E-05
14	5.12E-06	1.53E-01	4.96E-02	6.18E-02
15	3.01E+01	3.06E+01	3.01E+01	6.25E-02
16	6.10E-01	6.41E-01	6.17E-01	1.07E-02
17	5.00E+00	5.00E+00	5.00E+00	9.58E-06
18	5.00E+00	5.04E+00	5.00E+00	5.08E-03
19	6.01E-02	6.70E-01	1.22E-01	1.29E-01
20	3.50E-01	4.75E-01	3.53E-01	1.73E-02
21	4.27E-03	1.29E-01	6.96E-03	1.72E-02
22	1.41E-04	6.55E-04	1.94E-04	1.08E-04
23	3.09E+02	3.09E+02	3.09E+02	2.87E-02
24	1.56E+01	1.57E+01	1.56E+01	2.41E-02
25	6.73E-04	1.56E+02	5.01E+01	6.82E+01
26	4.00E+01	1.08E+02	5.12E+01	2.05E+01
27	3.87E+02	3.87E+02	3.87E+02	5.86E-02
28	3.09E-03	2.88E-01	9.61E-03	3.94E-02

One can observe that the best combination differs if the dimension varies and it is clear that COBRA should have at least three component-algorithms, otherwise it will not work efficiently. Experimental results demonstrated that the number of component-algorithms can be adjusted by a fuzzy controller, for example. In this case, the number of component-algorithms and the process of varying the population size will be controlled by some rule base.

5. Conclusions

In this paper new modifications of the meta-heuristic called COBRA (Co-Operation of Biology Related Algorithms), which was originally based on five nature-inspired algorithms (the Particle Swarm Optimization Algorithm, the Wolf Pack Search Algorithm, the Firefly Algorithm, the Cuckoo Search Algorithm and the Bat Algorithm), have been introduced. The proposed modifications involve variation of the number of component-algorithms and consequently variation of the whole composition of COBRA. Besides, the Fish School Search algorithm was added as a potential component of the COBRA approach. These modifications were validated and compared. Simulations and comparison showed that the COBRA-6 algorithm is superior to other modifications when the dimension is equal to 5, the COBRA-4 algorithm is superior to the others when the dimension is equal to 10 and the COBRA-3 algorithm is superior to the others when the dimension is equal to 30.

Experimental results showed that there is a necessity for some structure, for example, a fuzzy controller, which can be used to adjust the number of component-algorithms in an automated way. Although some modifications of the component algorithms that improve their performance are known, we used the original design of these algorithms, bearing in mind the aim of our investigation. Thus, in further work their modifications will also be included in COBRA.

Acknowledgments

The reported study was funded by RFBR and Government of Krasnoyarsk Territory according to the research project № 16-41-243064.

IOP Conf. Series: Materials Science and Engineering **173** (2017) 012001 doi:10.1088/1757-899X/173/1/012001

References

[1] Akhmedova Sh and Semenkin E 2013 Co-Operation of Biology Related Algorithms *Proceedings of the IEEE Congress on Evolutionary Computation* (IEEE CEC) pp 2207–2214
[2] Kennedy J and Eberhart R 1995 Particle Swarm Optimization *Proceedings of the IEEE International Conference on Neural networks* pp 1942–1948
[3] Yang Ch, Tu X and Chen J 2007 Algorithm of marriage in Honey Bees Optimization based on the Wolf Pack Search *Proceedings of the International Conference on Intelligent Pervasive Computing* pp 462–467
[4] Yang X S 2009 Firefly algorithms for multimodal optimization *Proceedings of the 5th Symposium on Stochastic Algorithms, Foundations and Applications* pp 169–178
[5] Yang X S and Deb S 2009 Cuckoo Search via Levy flights *Proceedings of the World Congress on Nature & Biologically Inspired Computing* pp 210–214
[6] Yang X S 2010 A new metaheuristic bat-inspired algorithm *Nature Inspired Cooperative Strategies for Optimization, Studies in Computational Intelligence* **284** pp 65–74
[7] Bastos Filho C and Lima Neto F et al 2009 Fish School Search: an overview *Nature-Inspired Algorithms for Optimisation. Series: Studies in Computational Intelligence* ed R Chiong **193** pp 261–277
[8] Molga M and Smutnicki Cz 2005 *Test finctions for optimization need 3 kwietnia*
[9] Liang J J, Qu B Y, Suganthan P N and Hernandez-Diaz A G 2012 Problem definitions and evaluation criteria for the CEC 2013 Special Session on Real-Parameter Optimization *Technical Report, Computational Intelligence Laboratory, Zhengzhou University, Zhengzhou China, and Technical Report, Nanyang Technological University, Singapore*
[10] Bastos Filho C and Lima Neto F et al 2009 On the influence of the swimming operators in the Fish School Search algorithm *Proceedings of the IEEE International Conference on Systems, Man, and Cybernetics (San Antonio)*

Computer simulation of the surface heating process by the movable laser

T A Akimenko, V A Dunaev, E V Larkin
Tula State University
92 "Lenina" prospect, Tula, 300012, Russia,

E-mail: tantan72@mail.ru

Abstract. The model of heating the target surface in air environment with moving laser beam is created. The equation was solved for the case of dynamically changing laser beam energy and constant parameters of the air flow.

1. Introduction

The task of heating the surface of the target moving laser beam is relevant, as is used in many industrial processes.

Practical implementation of this operation to ensure optimal heating and cooling parameters of the target is impossible without building a mathematical model that takes into account the specific conditions of the heating laser beam and cooling the surface by convection. The complexity and urgency of this problem increases when the beam heats the surface is moving.

The mechanism of target surface heating with a pulsating-periodic laser beam, moves over the surface of the plate in a predetermined manneris considered. [1-3].

The target is a flat plate with predetermined thickness δ. Spatial coordinate system, in which the problem is solved, is stationary. The laser beam moves over the surface of the plate in a predetermined manner with the speed $v(x,y,t)$ (fig.1).

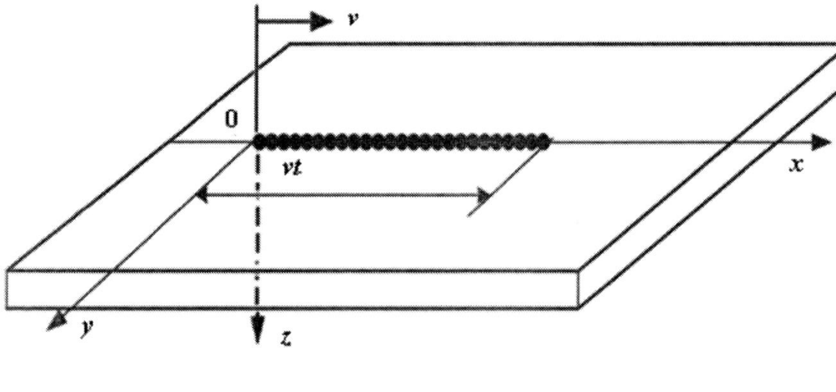

Figure 1. The scheme of the concentrated heat source movement on the plate surface.

2. Mathematical model

The system of equations describing the process of heating-cooling the target surface is as follows:

$$c\rho\frac{\partial T}{\partial t}=\frac{\partial}{\partial x}\left(\lambda_x\frac{\partial T_t}{\partial y}\right)+\frac{\partial}{\partial z}\left(\lambda_y\frac{\partial T}{\partial y}\right)+\frac{\partial}{\partial z}\left(\lambda_z\frac{\partial T}{\partial z}\right),\quad \forall x,y,z\in V,\ t>0, \tag{1}$$

$$q_l=-\lambda_z\frac{\partial T}{\partial z},\quad \forall x,y\in S,\ t>0, \tag{2}$$

$$\alpha(T_g-T)=-\lambda_z\frac{\partial T}{\partial z},\quad \forall x,y\in S,\ t>0, \tag{3}$$

$$T(x,y,z)\big|_{t=0}=T_0=\text{const},\ \forall x,y,z\in V, \tag{4}$$

where (1) – is the differential equation of the heat conductivity; (2) – is the boundary conditions on the laser hot spot surface *S*are of follows; (3) – is the boundary conditions of heat exchange with ambience on the surface *S*are the next; (4) – is the initial conditions for $t = 0$; T – is the target surface temperature; t – is the time; x, y, z – are spatial coordinates; λ_x, λ_y, λ_z– are heat conductivity coefficientsin the direction of axis*x*, *y*and *z*; V – target volume; $q_l = f(x,y,t,e)$ – theheat flux from the laser beam; α – is the heat exchange coefficient; T_g – ambient temperature; S – target surface; T_0– initial temperature.

To solve this problem a method of finite elements [4,5] is used. The decision comes down to the definition of steady-state value of the next functional:

$$\Phi[T(r,z)]=\int_x\left\{\frac{1}{2}\left[\lambda_x^*\left(\frac{\partial T}{\partial r}\right)^2+\lambda_y^*\left(\frac{\partial T}{\partial y}\right)^2+\lambda_z^*\left(\frac{\partial T}{\partial z}\right)^2\right]+q_C^*T\right\}dxdydz-$$
$$-\int_S\left[q_lT+\alpha\left(T_c-\frac{1}{2}T\right)T\right]dS, \tag{5}$$

where $q_C^*=c^*\rho^*\frac{\partial T^*}{\partial t}$ – the heat flux, conditioned by the heat capacity of the material, index " * " corresponds to a temperature distribution in the space fixed at the moment.

The problem reduces to the definition of change of the temperature distribution over the volume of the target as a function of time:

$$\delta\Phi\left[T\left(x,y,z,t^*\right)\right]=0. \tag{6}$$

As a finite element we use the cubic hybrid element, which is formed by the union of several tetrahedrons. A function of the temperature distribution is a linear function and the heat capacity of the element is concentrated in the nodes [4,5,6] (fig.2):

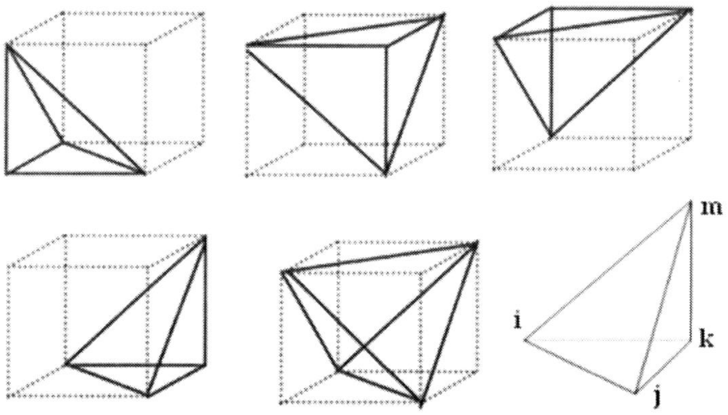

Figure 2. Scheme of the hybrid and tetrahedral finite element

The temperature at any point of finite element is determined by nodal temperatures quantity linearly:

$$T(x,y,z) = \left(N_1T_1 + N_2T_2 + N_3T_3 + N_4T_4\right), \tag{7}$$

where N_l – shape coefficient of tetrahedral finite element; T_l – the temperature of the nodal points of a finite element; l =1, 2, 3, 4.

The linear function of a temperature approximation (7) in this element is as follows:

$$T(x,y,z) = [1, x, y, z]\begin{bmatrix}\alpha_1\\ \alpha_2\\ \alpha_3\\ \alpha_4\end{bmatrix}; \tag{8}$$

where α_1, α_2, α_3, α_4 – are the coefficients of the approximating function.

Vector temperature values in nodes is as follows:

$$\boldsymbol{T} = \begin{bmatrix}T_1\\ T_2\\ T_3\\ T_4\end{bmatrix} = \begin{bmatrix}1 & x_i & y_i & z_i\\ 1 & x_j & y_j & z_j\\ 1 & x_k & y_k & z_k\\ 1 & x_m & y_m & z_m\end{bmatrix}\begin{bmatrix}\alpha_1\\ \alpha_2\\ \alpha_3\\ \alpha_4\end{bmatrix} = \boldsymbol{C\alpha}.$$

Inversion of matrix **C**permits to determine the coefficients of the tetrahedron N_i :

$$N_i = \frac{1}{6V_T}\sum_{i=1}^{4}\left(a_i + b_ix + c_iy + d_iz\right);$$

$$\frac{1}{6V_T}\begin{bmatrix}a_1 & a_2 & a_3 & a_4\\ b_1 & b_2 & b_3 & b_4\\ c_1 & c_2 & c_3 & c_4\\ d_1 & d_2 & d_3 & d_4\end{bmatrix} = \boldsymbol{C}^{-1};\ V_T = \frac{1}{6}\det \boldsymbol{C},$$

where V_T – volume of the tetrahedron; x_l, y_l, z_l – coordinates of the nodal points; indexes a_l, b_l, c_l, d_l – the coefficients of the approximating function of temperature in the amount of finite element; $l \in \{i, j, k, m\}$; $i, j, k, m = 1, 2, 3, 4$ are obtained by cyclic permutation in the sequence.

The heat flow, which is absorbed by the finite element due to the heat capacity of the material expressed through heat flows in nodes:

$$q_c = \int_v c\frac{\partial T}{\partial t}dv = \int_v c\left[\sum_{l=1}^{4} N_l \frac{\partial T_l}{\partial t}\right]dv = \sum_{l=1}^{4}\frac{\partial T_l}{\partial t}\int_v cN_l dv = \sum_{l=1}^{4} q_{cl},$$
$$q_{cl} = \frac{\partial T_l}{\partial t}C_l, \quad C_l = \int_v cN_l dv, \tag{9}$$

where C_l – reduced heat capacity of the node; v – the volume of finite element; c – the specific heat capacity of the target material.

The surface heat flux on the boundary surfaces are also represented in the form of concentrated nodal values:

$$q_l = \int_S q_l dS_l + \int_s \alpha(T_g - T)dS_g = \sum_l q_l + \sum_l q_{gl} - \sum_l q_{al},$$

where summation is performed over the nodal points of the boundary side of the finite element.

Let us consider the original functional in the class of linear functions of temperature (9). Minimization of this functional can be carried out over the nodal values [4], as they uniquely define the temperature at any point in the field of research.

Let us introduce into the original functional (5) conditions (7) – (9) and differentiate it respect to the nodal temperature of the tetrahedron T_l for the field of the finite element:

$$\frac{\partial \Phi}{\partial T_l} = \iint_v \left[\lambda_x \frac{\partial}{\partial x}\frac{\partial}{\partial T_l}\left(\frac{\partial T}{\partial x}\right) + \lambda_y \frac{\partial}{\partial y}\frac{\partial}{\partial T_l}\left(\frac{\partial T}{\partial y}\right) + \lambda_z \frac{\partial}{\partial z}\frac{\partial}{\partial T_l}\left(\frac{\partial T}{\partial z}\right)\right]dxdydz + q_{cl}. \tag{10}$$

To get the matrix form one should combine equations:

$$\left[\frac{\partial \Phi}{\partial T}\right] = \boldsymbol{\lambda T} + {}^4\boldsymbol{C}\frac{\partial \boldsymbol{T}}{\partial t}, \tag{11}$$

where $\boldsymbol{\lambda}$ – is the heat conduction matrix of the finite elements; ${}^4\boldsymbol{C}$ – is the heat capacity diagonal matrix of size 4×4; $\boldsymbol{T}$ – vector of temperature in all nodal points [4].

The final equation of the process of the functional minimization on the temperature at the nodal points are obtained by combining all the derivatives (11) over all finite elements on which the field of research is sampled [4].

Heat conduction matrix, heat capacity matrix and vectors of the nodal heat flow for the structure can be obtained by adding the respective members of heat conduction matrix, heat capacity matrix and finite elements heat flows. Resulting equation of the finite element method for this case takes the form:

$$\boldsymbol{\Lambda T} = -{}^n\boldsymbol{C}\left[\frac{\partial T}{\partial t}\right] + \boldsymbol{Q}_a + \boldsymbol{Q}_L, \tag{12}$$

where $\boldsymbol{\Lambda}$, ${}^n\boldsymbol{C}$ –is the global heat conductivity matrix and heat capacity matrix correspondingly of size $n \times n$; $\boldsymbol{Q}_L$, $\boldsymbol{Q}_a$ –are the global vectors of the nodal heat flow of radiant and convective heat exchange correspondingly;

IOP Conf. Series: Materials Science and Engineering **173** (2017) 012002 doi:10.1088/1757-899X/173/1/012002

$$A=\begin{bmatrix} A_{11} & A_{12} & \dots & A_{1n} \\ A_{21} & A_{22} & \dots & A_{2n} \\ \dots & \dots & \dots & \dots \\ A_{n1} & A_{n2} & \dots & A_{nn} \end{bmatrix};\ {}^{n}C=\begin{bmatrix} C_{11} & 0 & \dots & 0 \\ 0 & C_{22} & \dots & 0 \\ \dots & \dots & \dots & \dots \\ 0 & 0 & \dots & C_{nn} \end{bmatrix};Q_L=\begin{bmatrix} Q_{L1} \\ Q_{L2} \\ \dots \\ Q_{Ln} \end{bmatrix},\ Q_a=\begin{bmatrix} Q_{\alpha 1} \\ Q_{\alpha 2} \\ \dots \\ Q_{\alpha n} \end{bmatrix} \tag{13}$$

3. Numerical solution

For numerical calculations of the equation (12) the implicit difference scheme of solution was used.Usage of finite element with concentrated heat capacity shown on fig.2 allows to exclude oscillation of numerical values of temperature for sharply pronounced transient heating processes and get the calculating scheme for transient temperature field, suitable for constructing the method of numerical investigation of the heat state of the investigated target.

The finite-difference expression for the time derivative gives:

$$AT^{p+1}=-C\left(T^{p+1}-T^{p}\right)\frac{1}{\Delta t}+A\left(T_g^{p+1}-T^{p+1}\right)+Q_L^{p+1}, \tag{14}$$

where p – is the index corresponding to the time; T^{p+1}, T^{p} –are vectors of nodal temperatures in current and subsequent time; T_g^p – ambient temperature.

To combine in the left side of an equation all terms containing unknown quantities we pick out from the matrix Q_a unknown temperatures of structure nodal points:

$$Q_\alpha = A(T_g - T), \tag{15}$$

where A –is the diagonal matrix of convective heat exchange on surface of target;

$$A=\begin{bmatrix} A_{11} & 0 & \dots & 0 \\ 0 & A_{22} & \dots & 0 \\ \dots & \dots & \dots & \dots \\ 0 & 0 & \dots & A_{nn} \end{bmatrix}$$

Shifting to the left side of the equation (15) the terms containing the unknown T^{p+1}, one can get:

$$\left(A+\frac{1}{\Delta t}C+A\right)T^{p+1}=\frac{1}{\Delta t}CT^{p}+AT_g^{p+1}+Q_L^{p+1}, \tag{16}$$

The vector T of size n at the initial time is determined from the initial conditions. The solution of resulting linear system of algebraic equations is carried out by the method of adjoint gradient [4,8].

The algorithm for solving system (1) – (4) was worked out.

The results are shown on fig.3 as isotherms.

Computer experiment was held for target of 10*mm* thick, the material has the conductivity 0.5 W/(mK), grid of size 300×100×20 vas used.

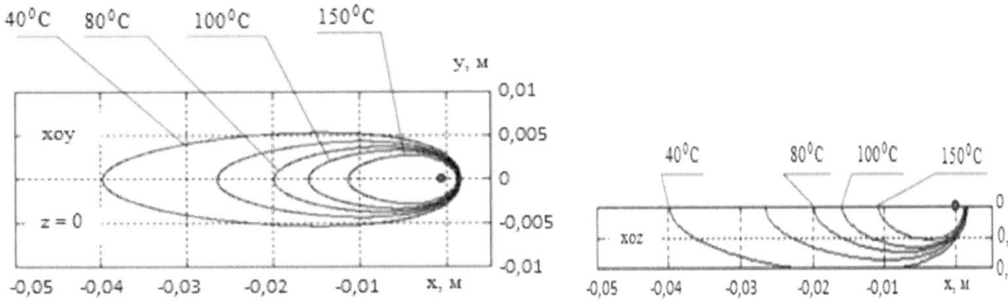

targetsurface target cross section

Figure 3. Isotherms in the plane x-y and x-z of the plate

Numerical simulation of the process is carried out by means of theoretical and computational complex of computer modeling and visualization of the heat exchange processes [9].

4. Conclusion

The developed method of numerical solution process of the surface heating by the movable laser beam allows to carry out a computer simulation of the target surface heating and cooling, and investigate the dynamics of target surface heating by pulsating-periodic laser.

Acknowledgments

This article was written with financial support of the Russian Ministry of Education under the Agreement of granting subsidies № 14.577.21.0084 on June 5, 2014. Unique identifier for Applied Scientific Research RFMEFI57714X0084.

References

[1] Sheich M A, Taylor S C, Hayhurst D R, Taylor R 2000 Measurement of thermal diffusivity of isotropic materials using a laser flash method and its validation by finite element analysis *J. Phys.D: Appl. Phys.* **33** pp 1536 – 1550

[2] Shuja S Z, Yilbas B S, Shazli S Z 2007 Laser repetitive pulse heating influence of pulse duty *Heat Mass Transf.* **43** pp 949–955

[3] Yilbas B S 1997 Analytical solution for time unsteady laser pulse heating of semi-infinite solid *Int. J. Mechanical Sciences* **39(6)** pp 671–682

[4] Zienkiewiez O C 1971 *The Finite Element Method in Engineering Science* 541 p

[5] Segerlind L 1976 *Applied Finite Element Analysis* 392p

[6] Fornberg B 1998 Calculation of weights in finite difference formulas **40** pp 685 –691

[7] Babuška I, Banerjee U, Osborn E 2004 Generalized Finite Element Methods: Main Ideas, Results, and Perspective *International Journal of Computational Methods* **1(1)** pp 67–103 DOI:10.1142/S0219876204000083

[8] Gerald C F, Wheatley P O 1997 *Applied Numerical Analysis* 680p.

[9] Dunaev V A 2001 Theoretical and computational complex computer modeling and visualization conjugate heat and mass transfer processes. Thermal measurements at the beginning of the XXI century *Abstracts of the Fourth International thermophysical school* pp 112-113

Attractor-based models for individual and groups' forecasting

N N Astakhova[1], L A Demidova[1, 2], A V Kuzovnikov[3] and R V Tishkin[1]

[1] Ryazan State Radio Engineering University, Gagarin Str., 59/1, Ryazan, Russian Federation, 390005

[2] Moscow Technological Institute, Leninskiy pr., 38A, Moscow, Russian Federation, 119334

[3] JSC Academician M.F. Reshetnev Information Satellite Systems, Lenin Street, 52, Zheleznogorsk, Krasnoyarsk region, Russian Federation, 662972

E-mail: asnadya@yandex.ru

Abstract. In this paper the questions of the attractors' application in case of the development of the forecasting models on the base of the strictly binary trees have been considered. Usually, these models use the short time series as the training data sequence. The application of the principles of the attractors' forming on the base of the long time series will allow creating the training data sequence more reasonably. The offered approach to creation of the training data sequence for the forecasting models on the base of the strictly binary trees was applied for the individual and groups' forecasting of time series. At the same time the problems of one-objective and multiobjective optimization on the base of the modified clonal selection algorithm have been considered. The reviewed examples confirm the efficiency of the attractors' application in sense of minimization of the used quality indicators of the forecasting models, and also the forecasting errors on 1 – 5 steps forward. Besides, the minimization of time expenditures for the development of the forecasting models is provided.

1. Introduction

Usually the forecasting models on the base of the strictly binary trees use the short time series (about 15 – 30 elements) as the training data sequence [1 – 4]. Therefore, it is necessary to allocate reasonably some part of the analyzed time series (TS), if its length is significantly bigger. Obviously, that application of the long training data sequences will be followed by increase in time expenditures for the development of the forecasting models. Also, it can lead to receive of the not quite adequate forecasting model which will yield the bad forecasting results. This fact can be explained, firstly, by the influence of already insignificant data on the forecasting results, and, secondly, by the difficulties of matching of the forecasting models on the base of the SBT.

In this paper it is offered to use the attractor-based approach to find the adequate length of the training data sequence for the development of the forecasting models on the base of the strictly binary trees. The offered approach to the choice of the training data sequence can be applied for the individual and groups' forecasting of TSs [1 – 6], that will allow minimizing the time expenditures for the development of the forecasting models and the obtaining of the forecasting values.

The rest of this paper is structured as follows. Section 2 describes the main ideas of the development of the forecasting models on the base of the strictly binary trees (SBT) and the modified clonal selection algorithm (MCSA). Herewith, the main principles of one-objective and multiobjective

IOP Conf. Series: Materials Science and Engineering **173** (2017) 012003 doi:10.1088/1757-899X/173/1/012003

optimization are considereted. Section 3 details the attractor-based approach to the choice of the training data sequence for the forecasting model on the base of the SBT. Besides, the main recommendations on the attractors' application for the individual and groups' forecasting of TSs are provided. The experimental results which clearly highlight the efficiency of the offered approach to the choice of the training data sequence for the forecasting model on the base of the SBT follow in Section 4. Finally, conclusions are drawn in Section 5.

2. Forecasting models on the base of the strictly binary trees and the modified clonal selection algorithm

The principles of the development of the forecasting models on the base of the one-objective MCSA were investigated in [2]. The main ideas of the multi-objective MCSA are described in [6]. The MCSA allows forming an analytical dependence on the base of the SBT at an acceptable time expenses. This analytical dependence in the one-objective MCSA describes the certain TS values and provides a minimum value of the average forecasting error rate ($AFER$):

$$AFER = \frac{\sum_{j=k+1}^{n} |(f^j - d^j)/d^j|}{n-k} 100\% , \quad (1)$$

where d^j and f^j are respectively the actual (fact) and forecasted values for the j-th element of the TS; n is the number of TS elements; k is the model order.

Also the $AFER$ (1) can be named as the affinity indicator Aff [16].

The multiobjective MCSA (MMCSA) uses two quality indicator: the mentioned above affinity indicator Aff and the tendencies discrepancy indicator $Tendency$:

$$Tendency = \frac{h}{n-k-1}, \quad (2)$$

where h is the number of negative multiplications $(f^{j-1} - f^j)\cdot(d^{j-1} - d^j)$; $j = \overline{k+2, n}$; d^j and f^j are respectively the actual (fact) and forecasted values for the j-th element of TS; n is the number of TS elements; k is the model order; $n-k-1$ is the total number of multiplications $(f^{j-1} - f^j)\cdot(d^{j-1} - d^j)$.

This indicator describes the rate of discrepancy between the tendencies of two time series (real TS and model TS). Both indicators must be minimized.

Possible options for analytical dependences are presented in the form of antibodies Ab which recognize antigens Ag (certain TS values). An antibody Ab is selected as "the best one". It provides the minimum value of the affinity indicator Aff [1, 2]. Coding of an antibody Ab is carried out by recording signs in a line. The signs are selected from three alphabets [1, 2]: the alphabet of arithmetic operations (addition, subtraction, multiplication and division) – $Operation = \{'+', '-', '\cdot', '/'\}$; the functional alphabet, where letters $'S', 'C', 'Q', 'L', 'E'$ define mathematical functions "sine", "cosine", "square root", "natural logarithm", "exhibitor", and the sign $'\_'$ means the absence of any mathematical function, – $Functional = \{'S', 'C', 'Q', 'L', 'E', '\_'\}$; the alphabet of terminals, where letters $'a', 'b', ..., 'z'$ define the arguments required analytical dependence and the sign $'?'$ defines a constant, $Terminal = \{'a', 'b', ..., 'z', '?'\}$. The use of these three signs alphabets provides a correct conversion of randomly generated antibodies into the analytical dependence. The structure of such antibodies can be described with the help of the SBT [1, 2].

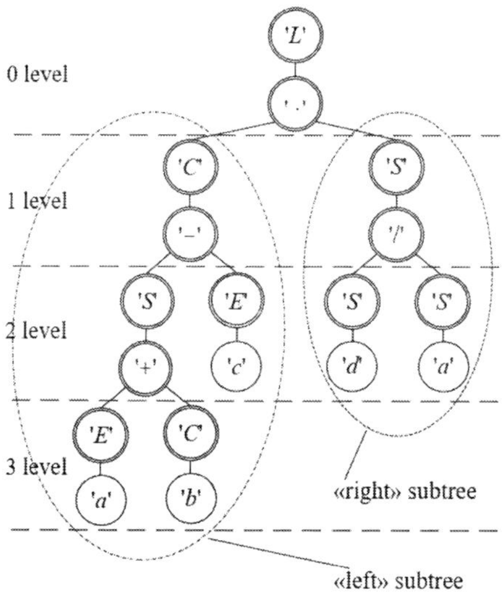

Figure 1. An example of a strict binary tree, which is used to form antibodies.

The number of signs in the alphabet of terminals *Terminal* in the antibody Ab determines maximal possible order K of the forecasting models with $K \geq k$, where k is the real model order), i.e. having the value of the element d^j in the forecasting TS at the j-th moment of time, K values of TS elements can be used as: d^{j-K},..., d^{j-2}, d^{j-1} [1 – 6].

The use of the SBT, illustrated in figure 1, allows building the complex analytical dependence and provides high accuracy of the forecasting TS [2]. Such SBT can be generated as a composition result of one "left" subtree of the maximum possible order $K = 3$ and some "right" subtrees of the maximum possible order $K = 2$. Thus the term "left" subtree ("right" subtree) is used for the branch (left or right) of SBT level in which it is necessary to include a new subtree. In this case it is rational to form antibodies by subdividing SBT into subtrees, then execute the subtree-walk of each vertex forming the ordered symbol lists on its vertices and then combining these lists consecutively [1 – 6].

Forming the symbol ordered list on the base of a subtree the consecutive double subtree-walk is carried out: at first moving the subtree bottom-up left to right we walk the vertices containing the alphabetic terminal signs *Terminal* in pairs and correspondingly above placed vertices containing the alphabetic functional symbols *Functional* and then moving in the same direction it is necessary to go around in pairs the vertices containing the alphabetic arithmetic operation signs *Operation* and correspondingly above placed vertices containing the alphabetic functional signs *Functional*. The first two signs in such an antibody contain the pair of zero level SBT from the functional alphabet *Functional* and arithmetic operation alphabet *Operation*.

Then there are the lists of the signs corresponding to the "right" maximum possible ordered subtrees $K = 2$ (moving the SBT bottom-up) and finally the symbol list of the "left" maximum possible ordered subtree $K = 3$. Using such a way of antibody formation we ensure the visualization of the SBT structure representation in the form of the subtrees union, and the antibody is easily interpreted in the analytical dependence.

For example, the antibody formed on the base of the SBT as shown in figure 1 is coded by the line of signs: $L \cdot S/SeSdC - S + EcCbEa$, which can be transformed into an analytical dependence:

$$f(a,b,c,d) = ln(cos(sin(exp(a) + cos(b)) - exp(c)) \cdot sin(sin(d) / sin(a))).$$

For the k-order forecasting model with $k = 4$ this analytical dependence can be written as:

$$f(d^{j-1}, d^{j-2}, d^{j-3}, d^{j-4}) = ln(cos(sin(exp(d^{j-1}) + cos(d^{j-2})) - exp(d^{j-3})) \cdot sin(sin(d^{j-4}) / sin(d^{j-1})).$$

Interpreting the antibodies into the analytical dependences it is rational to use the recursive procedure of interpretation [1, 2]. The MCSA applied to the searching for "the best" antibody defining "the best" analytic dependence includes the preparatory part (realizes the formation of the initial antibody population) and iterative part (presupposes the ascending antibodies ordering of affinity Aff the selection and cloning the part of "the best" antibodies, that are characterized by the least affine value Aff the hypermutation of the antibodies clones; self-destruction of the antibodies clones "similar" to the other clones and antibodies of the current population; calculating the affinity of the antibodies clones and forming the new antibodies population; suppression of the population received; generation of the new antibodies and adding them to the current population until the ingoing size; the conditional test of the MCSA completion.

The forecasting model on the base of the SBT can be applied for the individual and groups' forecasting of TSs [3 – 6]. In the second case this model is used as the general forecasting model for describing the clusters' centroids. Herewith, the general forecasting model can be specified for some individual TS during the additional iterations of the MCSA.

3. Attractors

In the mathematical field of dynamical systems, an attractor is a set of numerical values toward which a system tends to evolve, for a wide variety of starting conditions of the system. System values that get close enough to the attractor values remain close even if slightly disturbed [7]. Attractor can be used for the solution of many complex applied problems, including the forecasting problems [8 – 11].

In applied mathematics, the phase space method is a technique for constructing and analyzing solutions of dynamical systems, that is, solving time-dependent differential equations. The method consists of first rewriting the equations as a system of differential equations that are first-order in time, by introducing additional variables. The original and the new variables form a vector in the phase space. The solution then becomes a curve in the phase space, parameterized by time. The differential equation is reformulated as a geometrical description of the curve, that is, as a differential equation in terms of the phase space variables only, without the original time parameterization. Finally, a solution in the phase space is transformed back into the original setting [1].

One of approaches to creation of the phase curve suggests to pass from the differential equations to the difference ones and assumes that value d^j $(j=\overline{2,n})$ of element of the analyzed by TS is postponed on the abscissa axis in each concrete timepoint, and the corresponding chain pure gain:

$$\Delta d^j = d^j - d^{j-1} \quad (j=\overline{2,n}) \tag{3}$$

is postponed on the ordinate axis. It is not a derivative in the classical understanding, it is its discrete analog.

The attractors found thus can be used to form reasonably the training data sequence in the context of the forecasting problem. In this case, it is necessary to find all attractors and analyze the last attractor, if there are several attractors.

If attractor is single, it is necessary to consider it. Further, it is necessary to estimate the length of such attractor, i.e. the number of elements forming it.

If the attractor length corresponds to the condition imposed at the length of the training data sequence for the forecasting model on the base of the SBT, then such attractor can be recommended for the use.

If the attractor length is significantly less, than required, then it is necessary to refuse the use of the forecasting model on the base of the SBT. Also, it is possible to unite two or more attractors in one general training data sequence with the acceptable length. If the attractor length is significantly more, than required, then it is necessary to be ready to the possible side effects in case of its use as the training data sequence for the forecasting model on the base of the SBT. Besides, the attempt accomplishment of the reasonable reduction of length of the training data sequence created on the base of the attractor is possible. Also, the reasonable step is search of the alternative approaches to creation of the forecasting model.

In case of the individual forecasting it is necessary to use the attractor-based approach to the choice of the training data sequence for the forecasting model on the base of the SBT.

In case of the group's forecasting it is necessary, firstly, to clusterize the TS group by means of some clustering alrorithm (for example, by means of fuzzy *c*-means algorithm (FCM-algorithm) [3, 4] or *k*-means algorithm [5]) and, secondly, to use the attractor-based approach to the choice of the training data sequence for the forecasting model on the base of the SBT for the general forecasting model corresponding to the the cluster centroid of each formed cluster.

Also, it is possible to analyze all TS in the cluster to find attractors in them, to find lengths of all attractors in the cluster, to reveal the most often found attractors' lengths in the cluster and to use their maximum value (or their average value rounded to the next bigger integer) for determination of the length of the training data sequence. It is obviously, that the training data sequences for the general forecasting models on the base of the SBT in each cluster can have the different length.

4. Experimental studies

The experimental studies executed with the use of TSs, describing various socio-economic indexes, confirm the expediency and prospects of the attractor-based approach to creation of the training data sequence for the forecasting models on the base of the SBT.

The offered approach to creation of the training data sequence was applied for individual and groups' forecasting of TSs, describing the number of references of the E-Commerce systems in the requirements to vacancies posted on the websites of 2 famous recruiting network services – HeadHunter.ru (Russia) and Indeed.com (USA) [12].

In particular, the TS "Magenta (USA)" was considered (figure 2, a). This TS contains information on the number of vacancies which include a specific keyword "Magenta". This keyword defines the name of E-Commerce system for development of online stores. The obtained forecasting results can be used for the analysis of tendencies of the labour market.

The TS "Magenta (USA)" contains 64 elements (the supervision period: 03.02.2016 – 06.04.2016, a unit of measure – the number of references). Herewith, the values of 5 last elements must be predicted, and the others can be used for development of the forecasting model. It is possible to allocate two attractors on the phase curve (figure 2, b). The values by dates are designated by the sequence numbers of the TS elements.

The first attractor is determined during the period from 04.02.2016 to 14.03.2016. As the transition period it is possible to consider the period from 15.03.2016 to 16.03.2016, during which the transition to the new attractor has been executed. From 17.03.2016 to 06.04.2016 the second attractor is determined (from the 43-th value to the 64-th value in figure 2, b).

Let the TS values in the last 5 counting of time (from 02.04.2016 to 06.04.2016, that is from the 60-th value to the 64-th value in figure 2, b) are unknown and it is necessary to predict them.

In this case the length of the training data sequence received on the base of the second attractor with the length of 22 elements will be equal to 17 elements. Such length of the training data sequence can be considered as acceptable for development of the forecasting model on the base of the SBT.

In this example it is visible practically directly what data should be used as the training data sequence (though usually it isn't so obviously). Nevertheless, for confirmation of correctness of the decision on the choice of the training data sequence the attractor-based approach has been used.

Firstly, two forecasting models on the base of the SBT and the one-objective MCSA have been developed: the first model on the base of the elements' values of the initial TS with the length of 59 elements and the second model on the base of the elements' values of the second attractor with the length of 17 elements. Then, the same forecasting models on the base of the SBT and the multiobjective MCSA have been developed.

In case of the one-objective MCSA the average time of development of the required forecasting model in the first and second cases have constituted 1342.924 seconds and 1094.638 seconds respectively. Hence, the average time expenditures have been reduced more, than by 1.23 times. The average value of the affinity indicator in the first and second cases have constituted 2.619% and

1.893% respectively. The average value of the forecasting error on 5 steps forward in the first and second cases have constituted 2.136% and 1.043% respectively.

In case of the multiobjective MCSA the average time of development of the required forecasting model in the first and second cases have constituted 1377.805 seconds and 1107,07 seconds respectively. Hence, the average time expenditures have been reduced more, than by 1.24 times.

The average value of the affinity indicator the tendencies discrepancy indicator *Aff* in the first and second cases have constituted 2.411% and 1.676% respectively. The average value of the forecasting error on 5 steps forward in the first and second cases have constituted 2.007% and 0.988% respectively.

It is visible, that the one-objective MCSA works a little faster (as expected) than the multiobjective MCSA, but the last algorithm provides smaller values of errors of all types thanks to the corresponding correction (by means of the accounting of the second quality indicator – the tendencies discrepancy indicator *Tendency*) of the search direction of the forecasting model. Herewith, the use of attractor allows mimimizing the values of errors of all types, because the elements of TS, which lost the relevance were excluded from the consideration during the process of creation of the forecasting model. Also, the use of attractor allows mimimizing the time expenditures for the development of the forecasting model. For both MCSAs the execution of one run was considered as successful if the value of the affinity indicator *Aff* did not exceed 5%. It has as a result turned out that the share of the successful runs when using the elements' values of the initial TS as the training data sequence is equal to 32%, and the share of the successful runs when using the elements' values of the second attractor as the training data sequence is equal to 54%.

In the reviewed example 1000 iterations of the MCSA for population of 20 antibodies were executed. Coefficient of antibodies' cloning was equal to 0.3. Coefficient of clones' reproduction was equal to 0.8. Computer working under the 64-bit Windows 7 version with RAM of 2 Gb and the two-nuclear Pentium 4 processor with a clock frequency of 3.4 GHz was used for experiment.

Also, the attractor-based approach was applied for groups' forecasting on the base of the model data, which were generated with the use of the data about the references of the E-Commerce systems in the requirements to vacancies. Figure 3, a shows one of clusters, received during the clusterization of the group of TSs. Figure 3, b presents the TS "Cluster's centriod" for this cluster. Figure 3, c shows the phase curve of the TS "Cluster's centriod". It is possible to allocate two attractors on the phase curve (figure 3, c). As in case of the individual forecasting, it is expediently to use the second attractor, which is determined from the 43-th value to the 64-th value in figure 3, c.

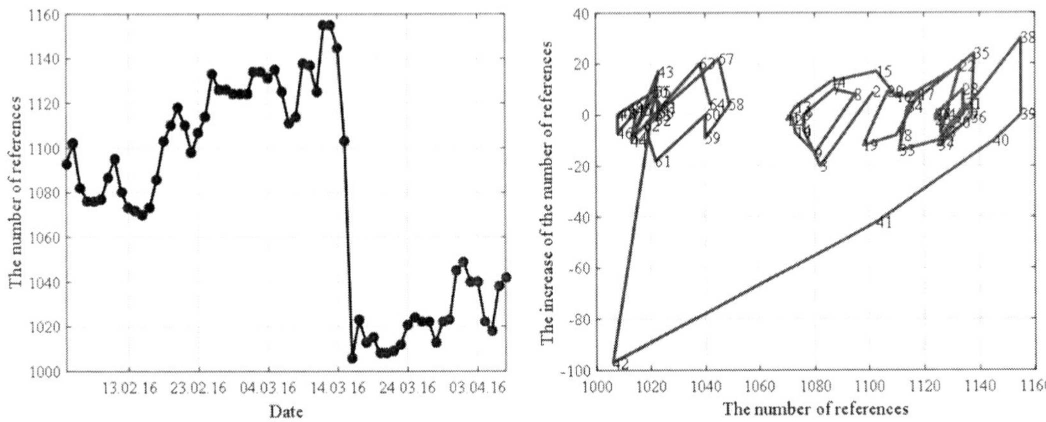

a – The TS "Magenta (USA)" b – The phase curve of the TS "Magenta (USA)"

Figure 2. Identification of attractors.

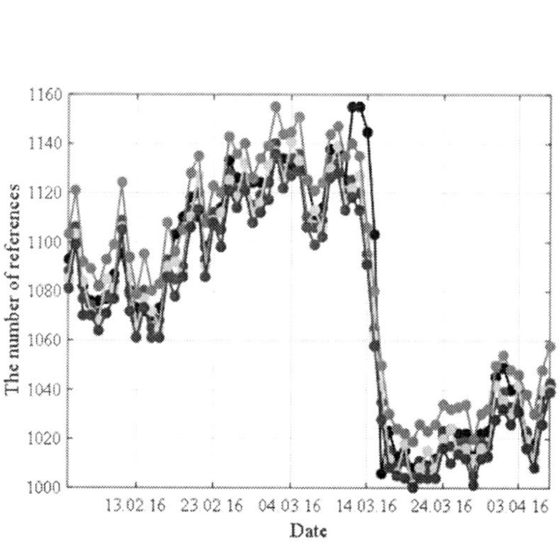

a – The cluster of TSs

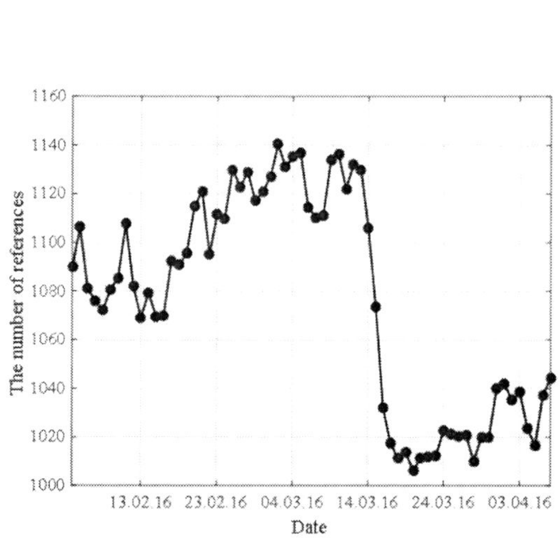

b – The TS "Cluster's centriod"

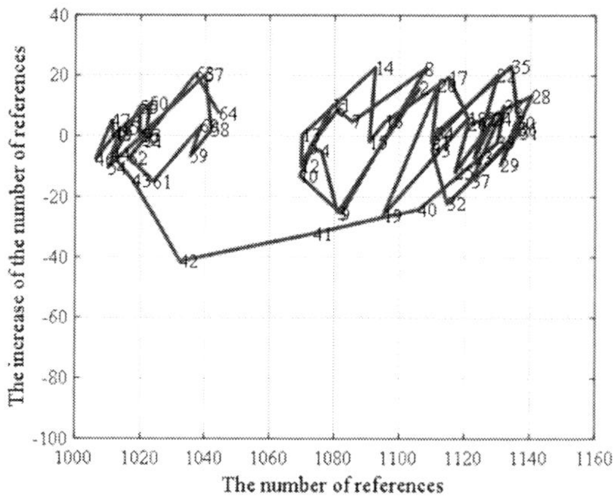

c – The phase curve of the TS "Cluster's centriod"

Figure 3. Identification of attractors for the cluster of TSs.

The rest data should be excluded from the analysis. It is visible (figure 3), that data from the first attractor have the absolutely other law of change. Therefore, their accounting would lead to forming the analytical dependences, not absolutely urgent for the present moment of time. Besides, it will increase the time expenditures for creation of the forecasting model on the base of the SBT. It is necessary to say, that the general forecasting model, which describes the cluster's centroid can be specified for the individual TS, if it is necessary.

The obtained results can be explained by the fact that the length of the training data sequence has been reduced, and the elements with the little actuality for the present moment of time have been excluded from the consideration by the use of the mathematically reasonable approach.

5. Conclusions

The experimental results confirm the expediency and prospects of the attractor-based approach to the choice of the training data sequence for the forecasting model on the base of the SBT, which can be applied for individual and groups' forecasting.

In case of the individual forecasting it is neccessary to find the attractor in the TS and use it to form the training data sequence. It allows reducing the time expenditures on the creation of the forecasting models and the obtaining of the forecasting results. In case of the group's forecasting it is necessary to apply the attractor-based approach to clusters which are formed on the base of the TS group. Herewith, the attractors for the different clusters' centroids can have the different length. As a result, it can lead to the additional reduction of the time expenditures on the creation of the forecasting models and the obtaining of the forecasting results. Herewith, it is possible to minimize at the same time the value of the affinity indicator of the forecasting model on the base of the SBT, the values of forecasting errors on 1 – 5 steps forward, and also the time expenditures on the creation of the forecasting models.

Further, it is planned to use the attractor-based approach for forecasting as the values of TS elements as the values' increments of TS elements with aim to create a technique for individual and groups' forecasting using the forecasting models on the base of the SBT and the MMCSA.

Acknowledgments
The reported study was funded by RFBR according to the research project № 16-08-00771.

References

[1] Demidova L A 2014 Time series forecasting models on the base of modified clonal selection algorithm *2014 International conference on computer technologies in physical and engineering applications* 33–34.

[2] Demidova L A 2014 Assessment of the quality prediction models based of the strict on binary trees and the modified clonal selection algorithm *Cloud of Science* pp 202–222

[3] Astakhova N N, Demidova L A and Nikulchev E V 2015 Forecasting Of Time Series' Groups With Application Of Fuzzy C-Mean Algorithm *Contemporary Engineering Sciences* **8 (35)** pp 1659–1677.

[4] Astakhova N N, Demidova L A and Nikulchev E V 2015 Forecasting Method For Grouped Time Series With The Use Of K-Means Algorithm *Applied Mathematical Sciences* **9 (97)** pp 4813–4830.

[5] Astakhova N, Demidova L and Konev V 2015 The Description Problem Of The Clusters' Centroids 2015 *International Conference "Stability and Control Processes" in Memory of V.I. Zubov (SCP)* pp 448–451.

[6] Astakhova N and Demidova L 2016 Using of the notion "Pareto set" for development of the forecasting models based on the modified clonal selection algorithm *6th Seminar on Industrial Control Systems: analysis, modeling and computation* art 02001.

[7] Kolmogorov A, Petrovskii I and Piscounov N 1937 A study of the diffusion equation with increase in the amount of substance, and its application to a biological problem *In V.M. Tikhomirov editor, Selected Works of A.N. Kolmogorov I* pp 248-270.

[8] Palmer T N 2003 Predicting uncertainty in forecasts of weather and climate *Meteorological Training Course Lecture Series, ECMWF* pp 1-48.

[9] Marcelo Espinoza, Suykens Johan A K and Bart De Moor 2005 Short Term Chaotic Time Series Prediction using Symmetric LS-SVM Regression *2005 International Symposium on Nonlinear Theory and its Applications (NOLTA2005) (Belgium: Bruges)* pp 606-609.

[10] Nikulchev E 2014 Robust Chaos Generation on the Basis of Symmetry Violations in Attractors *2nd International Conference on Emission Electronics (ICEE)* pp 59-61.

[11] Nikulchev E V, Kondratov A P 2015 Method of Generation of Chaos Map in the Centre Manifold *Advanced Studies in Theoretical Physics* **9 (16)** pp 787-792.

[12] Nikulchev E, Ilin D, Biryukov D and Bubnov G 2016 Monitoring of Information Space for Professional Skills *Demand Contemporary Engineering Sciences* **9 (14)** pp 671-678.

[13] Deb K, Pratap A, Agarwal S, and Meyarivan T 2002 A Fast and Elitist Multiobjective Genetic Algorithm: NSGA II *IEEE Transactions on Evolutionary Computation* **6 (2)** pp 182–197

[14] Deb K 2001 Multi-objective Optimization using Evolutionary Algorithms *Chichester* pp 221-232.

[15] Seada H, and Deb K 2015 U-NSGA-III: A Unified Evolutionary Optimization Procedure for Single, Multiple, and Many Objectives: Proof-of-Principle Results Evolutionary Multi-Criterion Optimization *Lecture Notes in Computer Science* **9019** pp 34-49.

IOP Conf. Series: Materials Science and Engineering **173** (2017) 012004 doi:10.1088/1757-899X/173/1/012004

Multiobjective clonal selection algorithm for solving forecasting problems

N N Astakhova[1,4], L A Demidova[1,2], A V Kuzovnikov[3] and R V Tishkin[1]

[1] Ryazan State Radio Engineering University, Gagarin Str., 59/1, Ryazan, Russian Federation, 390005

[2] Moscow Technological Institute, Leninskiy pr., 38A, Moscow, Russian Federation, 119334

[3] JSC Academician M.F. Reshetnev Information Satellite Systems, Lenin Street, 52, Zheleznogorsk, Krasnoyarsk region, Russian Federation, 662972

E-mail: asnadya@yandex.ru

Abstract. The multiobjective modified clonal selection algorithm based on the use of the notion "Pareto dominance", which can be applied for the development of the forecasting models on the base of the strictly binary trees has been offered. It is suggested to use the affinity indicator based on the average forecasting error rate, and the tendencies discrepancy indicator in the role of the objective functions of this algorithm. The multiobjective modified clonal selection algorithm can be applied for solving problems of individual and groups' forecasting. The experimental results which confirm the efficiency of the offered algorithm in comparison with the basic modified clonal selection algorithm have been given.

1. Introduction

The problem of the right choice of the best forecasting model is the main problem of forecasting [1 – 3]. In particular, in the forecasting model on the base of the strict binary trees (SBT) and the basic modified clonal selection algorithm (MCSA) [1, 2] this model is presented in the form of antibodies. The MCSA allow generating the different variants of the forecasting models to find among them the best forecasting model according with the chosen quality indicator. The antibody is coded by a sequence of symbols randomly selected from the corresponding alphabets. Such sequence will be transformed to the analytical dependence which represents some function [1, 2]. This function will be applied to obtaining the predicted values of a time series (TS).

The best forecasting models must be determined at each step of the iterative process in the MCSA. Such models become parents for the next generation of models at the next step [1, 2, 4 – 7]. Obviously, the correct selection of antibodies is very importent for the effective use of the MCSA.

The MCSA possesses a number of benefits in comparison with analogic optimization algorithms. It allows weakening the requirements to data representation and establishing the optimum settings. Also, the MCSA provides high speed and high efficiency of calculations. This algorithm applies the evolutionary principles to modification and selection of antibodies constructed on the base of the SBT. As a result, it is possible to find the best antibody and the best forecasting model corresponding to this antibody. The MCSA is applicable for forecasting of TSs with a short length of an actual part which are characteristic for a large number of real processes and statistical indicators.

Nowdays, the one-factor forecasting models and multiple-factor forecasting models on the base of the SBT and the MCSA which use the values of elements of TSs and the values of their increments for creation of the training data sequence are developed by authors of this paper [1, 2]. Also, we have offered the forecasting method for the groups of TSs which uses the MCSA to create the general forecasting models for each cluster of TSs [4 – 6].

Usually, the average forecasting error rate, calculated for the training data sequence, is used as the main quality indicator of the forecasting model. Herewith, the average forecasting error rate should be minimized [1 – 7]. The MCSA steadily shows the low value of this quality indicator for data of various subject domains.

However, the use of the average forecasting error rate as the single quality indicator of the forecasting model is not always sufficient to determine the best forecasting model. Often it is required to consider the additional quality indicators of the forecasting model, such as the compliance to the seasonal tendencies of TS, the compliance to the trend of TS, lack of emissions, complexity of the forecasting model, etc. [2]. But it isn't possible to choose the single quality indicator correctly. Therefore, the problem of the choice of the quality indicators is very actual. It is expedient to use the additional quality indicator which will allow estimating the general tendency of values' change of the known elements of TS (for example, the tendencies discrepancy indicator) along with the average forecasting error rate [2].

Now, a number of multiobjective evolutionary algorithms (MOEAs) have been suggested. First of all, it is neccassary to say about the multiobjective genetic algorithms [8 – 17] and the multiobjective clonal selection algorithms [18 – 25]. These algorithms provide a solution of the account's problem of the several objective functions (criteria, quality indicators) at the analysis of various applied problems.

Obviously, it is possible to increase the efficiency of the forecasting models on the base of the SBT, using the multiobjective MCSA (MMCSA) at the solution of the problem of the medium-term forecasting. The MMCSA has to borrow the best ideas of the MOEAs in the context of forecasting problems. It is suggested to use in this algorithm the affinity indicator based on the average forecasting error rate and the tendencies discrepancy indicator in the role of the objective functions.

The rest of this paper is structured as follows. Section 2 presents the main ideas of the MCSA. Then, Section 3 details the proposed approach for solving the problem of multiobjective optimization for the MCSA. Here the review of the multiobjective evolutionary optimization algorithms and the choice of the prototype for the realization of the MMCSA have been carried out. The experimental results comparing the original MCSA to the MMCSA follow in Section 4. Finally, conclusions are drawn in Section 5.

2. Forecasting models on the base of the strictly binary trees and the modified clonal selection algorithm

The main ideas of development of the forecasting models on the base of the SBT and the MCSA were investigated in [1]. The MCSA allows forming an analytical dependence on the base of the SBT, which describes the certain TS values and provides a minimum value of the average forecasting error rate ($AFER$):

$$AFER = \frac{\sum_{j=k+1}^{n} |(f^j - d^j)/d^j|}{n-k} 100\% , \qquad (1)$$

where d^j and f^j are respectively the actual (fact) and forecasted values for the j-th element of the TS; n is the number of TS elements; k is the model order.

The possible options for analytical dependences are presented in the form of antibodies Ab which recognize antigens Ag (certain TS values). An antibody Ab is selected as "the best one". It provides the minimum value of the affinity indicator *Aff* [1, 2]. Coding of an antibody *Ab* is carried out by recording signs in a line. The signs are selected from three alphabets [1, 2]:

- the alphabet of arithmetic operations (addition, subtraction, multiplication and division) – $Operation = \{'+','-','\cdot','/'\}$;
- the functional alphabet, where letters $'S','C','Q','L','E'$ define mathematical functions "sine", "cosine", "square root", "natural logarithm", "exhibitor", and the sign $'\_'$ means the absence of any mathematical function, – $Functional = \{'S','C','Q','L','E','\_'\}$;
- the alphabet of terminals, where letters $'a','b',...,'z'$ define the arguments required analytical dependence and the sign $'?'$ defines a constant, $Terminal = \{'a','b',...,'z','?'\}$.

The use of these three signs alphabets provides a correct conversion of randomly generated antibodies into the analytical dependence. The structure of such antibodies can be described with the help of the SBT [1, 2]. The use of the SBT allows building the complex analytical dependence and provides high accuracy of the forecasting TS [2].

The MCSA applied to the searching for "the best" antibody defining "the best" analytic dependence includes the preparatory part (realizes the formation of the initial antibody population) and iterative part (presupposes the ascending antibodies ordering of affinity *Aff* the selection and cloning the part of "the best" antibodies, that are characterized by the least affine value *Aff* the hypermutation of the antibodies clones; self-destruction of the antibodies clones "similar" to the other clones and antibodies of the current population; calculating the affinity of the antibodies clones and forming the new antibodies population; suppression of the population received; generation of the new antibodies and adding them to the current population until the ingoing size; the conditional test of the MCSA completion.

The forecasting model on the base of the SBT can be applied for for the individual and groups' forecasting of time series [3 – 5]. In the second case this model is used as the general forecasting model for describing the clusters' centroids. Herewith, the general forecasting model can be specified for some individual TS during the additional iterations of the MCSA.

3. The multiobjective modified clonal selection algorithm

In the context of working with the MSCA the average forecasting error rate *AFER* can be considered as the affinity indicator *Aff* [1, 2]. It can be used as the first quality indicator for the forecasting model.

The rate of discrepancy between the tendencies of two TSs (the tendencies discrepancy indicator *Tendency*) can be used as the second quality indicator for the forecasting model [2]. The tendencies discrepancy indicator *Tendency* can be calculated as:

$$Tendency = \frac{h}{n-k-1}, \tag{2}$$

where h is the number of negative multiplications $(f^{j-1} - f^j)\cdot(d^{j-1} - d^j)$; $j = \overline{k+2, n}$; d^j and f^j are respectively the actual (fact) and forecasted values for the j-th element of TS; n is the number of TS elements; k is the model order; $n-k-1$ is the total number of multiplications $(f^{j-1} - f^j)\cdot(d^{j-1} - d^j)$.

This indicator allows adapting the forecasting models on the base of the SBT and the MCSA for the medium-term forecasting.

Both indicators (*Aff* and *Tendency*) determine the similarity of the predicted values of the TS with the real ones. However, these indicators use the different principles of assessment. The affinity indicator *Aff* is used in the implementation of the MCSA to define value of "adaptability" (quality) of the antibody *Ab* on the base of elements' values of the TS. The tendencies discrepancy indicator

Tendency allows estimating the quality of the antibody *Ab* taking into account the coincidence with the trend of the TS. Herewith both indicators must be minimized. These indicators are based on the various principles of the quality assessment of the forecasting model. The affinity indicator *Aff* estimates the similarity and difference between the predicted and actual values of the known elements of the TS. The tendencies discrepancy indicator *Tendency* estimates the similarity and difference of the change's directions between the predicted and actual values of the known elements of the TS. This indicator helps to analyze tendencies in the TS and the presence of seasonal fluctuations.

These indicators (*Aff* and *Tendency*) must be used simultaneously at the quality assessment of the forecasting models on the base of the SBT to solve the problem of the medium-term forecasting. Now the problem of the simultaneous accounting of the several quality indicators can be successfully solved using the multiobjective optimization algorithms, including, evolutionary algorithms.

In recent years, a number of multiobjective evolutionary algorithms (MOEA) have been suggested. The main reason for this is their ability to find the multiple Pareto-optimal solutions [8 – 25] in one single simulation run. These algorithms work with a population of solutions. Therefore the primary attention has to be paid to maintaining the diversity and spread of solutions. Such multiobjective evolutionary optimization algorithms provide a solution of the account's problem of the several objective functions (criteria, quality indicators) at the analysis of various applied problems.

Nowdays the multiobjective genetic algorithms (MOGA) [8 – 17] are the most known algorithms of the multiobjective optimization. It is necessary to say about the multiobjective clonal selection algorithms (MOCSA) [17 – 24]. However these algorithms are less designed and, in the majority, borrow the principles of multiobjective optimization underlain in the genetic algorithms. The possibility of this loan can be explained with many similar mechanisms of the evolutionary process realization in the MOGA and MOCSA. At the beginning of the 1990s, a number of different multiobjective genetic algorithms were suggested. Such algorithms as Fonseca and Fleming's FFGA (Fonseca and Fleming's Genetic Algorithm) [10], Horn, Nafpliotis and Goldberg's NPGA (Niched Pareto Genetic Algorithm) [11], Srinivas and Deb's NSGA (Non-dominated Sorting Genetic Algorithm) [15, 16] are the most known among them.

The main common features of these multiobjective evolutionary optimization algorithms are the following: i) assigning fitness to the population members based on the nondominated sorting and ii) preserving the diversity among the solutions of the same nondominated front. These algorithms have been shown to find multiple nondominated solutions. The essential interest has been to introduce elitism to enhance the convergence properties of a MOEA. Among the existing elitist MOEAs, Zitzler and Thiele's SPEA (Strength Pareto Evolutionary Algorithm) [12], Knowles and Corne's PAES (Pareto Archived Evolution Strategy) [13], and Rudolph's elitist GA [14] are well studied. They appeared at the end of 1990s.

The fast and elitist MOEA, called the NSGA-II, has been suggested in 2002 by Deb K., Agrawal S., Pratap A. and Meyarivan T. This algorithm uses a fast nondominated sorting procedure with $O(MN^2)$ computational complexity (where M is the number of objectives and N is the population size), an elitist-preserving approach, and a parameterless niching operator [15, 16]. As result, the NSGA-II alleviates the following difficulties of the MOEAs: $O(MN^3)$ computational complexity, nonelitism approach, the need for specifying a sharing parameter. This algorithm creates a mating pool by combining the parent and offspring populations and selecting the best (with respect to fitness and spread) solutions [15, 16]. The NSGA-II, in most problems, is able to find much better spread of solutions and better convergence near the true Pareto-optimal front compared to the PAES and SPEA, which are the elitist MOEAs that pay special attention to creating a diverse Pareto-optimal front. Also, the NSGA-II uses new the modified definition of dominance in order to solve constrained multiobjective problems efficiently. In the NSGA-II the sharing function approach is replaced with a crowded-comparison approach (on the base of the notion of the "crowding distance") that eliminates the following difficulties: i) the performance of the sharing function method in maintaining a spread of solutions depends largely on the chosen value of the sharing parameter; the overall complexity of the sharing function approach is $O(N^2)$, since each solution must be compared with all other solutions in the population. The NSGA-II does not require any user-defined parameter for maintaining diversity among population members. The basic operations of the NSGA-II and their worst-case complexities are as follows: i) nondominated sorting is $O(M(2N)^2)$; ii) crowding-distance assignment is $O(M(2N)log(2N))$; iii) sorting on the base of the ranks and the crowing distances is $O(2N\,log(2N))$. The overall complexity of the NSGA-II is $O(MN^2)$, which is governed by the nondominated sorting part of the algorithm. It should be noted that the diversity preserving mechanism used in the NSGA-II is the best than in the PAES and the SPEA [15, 16].

Later, in 2012 the reference-point based multiobjective NSGA-II, called the NSGA-III, with $O(MN^2)$ computational complexity was been suggested [17]. The NSGA-III uses a predefined set of the reference points to ensure the diversity in the obtained solutions. The chosen reference points can either be predefined in a structured manner or supplied preferentially by the user. In the absence of any preference information, any predefined structured placement of the reference points can be adopted. The NSGA-III uses Das and Dennis's [17] systematic approach, that places the points on a normalized hyperplane – a ($M-1$)-dimensional unit simplex – which is equally inclined to all objective axes and has an intercept of one on each axis. If the p divisions are considered along each objective, the total number H of the reference points in the M-objective problem is given by the $\tilde{N}^{p}_{M+p-1}$. This algorithm emphasizes the population members which are in some sense associated with each of these reference points. Since the reference points are widely distributed on the entire normalized hyperplane, the obtained solutions are also likely to be widely distributed on or near the Pareto-optimal front. In the case of a user-supplied set of the preferred reference points, ideally the user can mark H points on the normalized hyperplane or indicate any H vectors for the purpose (a herewith these vectors must be M-dimensional). For the rest the NSGA-III is similar to NSGA-II at the realization. The NSGA-III works better on the different classes of multiobjective optimization problems, than others MOEAs, but demand to use a predefined set of the reference points.

At the beginning of the 2000s, the first multiobjective clonal selection algorithms were suggested. The most known MOCSAs are the following: MISA (Multiobjective Immune System Algorithm) [18], MOIA (Multiobjective Immune Algorithm) [19], MOCSA (Multiobjective Clonal Selection Algorithm) [20], IDCMA (Immune Dominance Clonal Multiobjective Algorithm) [21], ACSAMO (Adaptive Clonal Selection Algorithm for Multiobjective Optimization) [22], CNMOIA (Constrained Nonlinear Multiobjective Optimization Immune Algorithm) [23] and NNIA (Nondominated Neighbor Immune Algorithm) [24]. All these algorithma are based on the principles of the Pareto dominance. Such algorithms as MISA, IDCMA, ACSAMO and NNIA use elitism in their realization. A herewith, the NNIA is the modification of the IDCMA. The NNIA as the NSGA-II and NSGA-III uses the notion of the crowding distance as the measure of density of solutions in the neighborhood. It allows maintaining the diversity among the solutions in the population. The computational complexity of the MOCSAs depends on the number of objectives M, the population size N, the clone population size N_c and so on. For example, The computational complexity of the IDCMA and the NNIA is $O((N_d+N_c)^2)$, where N_d is the size of immune dominance antibody population, N_c is the size of generic antibody population (the clone population size) [25].

The analisys of merits and demerits of the MOEAs shows that such the MOGAs as the NSGA-II and the NSGA-III are significantly better than othes because they can successfully solve more difficult problems of the multiobiective optimization [15 – 17]. Besides, the MOGAs have more successful development and application in the solution of many real difficult problems. A herewith the MOCSAs borrows some ideas of the MOGAs (for example, the NNIA uses the notion of the crowding distance for the first time introduced in the NSGA-II).

In this regard the decision on expediency of the adaptation of the ideas put in the NSGA-II at the realization of the MMCSA which is applied for the selection of the forecasting models on the base of the SBT had been made. In this case, it is necessary to understand the forecasting model (and the antibody corresponding to it) as the decision, and the quality indicator of the forecasting model as the objective function at the realization of the MMCSA when developing the forecasting models on the base of the SBT. All forecasting models with use of the notion "Pareto dominance" can be divided to dominated and nondominated models.

Let $Q_{s,v}$ be a value of the v-th quality indicator for the s-th forecasting model; $v=\overline{1,V}$; $s=\overline{1,S}$; let V be a quantity of the quality indicators of the forecasting model; let S be a quantity of the forecasting models (for example, in the population, that is the population size).

The s-th forecasting model is dominated by the z-th forecasting model ($s=\overline{1,S}$; $z=\overline{1,S}$), if the following conditions are satisfied: the s-th forecasting model is dominated by the z-th forecasting model ($s=\overline{1,S}$; $z=\overline{1,S}$), if the following conditions are satisfied for all quality indicators:

$$Q_{s,v} \geq Q_{z,v} \ (v=\overline{1,V}), \tag{3}$$

and there is at least one the v^*-th ($1 \leq v^* \leq V$) indicator for which the condition

$$Q_{s,v^*} > Q_{z,v^*} \tag{4}$$

is satisfied.

Herewith, all quality indicators must be minimized (the smaller the values of the quality indicators, the better the forecasting model).

The rank R_s must be calculated for every s-th forecasting model ($s=\overline{1,S}$). The rank R_s is equal to the quantity of the forecasting models in the population which dominate over the s-th forecasting model. The rank R_s of the s-th nondominated forecasting model is equal to zero [15 – 16].

Let $V=2$, $Q_{s,1}=Aff_s$, $Q_{s,2}=Tendency_s$, where Aff_s and $Tendency_s$ are the values of the affinity indicator (1) and the tendencies discrepancy indicator (2) for the s-th forecasting model ($s=\overline{1,S}$) accordingly. Then the s-th forecasting model is dominated by the z-th forecasting model ($s=\overline{1,S}$; $z=\overline{1,S}$), if the following conditions are satisfied:

($Q_{s,1} \geq Q_{z,1}$ and $Q_{s,2} > Q_{z,2}$) or ($Q_{s,1} > Q_{z,1}$ and $Q_{s,2} \geq Q_{z,2}$), that is

($Aff_s \geq Aff_z$ and $Tendency_s > Tendency_z$) or ($Aff_s > Aff_z$ and $Tendency_s \geq Tendency_z$).

The crowing distances τ_s ($s=\overline{1,S}$) can be calculated using the following algorithm [15 – 17].

Step 1. To calculate ranks for all forecasting models in the population. To unite the models with identical values of the rank into one group.

Step 2. For every group of the forecasting models:

- to sort the forecasting models according to each quality indicator value in ascending order of magnitude;
- to assign infinite distance to boundary values of the forecasting models in the group, i.e. $\tau_1=\infty$ and $\tau_{G_w}=\infty$, where G_w is the quantity of the forecasting models in the w-th group ($w=\overline{1,W}$); W is the groups' quantity; to assign $\tau_s=0$ for $s=\overline{2,G_w-1}$;
- to calculate the the crowding distance τ_s as:

$$\tau_s = \sum_{v=1}^{V} \frac{Q_{s-1,v} - Q_{s+1,v}}{Q_v^{\max} - Q_v^{\min}}, \tag{5}$$

where $Q_{s-1,v}$ and $Q_{s+1,v}$ are the values of the v-th quality indicator ($v=\overline{1,V}$) for the forecasting models with the numbers ($s-1$) and ($s+1$), which are the nearest "neighbors" for the s-th model; $Q_v^{\min}$ and $Q_v^{\max}$ are the minimum and maximum values of the v-th quality indicator ($v = \overline{1,V}$) accordingly.

Figure 1 shows, how we can calculate the crowding distance on the base of two quality indicators. The points, marked with solid circles, correspond to the models with the minimum (zero) value of the rank (i.e. these points correspond to the Pareto front with the zero rank).

To calculate the crowing distance for the s-th forecasting model it is required to define values of both quality indicators for the ($s-1$)-th and the ($s+1$)-th models, which are the nearest "neighbors" for the s-th model and have the same rank. Also, it is necessary to define the best and worst values of each quality indicator (regardless of the decision rank). The crowing distances τ_s ($s = \overline{1,S}$) for the s-th forecasting model on the base of two quality indicators can be calculated as:

$$\tau_s = (Q_{s-1,1} - Q_{s+1,1})/(Q_1^{max} - Q_1^{min}) + (Q_{s-1,2} - Q_{s+1,2})/(Q_2^{max} - Q_2^{min}) \quad (6)$$

At the realization of the MMCSA the s-th forecasting model will be better than the z-th forecasting model, if: ($R_s < R_z$) or ($R_s = R_z$ and $\tau_s > \tau_z$). If the s-th forecasting model is better, than the z-th forecasting model, the s-th forecasting model is the candidate for transfer into the new generation.

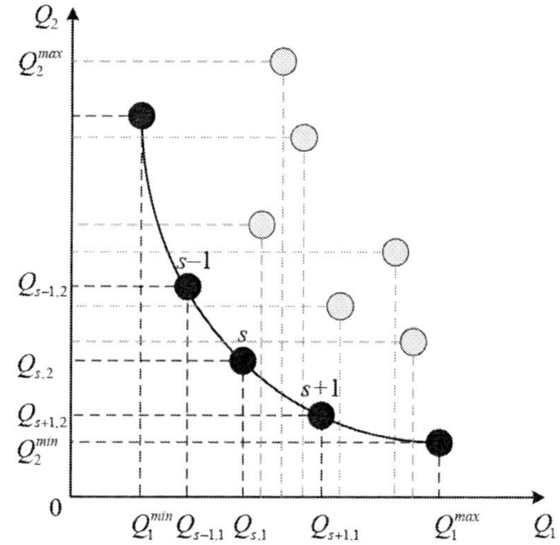

Figure 1. The points using for calculation of the crowding distance.

It should be noted that the forecasting models at the realization of the MMCSA will be exposed twice to the selection procedure providing the maintenance of a variety in the population: at first, when the suppression of antibodies has been carried out, and then, when the simultaneous analysis of values of ranks and values of the crowding distance has been executed.

The MMCSA can be described by the following sequence of steps.

Step 1. To generate initial population of antibodies. Each antibody is coded on the base of the SBT and represents some forecasting model.

Step 2. To perform the nondominated sorting to population of antibodies on the base of two quality indicators for the forecasting model (*Aff* and *Tendency*).

Step 3. To choose the parents-antibodies for the next generation of the clones-antibodies based on the values of the rank and crowding distance.

Step 4. To pass to step 5 if desirable values of the quality indicators are reached or the quantity of generations in the MMCSA is settled. Otherwise to pass to step 2.

Step 5. To accept the antibody with the minimum value of the affinity indicator (1) in the last population as the optimum decision. To use the forecasting model corresponding to this antibody for the forecasting.

As a result of application of the offered MMCSA the Pareto set of the nondominated forecasting models will be received. These models provide the best combinations of values of the used quality indicators for the forecasting models. The received forecasting models can be applied at the solution of a problem of the medium-term forecasting.

4. Experimental studies

To confirm the prospects of the MMCSA the experiments for 5 following TSs were conducted:

T1 (TS "The Brent crude oil price"; range of values: 15.12.2015 – 02.01.2016; unit of measure: ruble) = [46.04; 45.11; 44.46; 44.38; 44.30; 43.97; 44.70; 44.70; 44.08; 45.92; 47.08; 47.32; 47.46; 48.05; 48.84; 50.25; 48.90; 49.50];

T2 (TS "Total number of the made energy resources"; range of values: 1992 – 2012 years; unit of measure: quadrillion Btus) = [47.99; 44.69; 42.30; 41.42; 39.35; 38.74; 39.07; 40.81; 41.70; 42.63; 44.16; 47.16; 49.86; 51.05; 52.06; 52.52; 52.52; 50.01; 53.74; 54.63; 55.30];

T3 (TS "Number of Internet users per 100 people"; range of values: 1990 – 2014 years; unit of measure: percent) = [0.00000; 0.00067; 0.01345; 0.05382; 0.14815; 0.26975; 0.47296; 0.81274; 1.01899; 1.97723; 2.94437; 4.12827; 8.29886; 12.85939; 15.22667; 17.02328; 24.66000; 26.83000; 29.00000; 43.00000; 49.00000; 63.80000; 67.97000; 70.52000];

T4 (TS "Passenger traffic"; range of values: 1989 – 2012 years; unit of measure: millions passenger-kilometers) = [3669.202; 3635.441; 3305.227; 3145.413; 3087.016; 3251.615; 2714.314; 3268.434; 3743.247; 4249.406; 4468.123; 4489.569; 4651.419; 4726.832; 4986.468; 5054.200; 5482.314; 5655.296; 5129.103; 5987.268; 5933.041; 5973.308];

T5 (TS "Value added in agriculture"; range of values: 1992 – 2013 years; unit of measure: dollar) = [270112; 274000; 255000; 253200; 272475; 227102; 192117; 181200; 170300; 152900; 141042; 167100; 158000; 152900; 155573; 164272; 164262; 173699; 173411; 175800; 153500; 139028; 139842; 144612].

These TSs describe the macroeconomic indicators of the Russian Federation.

Values of the forward forecasting errors (for 1 – 5 steps forward) specify that the offered MMCSA is effective as for the solution of problems of the short-term forecasting (for 1 – 3 steps forward) as for the solution of problems of the medium-term forecasting (for 4 and 5 steps forward). It should be noted that use of the additional quality indicator of the forecasting model (the indicator *Tendency*) allowed carrying out the "search" of the forecasting model in the necessary (correct) direction. As a result, for all reviewed examples of time series for the small number of generations of the MCSA the smaller values of the affinity indicator *Aff* (training error (1)) and in most cases the smaller values of the forecasting errors for 1 – 5 steps forward were received (table 1). These errors were calculated on the base of 1000 program runs for 100 generations of population. The size of population is set by 20 antibodies. Results of the forecasting for TS "The Brent crude oil price" have been presented in figures 2 and 3. These results are received with use of the forecasting models on the base of one and two quality indicators respectively.

The presented graphic dependences show that the second model repeats the mathematical law of initial TS for the training set of data better than the first one. Moreover, this feature is kept also for the test set (when the forecasting for 5 steps forward is carried out). Herewith, the forecasting model on the base of two quality indicators (*Aff* and *Tendency*) demonstrates the best survival for the extended forecasting horizon than the forecasting model on the base of one quality indicator (*Aff*). Besides, this model is more effective for the short-term forecasting. The same results had been obtaited for all other TSs.

Also, both variants of optimization have been applied for the development of the forecasting models intended for forecasting of the names' references' quantity of the E-Commerce systems in the requirements to vacancies posted on the websites of 2 famous recruiting network services –

HeadHunter.ru (Russia) and Indeed.com (USA). Each of the TSs contains the information on the number of vacancies which include a specific keyword (Magento, OpenCart, PrestoShop, Hybris, Demandware). This keyword defines the name of E-Commerce system for development of online stores. Herewith, 7 TSs have been considered: 4 TSs with information on vacancies in Russia and 3 TSs with information on vacancies in USA.

The chosen TSs have the different number of elements, the scale level of elements values, sampling points, amplitude and period. It allows considering the further conclusions universal.

Table 1. The forecasting errors.

Error type	Forecasting error with use of *Aff* (%)	Forecasting error with use of *Aff* (%) and *Tendency*
TS #1 ("The Brent crude oil price")		
Training error (1)	3,12	1,10
Error for 1 step	1,43	0,22
Error for 2 step	0,85	0,01
Error for 3 step	0,67	0,38
Error for 4 step	1,06	0,22
Error for 5 step	0,45	0,57
Average error	0,89	0,28
TS #2 ("Total number of the made energy resources")		
Training error (1)	5,23	2,89
Error for 1 step	1,14	0,62
Error for 2 step	0,87	0,20
Error for 3 step	1,53	0,73
Error for 4 step	0,29	0,29
Error for 5 step	0,97	0,68
Average error	0,96	0,50
TS #3 ("Number of Internet users per 100 people")		
Training error (1)	5,49	3,29
Error for 1 step	1,48	0,57
Error for 2 step	3,75	0,13
Error for 3 step	7,41	0,65
Error for 4 step	5,86	0,12
Error for 5 step	9,14	0,42
Average error	5,53	0,38
TS #4 ("Passenger traffic")		
Training error (1)	5,47	2,54
Error for 1 step	1,58	1,75
Error for 2 step	2,84	1,42
Error for 3 step	3,54	1,98
Error for 4 step	3,17	1,10
Error for 5 step	3,69	1,38
Average error	2,96	1,52
TS #5 ("Value added in agriculture")		
Training error (1)	3,98	2,75
Error for 1 step	0,82	0,82
Error for 2 step	1,37	0,54
Error for 3 step	1,89	0,86

Error for 4 step	2,29	1,09
Error for 5 step	3,97	0,22
Average error	2,27	0,71

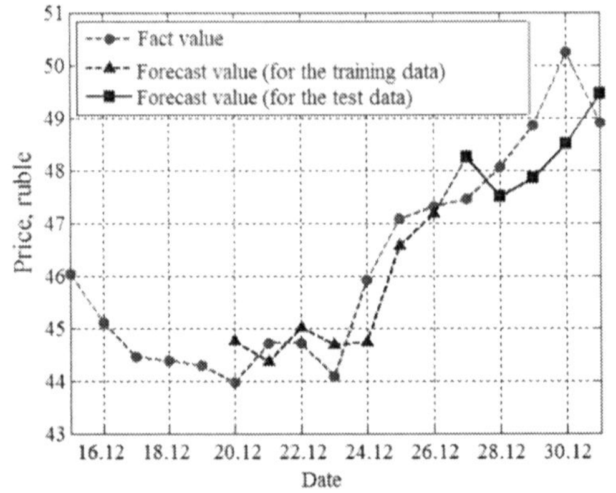

Figure 2. Forecasting of the TS "The Brent crude oil price" (with one quality indicator (*Aff*)).

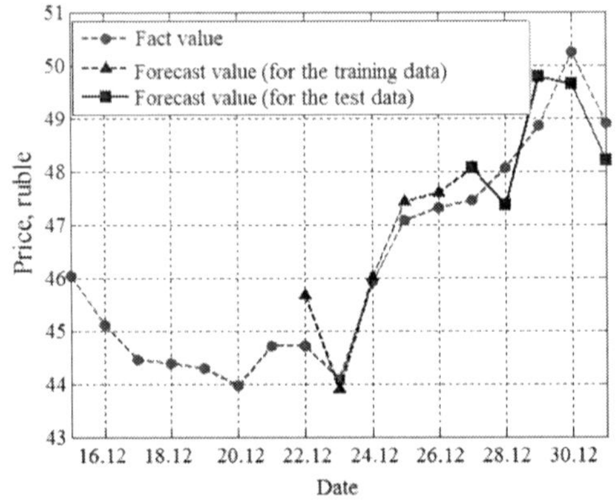

Figure 3. Forecasting of the TS "The Brent crude oil price" (with two quality indicators (*Aff* and *Tendency*)).

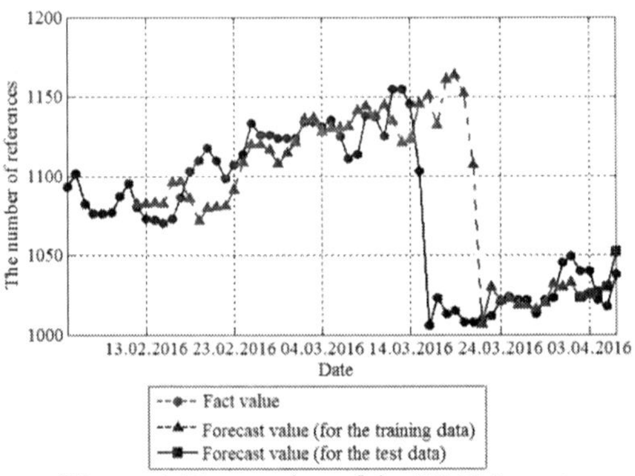

Figure 4. Forecasting of the TS "Magenta (USA)" (with one quality indicator (*Aff*)).

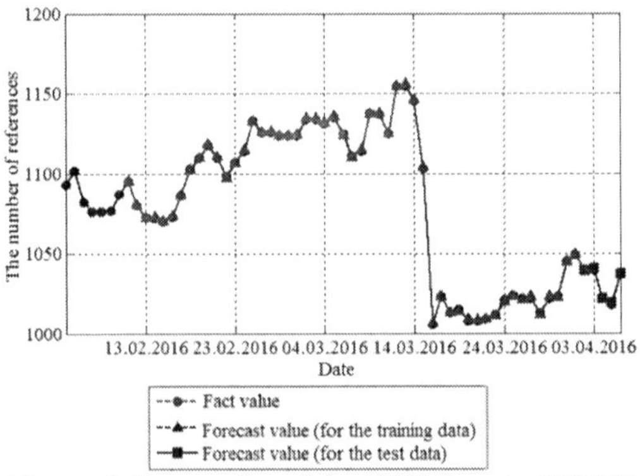

Figure 5. Forecasting of the TS "Magenta (USA)" (with two quality indicators (*Aff* and *Tendency*)).

Forecasting for each TS was executed with use of one (*Aff*) and two (*Aff* and *Tendency*) quality indicators of the forecasting model. The average errors values of the forecasting are presented in table 1. Figure 4 shows the forecasting results for the TS "Magenta (USA)" on the base of the MCSA. Figure 5 shows the forecasting results for the TS "Magenta (USA)" on the base of the MMCSA. These results have been received under the same conditions that were in the experiment considered above. It is obvious that in this case the MMCSA has worked also better, than the MCSA. Herewith, the time expenditures have not increased practically (they are equal to 133.483 seconds and 136.984 seconds respectively).

The same results had been obtaited for all other TSs. It should be noted a problem which was revealed for this this group of TSs. It is the problem of the choice of urgent length of the training data sequence. This problem undoubtedly requires the fastest solution.

The obtained forecasting results allow making the following conclusion: the use of the second quality indicator of the forecasting model provides a way to increase the life time of the forecasting model.

The efficiency of the forecasting model for performing of the short-term forecasts is saved, that, in general, confirms the success of the offered MMCSA based on the ideas of the "Pareto dominance". Herewith, the computing complexity of the MMCSA is $O(M(N+N_c)^2)$, where M is the number of the quality indicators ($M=2$), N is the population size, N_c is the clone population size.

5. Conclusions

Initially, the basic MCSA was developed for solving the short-term forecasting problems. However, the fulfilled researches showed the possibility of application of the MMCSA for solving the medium-term forecasting problems. Herewith, the computing complexity of the MMCSA increases slightly.

It has been proved during the experiments on forecasting of the macroeconomic indicators of the Russian Federation and others, that the application of the principles of the Pareto dominance is the effective solution of the account's problem of the several quality indicators at the development of the forecasting models, which represent the analytical dependences on the base of the SBT.

Further, it is planned to use the offered MMCSA at the realization of the forecasting method for the groups of TSs, when it is necessary to create the general forecasting models for each cluster of TSs. The k-means algorithm, the FCM algorithm and some others will be involved in an experiment.

The application of the MMCSA will allow giving the more exact description of the TS centroids of clusters, and, as a result, to increase the forecasting accuracy. Herewith, any general forecasting model can be specified for some private TS in the cluster (if it is necessary) with the use of the same multiobjective optimization algorithm.

Acknowledgments

The reported study was funded by RFBR according to the research project № 16-08-00771.

References

[1] Demidova L A 2014 Time series forecasting models on the base of modified clonal selection algorithm *2014 International conference on computer technologies in physical and engineering applications* pp 33–34

[2] Demidova L A 2014 Assessment of the quality prediction models based of the strict on binary trees and the modified clonal selection algorithm *Cloud of Science* pp 202–322

[3] Demidova L A 2013 Genetic Algorithm For Optimal Parameters Search In The One-Factor Forecasting Model Based On Continuous Type-2 Fuzzy Sets *Automation and Remote Control* **74** **(2)** pp 313–320

[4] Astakhova N N, Demidova L A and Nikulchev E V 2015 Forecasting Of Time Series' Groups With Application Of Fuzzy C-Mean Algorithm *Contemporary Engineering Sciences (2015)* **8** **(35)** pp 1659–1677

[5] Astakhova N N, Demidova L A and Nikulchev E V 2015 Forecasting Method For Grouped Time Series With The Use Of K-Means Algorithm *Applied Mathematical Sciences* **9** **(97)** pp 4813–4830

[6] Astakhova N, Demidova L and Konev V 2015 The Description Problem Of The Clusters' Centroids *2015 International Conference "Stability and Control Processes" in Memory of V.I. Zubov (SCP)* pp 448–451

[7] Astakhova N and Demidova L 2016 Using of the notion «Pareto set» for development of the forecasting models based on the modified clonal selection algorithm *6th Seminar on Industrial Control Systems: analysis, modeling and computation* art 02001

[8] Michalewicz Z 1995 Genetic algorithms, numerical optimization and constraints *Proc. of the Sixth Int. Conf. on Genetic Algorithms and their Applications* pp 239–247

[9] Goldberg D E 1989 Genetic Algorithms in Search Optimization, and Machine *Learning Reading, Massachusetts: Addison-Wesley* p 372

[10] Fonseca C M and Fleming P J 1995 Multiobjective optimization and multiple constraint handling with evolutionary algorithms – Part I: A unified formulation *Technical report 564, University of Sheffield* pp 1–16

[11] Horn J, Nafpliotis N and Goldberg D E 1994 A niched Pareto genetic algorithm for multiobjective optimization *In Proceedings of the First IEEE Conference on Evolutionary Computation* **1** pp 82–87

[12] Zitzler E and Thiele L 1998 Multiobjective optimization using evolutionary algorithms – A comparative case study *In Parallel Problem Solving From Nature, V, A. E. Eiben, T. Back, M. Schoenauer and H.-P. Schwefel, Eds* pp 292–301

[13] Knowles J and Corne D 1999 The Pareto archived evolution strategy: A new baseline algorithm for multiobjective optimization *Proceedings of the 1999 Congress on Evolutionary Computation* pp 98–105

[14] Rudolph G 1999 Evolutionary search under partially ordered sets *Dept. Comput. Sci./LS11 Tech. Rep. CI-67/99*

[15] Deb K, Pratap A, Agarwal S and Meyarivan T 2000 A Fast and Elitist Multiobjective Genetic Algorithm: NSGA II *KanGAL Report No 200001 Indian Institute of Technology* pp 182–197

[16] Deb K 2001 Multi-objective Optimization using Evolutionary Algorithms pp 221–232

[17] Seada H and Deb K 2015 U-NSGA-III: A Unified Evolutionary Optimization Procedure for Single, Multiple, and Many Objectives: Proof-of-Principle Results *Lecture Notes in Computer Science: Evolutionary Multi-Criterion Optimization* **9019** pp 34-49

[18] Coello P, Coello C A, Cruz Cortés N 2012 An approach to solve multiobjective optimization problems based on an artificial immune system *Proceedings of the First International Conference on Artificial Immune Systems* pp 212–221

[19] Luh G C, Chueh C H and Liu W W 2004 MOIA: Multi-Objective Immune Alghorithm *Computers and Structures* **82** pp 829–844

[20] Campelo F, Guimarães F G, Saldanha R R, Igarashi H, Noguchi S, Lowther D A and Ramirez J A 2004 A novel multiobjective immune algorithm using nondominated sorting *In 11th International IGTE Symposium on Numerical Field Calculation in Electrical Engineering.*

[21] Jiao J, Gong M, Shang R, Du H, Lu B 2005 Clonal selection with immune dominance and anergy based multiobjective optimization *3rd International Conference on Evolutionary Multi-Criterion Optimization* pp 474–489

[22] Wang X L, Mahfouf M 2006 2006 ACSAMO: An Adaptive Multiobjective Optimization Algorithm using the Clonal Selection Principle *2nd European Symposium on Nature-Inspired Smart Information Systems* pp 1–12

[23] Zhang Z 2006 Constrained Multiobjective Optimization Immune Algorithm: Convergence and Application *Computers and Mathematics with Applications* vol **52(5)** pp 791–808

[24] Jiao L, Gong M, Du H and Bo L 2008 Multiobjective immune algorithm with nondominated neighbor-based selection *Evolutionary Computation* **16 (2)** pp 225–255

Multistable states in the biosphere-climate system: towards conceptual models

S. Bartsev[1], P. Belolipetskii[2], A. Degermendzhi[1]

[1] Institute of Biophysics SB RAS, Krasnoyarsk Scientific Center
[2] Institute of Computational Modelling SB RAS, Krasnoyarsk Scientific Center

E-mail: bartsev@yandex.ru

Abstract. Forecasting response of the biosphere and regional ecosystems to observed and expected climate change is the fundamental problem with obvious practical significance. Fundamental non-linearity of the climate system and biosphere makes feasible implementing multiple states and threshold processes in the biosphere-climate system (BCS) in response to gradually increasing influence factor (greenhouse gas concentrations growth). Really time series analysis of global temperature and other global and local parameters indicates the presence of abrupt transitions between stationary states. Identification of the switching mechanisms using general circulation models of the atmosphere and the ocean is associated with the obvious difficulties due to their complexity. Understanding the nature of such switches at qualitative level can be achieved by using a conceptual small-scale models. Some variants of possible mechanisms capable of generating these shifts and simultaneously supporting quasi-stationary periods between them are discussed.

1. Introduction

Prediction of the biosphere and regional ecosystems reaction to observed and anticipated climate change under different scenarios of anthropogenic influence is a fundamental problem which is of obvious practical value. This prediction of the state of "biosphere-climate" system (BCS) is complicated by its fundamental non-linearity that can produce multiple steady states of the system and the threshold switching between these states.

A good example of the manifestation of non-linear properties of the BCS is so-called "pause" or "hiatus" of global temperature rise which is interpreted as a result of the transition of the climate system to another state [1]. Moreover it was observed [2], [3] that Earth temperatures on both local and global scales exhibit sharp shifts at 1976/1977, 1987/1988 and 1997/1998 years. These shifts are more or less visible on the background of natural climate variability. Changes of atmospheric [4], [5], hydrological [6], [7] and ecological [8], [9], [10] parameters were observed together with temperature shifts.

These fast shifts between stationary states cause particular attention since the time of shift is difficult to predict, and in addition, there is a chance that the next stationary temperature level can significantly impair the quality of life of the significant part of human population.

There are two points of view on these environmental shifts. The first one is the shifts are the result of superposition of natural oscillations and continuous anthropogenic warming [11], [12]. The second one is that these shifts are the manifestations of the properties of the climate system itself. They are not simply acceleration of temperature rise, caused by coincidence of various processes, but these shifts represent some structural changes in climate system masked by variability.

According to the first point of view there are no special mechanisms of shifts, and since they are the result of natural oscillations superposition, they do not need special attention. Accepting the second point of view and accounting that the most portion of observed recent warming is caused by the shifts we come to importance of the phenomenon investigation. In this case mechanics of shifts on the one hand and processes producing quasi-stationary periods (like recent temperature pause) on the other hand are to be in the focus of our attention.

The cessation of global temperature growth observed in the past 13 years contradicts the prediction based on the IPCC (Intergovernmental Panel on Climate Change) models of general atmosphere and ocean circulation [1]. The absence of a direct relationship between the growing concentration of green-house gases in the atmosphere and the global temperature can be interpreted as a manifestation of the own dynamics of the Earths climate system. So prediction of abrupt changes seems to be very serious challenge to investigators in comparison with the prediction of gradual dynamics since nonlinearity plays key role in the shifts origin.

Climate and biosphere indices are highly variable in space and in time. Commonly used technique for identifying patterns of variability in climate science is empirical orthogonal function (EOF) analysis [15]. Conceptually, EOF analysis determines a spatio-temporal pattern of variability that accounts for the maximum covariance between the field anomaly time series at all pairs of grid points in the data. Each EOF pattern is associated with corresponding principal component (PC) of time series [16], which describes the temporal evolution of the EOF pattern. The PC time series may be obtained by projecting the EOF pattern onto the original anomaly field at each time step so as to find the sign and amplitude of the pattern at any given time.

Early studies [13], [14], [15] on short and long timescales global temperature variability and trends showed that first principal component (PC_1) of global sea surface temperature (SST) is almost the same as various ENSO (El Nino South Oscillation) indexes (Nino34, SOI, MEI). Linear adjustment on ENSO by means of subtracting PC_1 projection from surface temperature anomalies revealed that residual temperature dynamics looks like staircase dynamics [13], [14]. After this adjustment 1976/1977 shift almost disappeared, that suggests its relation to ENSO. In contrast 1987/1988 and 1997/1998 shifts, which were previously detected in many regional processes, became obviously visible in global temperature anomalies (Fig. 1).

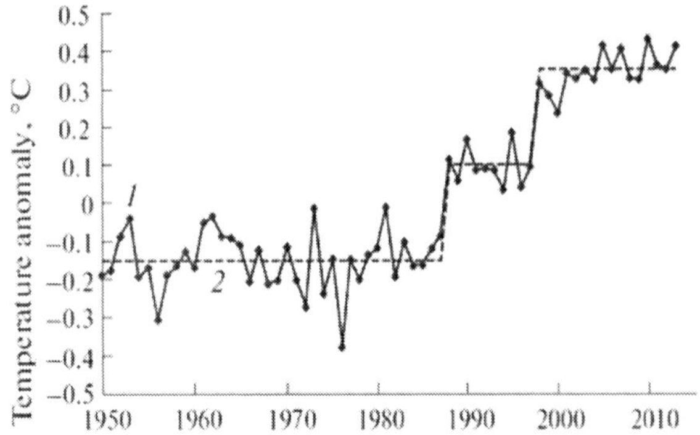

Fig. 1. Residual annual average temperature variations. (1)

Average annual global temperature anomaly, (2) approximation by the model step function.

IOP Conf. Series: Materials Science and Engineering 173 (2017) 012005 doi:10.1088/1757-899X/173/1/012005

Verification of these results [13], [14] on satellite-based temperature datasets and consideration of potential mechanisms of the shifts can serve as an additional argument in favor of physical reality of global climate structural shifts.

2. Abrupt climate shifts and observations

In this study we considered three global satellite-based temperature datasets - NCEP OI v2, UAH MSU v5.6 and RSS MSU v3.3. Satellite-based measurements period starts approximately at 1980 year. These measurements have some advantages over surface measurements e.g. better spatial coverage and less discrepancies between each other if compare to the surface ones. Since we are interested in the dynamics of climate and due to calculation convenience, all the datasets considered not as absolute values but as anomalies.

As was mentioned above application of EOF technique to detrended sea surface temperatures (HadSST3) data revealed [13], [14], [15] the leading EOF of monthly SST anomalies over the globe is the ENSO mode. These mode is persistent in most of climate datasets with global coverage. The major ENSO indexes are nearly identical to the PC_1 time series obtained from SST and temperature of lower troposphere (TLT) fields. Correlation coefficients between them is near 0.9.

In this study we used first principal components of detrended NCEP OI SST, tropical UAH MSU TLT and RSS MSU TLT in order to account ENSO influence on corresponding field. Then linear projections of PC_1 time series on corresponding fields were subtracted from initial fields. This procedure allows to adjust for main ENSO effects. All the datasets were achieved and manipulated at Climate Explorer site (climexp.knmi.nl).

Major features of the global atmospheric circulation, averaged over timescales longer than a month or two, are largely determined by (SST). The dynamics of global SST monthly mean temperature anomalies adjusted for linear ENSO effects shows three major perturbation events during 1980-2015 period. One event is associated with well-known Pinatubo eruption at 1991 [17]. Other evident events having outstanding amplitude are 1987/1988 and 1997/1998 shifts (Fig. 2). On the figure the contribution of volcanoes is not accounted in the SST graph since estimations of aerosols effect are very uncertain. Aerosols curve is shown for illustration of possible cause of SST steady state deviations.

All time-series have short-term fluctuations, which we consider as noise. In contrast to them 1987 and 1997 shifts and 1991 Pinatubo eruption produced more significant changes. Linear fit to constant temperature during 1980-1987 and 1998-2015 is quite good for all considered data, and Fig. 2 is shown as an typical example. And it seems that during 1988-1997 years temperatures will also be near constant after adjustment for Pinatubo eruption. It is seen in ENSO adjusted datasets that in the periods before, between, and after the shifts warming was absent or very small. In other words during 24 years of total 36 years of satellite measurements there were no warming if ENSO variability removed.

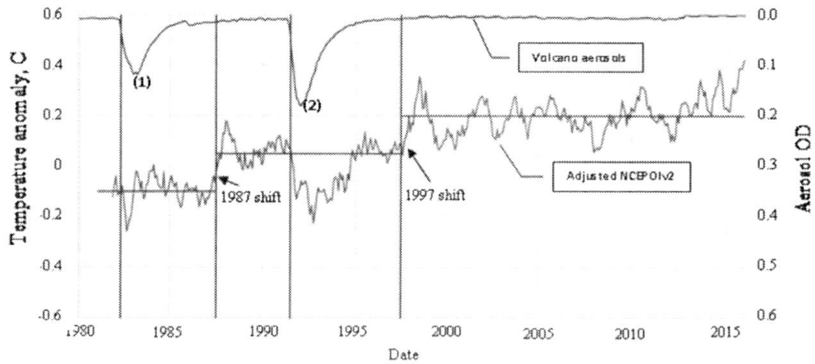

Fig. 2. Adjusted for ENSO NCEP OI v2 global SST, major perturbation events and quasi-stationary periods.

On the graph: (1) El Chichon eruption; (2) - Pinatubo eruption.

As a result, our analysis indicates that the dynamics of satellite-based temperature anomalies can be interpreted as stationary periods partitioned by sharp steps with superimposed ENSO variability on them. Thus the results of the study the temporal dynamics of reanalysis, SST and TLT data are in agreement with previous results on the which is additional argument in favor of physical reality of global climate structural 1987/1988 and 1997/1998 shifts.

The synthesis of satellite-based global temperature datasets and widely used EOF analysis conducted in the present study has provided robust evidence that previously found staircase signal [15] is a real climate signal. Forcing from anthropogenic greenhouse gases continuously rose from 1980 to 2015. From the other hand our analysis suggests that most of warming occurred during 1987/1988 and 1997/1998 events. Thus 1987/1988 and 1997/1998 shifts are outstanding events which underlying physical mechanics remain unresolved.

3. On possible mechanisms of abrupt climate shifts

Understanding that gradual global temperature changes can hardly be expected due to natural variability modulating the gradually increasing concentration of anthropogenic greenhouse is presented in literature [18]. Breach of graduality in the form of abrupt shifts can be attributed to the superposition of natural oscillations and monotonic trends however long-lasting quasi-stationary periods rise some doubts on this issue.

Let's consider possible mechanisms capable of generating these shifts and simultaneously supporting quasi-stationary periods between them. There are several pure versions: 1) there is some yet unknown parameter of biosphere-climate system possessing multiple (not less than three) stable stationary states, and observed stochastic-like variability is of external nature; 2) there is a set of more simple bistable systems and each shift is a manifestation of switch of only one system (variability is also of external nature); 3) variability is an inherent property of the system which is chaotic or near chaotic oscillator (the simplest classical example is Lorenz system) and shifts are transitions between attractors; 4) there is a system of interconnected (coupled) oscillators and shifts are transitions between attractors of this super system.

To these variants we can add spatial coordinate for example a set of bistable systems can be considered as spatially distributed systems of similar nature, or as locally combined system of different nature. Later some of them will be considered. In addition nobody can exclude from the consideration some mixed cases, which makes the system more complex.

Let's consider the first case. Formally three quasi-stable states correspond to the existence of at least five stationary states of considered parameter of BCS. And if we presume possibilities of new shifts then we have to assume the existence of additional stationary states, what corresponds to the existence of the number of roots of function describing time derivative of the parameter. Literature searching for BCS parameter possessing such property was unsuccessful.

Completely another situation is with the second case. Many papers describing models and observation data on bistable indices can be found. Some of them are listed here just for instance: bistable regime is possible due to interactions between SST anomaly and thermocline depth anomaly [19]; between SST anomaly and salinity [20], [21]; between vegetation and monsoon precipitation [22]; between sea level pressure and eddy heat flux [3]; between CO_2 and convection [23]; temperature and soil moisture [24]; between temperature gradient and precipitation [25]. However all these examples consider only bistable systems and each of them can not describe two shifts observed. Possible way to explain the phenomenon is in presuming similar bistable systems located in different Earth's regions, which are autonomous and due to different thresholds of switching undergoes transitions at different time moments. However statistical analyses does not support this versions, no well-marked distinct spatial anomalies were detected yet.

IOP Conf. Series: Materials Science and Engineering **173** (2017) 012005 doi:10.1088/1757-899X/173/1/012005

An alternative idea, that each observed shift is the result of state transition in one of feedback interconnected systems including listed above. Observations support this version since found 1987/1988 and 1997/1998 shifts have different physical mechanics. So 1987/1988 shift is associated with expansion of Hadley cell and a pole ward broadening and intensification of the Ferrel cell [3]. McLean [4] observed 7 percents reduction in total cloud cover followed by 1987 shift till 1997 shift. These features were not observed after 1997/1998 shift. Also in reanalysis parameters revealing these shifts [14] looks different. The shift of 1987/1988 is more sharp than 1997/1998 shift in global average meridional wind at 300 mb and in tropical near surface vertical wind speed.

Local manifestations of these shifts are also very different. In the southwestern Japan sea 1987/1988 shift is seen in sharp rise of winter temperatures and 1997/1998 shift is seen in summer temperature rise [6]. According to Chavez and coworkers [5] 1987/1988 shift is associated with maximum of Pacific Decadal Oscillation warm phase and 1997/1998 shift with transition to cold phase.

To illustrate possible interrelation between two bistable systems producing two shifts let's consider two pairs of parameters: SST anomaly and thermocline depth anomaly [19] and SST difference anomaly and salinity [20]. The first system can be described by the following model

$$\frac{dT_E}{dt} = RT_E + \gamma h_W - e_n(h_W + bT_E)^3$$
$$\frac{dh_W}{dt} = -rh_W - abT_E,$$

where h_W denotes the thermocline depth anomaly in the western Pacific; T_E denotes the SST anomaly, averaged over the central to eastern equatorial Pacific; b is coupling coefficient between the wind stress and SST anomalies; γ is thermocline feedback coefficients; r collectively represents the damping of the upper ocean system through mixing; α - coupling coefficient between wind forcing and zonally integrated wind stress; R collectively describes the positive feedback hypothesis of the tropical ocean-atmospheric interaction; e_n coefficient describes a nonlinear dependence of subsurface temperature on the thermocline depth.

The second system is finally described by only one equation on normalized salinity obtained after normalizing and limiting transition due to estimating temperature difference as fast variable [20]:

$$\frac{dy}{dt} = -[1 + \mu^2(y-1)^2]y + p_0 + p(t),$$

where the parameter μ corresponds to the ratio of diffusive timescale; p_0 and $p(t)$ denotes constant and stochastic components of fresh water flux depending on the difference between evaporation and precipitation.

One of possible interactions among these systems is the influence of temperature on the precipitation and consequently on fresh water flux, in its turn salinity affects thermocline depth via changes in mixing intensity. So GHG forcing can initialize changes in thermocline depth backward causing temperature change which in turn causes salinity changes affecting thermocline and temperature. An illustrative model of mentioned system interrelations can be described as:

$$\frac{dT_E}{dt} = F + (\sigma - 1)T_E - e_n\sigma T_E^{\,3} + Rnd(t)$$
$$\frac{dy}{dt} = P - y - \mu^2 y(y-1)^2 + Rnd(t)$$
$$\frac{dP}{dt} = k_1(P_0 + k_2 T_E - P),$$

where P_0 denotes some constant level of precipitation; k_2 denotes the influence of temperature to cloud formation and precipitation; k_1 represents time scale of precipitation formation; *Rnd(t)* imitate external stochastic influence.

The dynamics of this system at increasing forcing is shown on Fig. 3.

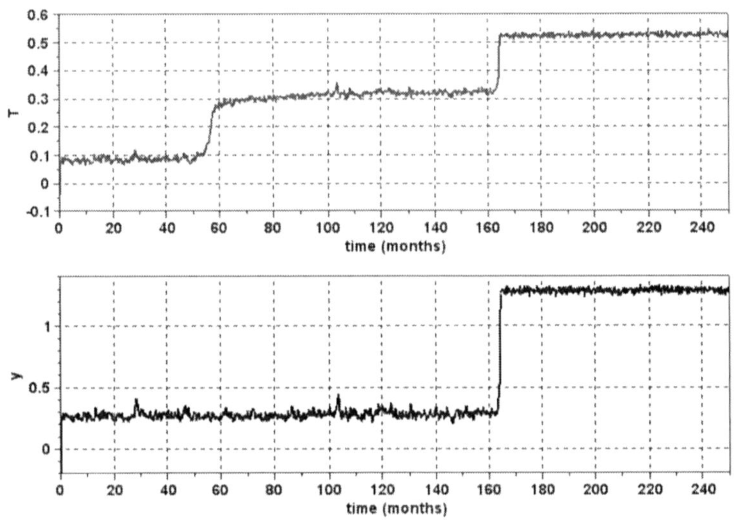

Fig. 3. Dynamics of "temperature-thermocline-salinity-temperature" model demonstrating two temperature shifts causing by thermocline and salinity transitions.

The abundance of potentially bistable systems is some kind of challenge for researchers since selecting and predicting actual interactions capable of initiating the next climate shift can not be conducted analytically and require huge amount simulations of uncertain credibility due to our uncertainty concerning what interaction is actual now and what will be the next one.

Let's return to the list of possible mechanism of multiple shifts. Third case presumes that observed climate variability is an inherent property of the system and observed shifts are manifestations of transitions over a set of attractors belonging chaotic or near chaotic oscillator responsible for this variability. The most probable candidate to this role is ENSO.

ENSO originates in the tropical Pacific through interactions between the ocean and the atmosphere, but its environmental impacts are felt worldwide. While air-sea interactions responsible for ENSO are centered over the equatorial Pacific Ocean, changes in tropical precipitation from deep convection associated with ENSO influence the global atmospheric circulation [19]. There are mathematical models of ENSO [26], [27] and atmospheric general circulation [28] based on some modifications of classic Lorenz model:

$$\frac{dx}{dt} = \sigma(y - x)$$
$$\frac{dy}{dt} = rx - y - xz$$
$$\frac{dz}{dt} = xy - bz,$$

Vallis [26] used the following transformation

$$x \to -x$$
$$y \to -\frac{C}{B}y$$
$$z \to -\frac{C}{B}z + 1$$

and obtained a model for El Nino:

$$\frac{dx}{dt} = By - C(x + p)$$
$$\frac{dy}{dt} = xz - y$$
$$\frac{dz}{dt} = -xy - z + 1,$$

where variable x represents velocity of the westeast ocean current; variable $y \sim (T_E - T_W)$ where T_E and T_W are averaged temperatures of the upper ocean layer in the western and eastern Pacific; $z \sim (T_E + T_W - 2T_d)$, where T_d is deep ocean temperature under the thermocline and is taken as a constant. The ocean current is driven by the surface wind which, in turn, is generated by the temperature gradient.

Very interesting properties of another modification of Lorenz model named Lorenz84 were discovered [28]:

$$\frac{dx}{dt} = -y^2 - z^2 - ax + aF$$
$$\frac{dy}{dt} = xy - y - bxz + G$$
$$\frac{dz}{dt} = bxy + xz - z,$$

where x represents the intensity of the symmetric globe encircling westerly wind current, and also the poleward temperature gradient, which is assumed to be in permanent equilibrium with it. The variables y and z represent the cosine and sine phases of a chain of superposed large-scale eddies, which transport heat poleward at a rate proportional to the square of their amplitude. The constant terms aF and G in the model represent symmetric and asymmetric thermal forcing.

This model demonstrates that there are areas of parametric space where four stable attractors coexist in phase space. So this is a rather nontrivial low-order system and it can be complex, useful tool for investigating problems of climate dynamics in a simple setting. A striking result is that the final climate scenario crucially depends on subtle and minute tuning of parameters.

In the context of this model the shifts can be the result of the model transitions from one attractor to another. Fig. 4 demonstrates the coexistence of two attractors corresponding to periodic solutions.

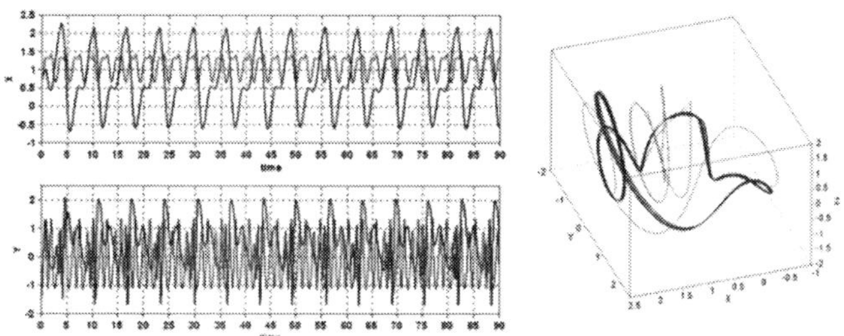

Fig. 4. Two attractors of Lorez84 model.

However addition of relatively weak external periodic forcing makes the behavior of the system hardly predictable (Fig. 5).

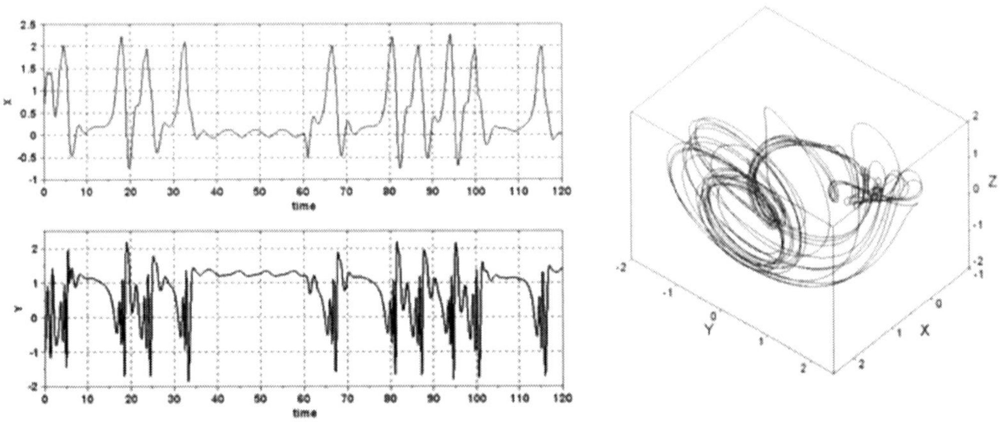

Fig. 5. Dynamics of Lorenz84 model under weak periodic external forcing.

Short impacts can switch the system from one attractor to another (Fig. 6). Certainly this example looks very simple however our objective is to consider potential mechanisms of shifts without claim to an accurate description.

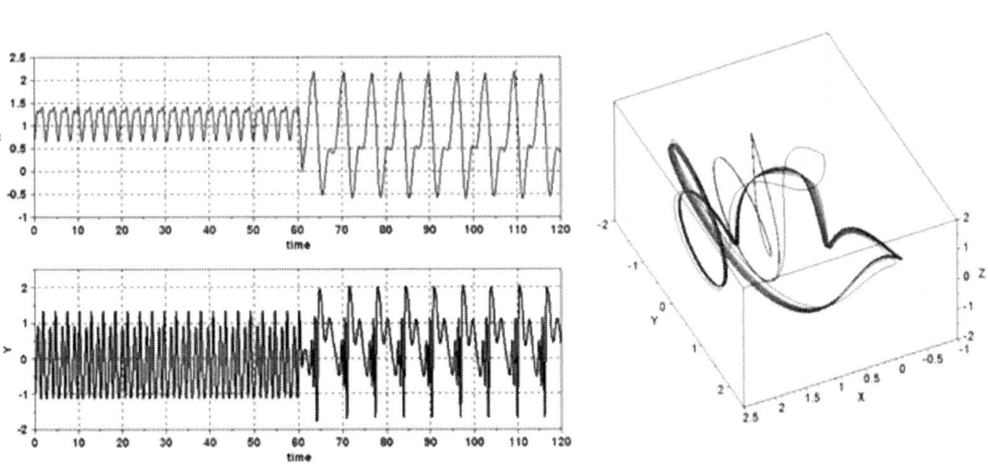

Fig. 6. Response of Lorenz84 model to weak shot impact.

Investigating of the fourth version of potential mechanism of the shifts generations (system of coupled stochastic or near stochastic oscillators) and mixed variants of all these mechanisms including spatial distribution of oscillators are extremely interesting area and many pictures of complex dynamics can be obtained [29], [30], [31].

4. Conclusions

Principal feasibility of complex regimes in BCS including abrupt shifts is not in doubt. We hardly will be able to know all about interactions and exact values of BCS indices. However predicting expected time of abrupt shifts by means small-scale conceptual models involving key variables and parameters seems to be rather possible.

The analysis conducted in this paper demonstrates sizable set of potential mechanisms capable of providing multi stable dynamics of BCS. In this context the primary objective is to reveal some rules and constrains (for instance mass and energy conservation lows), which reduce the set of possible combinations of multi stable systems or/and oscillators.

Further statistical searching for interactions between different global and local BCS variables for detecting key variables for conceptual mathematical model combined with parallel testing of different variants of the model are expected to provide consistent and explainable representation of shift generating mechanics.

Acknowledgments. This work was supported by grant RFBR-KKFN No 15-41-04300 and Complex Program of SB RAS No II.2. No 0360-2015-0002.

References

[1] Tollefson, J. 2014 Climate change: The case of the missing heat *Nature* 505 pp. 276–278.
[2] Hare, S. R., Mantua, N. J. 2000 Empirical evidence for North Pacific regime shifts in 1977 and 1989 *Prog. Oceanogr.* 47 pp. 104–146.
[3] Reid, P. C., Hari, R. E., Beaugrand, G., Livingstone, D. M. 2016 Global impacts of the 1980s regime shift *Global Change Biology* 22(2) pp. 682–703.
[4] Kim, Y.-H., Kim, M.-K., Lau, W. K. M., Kim, K.-M., Cho, C.-H. 2015 Possible mechanism of abrupt jump in winter surface air temperature in the late 1980s over the Northern Hemisphere *J. Geophys. Res. Atmos.* 120 pp. 12,474-12,485.
[5] McLean, J. 2014 Late Twentieth-Century Warming and Variations in Cloud Cover *Atmospheric and Climate Sci.* 4 pp. 727–742.
[6] Chavez, F. P., Ryan, J., Lluch-Cota, S.E., Niquen, C. M. 2003 From Anchovies to Sardines and Back: Multidecadal Change in the Pacific Ocean *Science* 299 pp. 217–221.
[7] Tian, Y., Kidokoro, H., Watanabe, T., Iguchi, N. 2008 The late 1980s regime shift in the ecosystem of Tsushima warm current in the Japan/East Sea: Evidence from historical data and possible mechanisms *Progress in Oceanography* 77 pp. 127-145.
[8] Beaugrand, G., Reid, P.C. 2003 Long-term changes in phytoplankton, zooplankton and salmon related to climate *Global Change Biology* 9 pp. 801–817.
[9] Reid, P. C., de Ftima Borges, M., Svendsen, E. 2001 A regime shift in the North Sea circa 1988 linked to changes in the North Sea horse mackerel fishery *Fisheries Research* 50(1/2) pp. 163–171.
[10] Flint, P.L. 2013 Changes in size and trends of North American sea duck populations associated with North Pacific oceanic regime shifts *Mar. Biol.* 160 pp. 59-65.
[11] Lean, J. L., Rind, D. H. 2008 How natural and anthropogenic influences alter global and regional surface temperatures: 1889 to 2006 *Geophys. Res. Lett.* 35 L18701.
[12] Douglass, D. H. 2011 The Pacific sea surface temperature *Physics Letters* A 376(2) pp. 128–135.
[13] Belolipetsky, P. V. 2014 The Shifts Hypothesis - an alternative view of global climate change. *Preprint.* http://arxiv.org/ftp/arxiv/papers/1406/1406.5805.pdf.
[14] Belolipetsky, P., Bartsev, S., Ivanova, Y., Saltykov, M. 2015 Hidden Staircase Signal in Recent Climate Dynamic *Asia-Pac. J. Atmos. Sci.* 51(4) pp. 323–330.
[15] Deser, C., Alexander, M. A., Xie, S-P., Phillips, A. S. 2010 Sea surface temperature variability: Patterns and mechanisms *Ann. Rev. Mar. Sci.* 2 pp. 115-143.
[16] Jolliffe I.T. 2002 Principal Component Analysis, *Springer-Verlag, New York, Inc. (2nd edition)*, 487 p.
[17] Robock, A. 2000 Volcanic eruptions and climate *Rev. Geophys.* 38 pp. 191–219.

[18] Trenberth, K. E., Hurrell, J. W. 1994 Decadal atmosphere-ocean variations in the Pacific *Clim. Dynam.* 9 pp. 303–319.
[19] Jin, F.-F. 1997 An Equatorial Ocean Recharge Paradigm for ENSO. Part I: Conceptual Model. *Journal of the Atmospheric Sciences* 54 pp. 811–829.
[20] Cessi, P. 1994 A simple box model of stochastically forced thermohaline flow *Journal of Physical Oceanography* 24 pp. 1911–1920.
[21] Marotzke, J. 2000 Abrupt climate change and thermohaline circulation: Mechanisms and predictability *PNAS* 97(4) pp. 1347-1350.
[22] Brovkin, V. 2002 Climate-vegetation interaction *J. Phys. IV France* 12 pp. 52–57.
[23] Abbota, D. S., Tzipermana, E. 2008 A high-latitude convective cloud feedback and equable climates *Q. J. R. Meteorol. Soc.* 134 pp. 165-185.
[24] DAndrea, F., Provenzale, A., Vautard, R., De Noblet-Decoudre, N. 2006 Hot and cool summers: Multiple equilibria of the continental water cycle *Geophysical Research Letters* 33 L24807.
[25] Sessions, S. L., Sugaya, S., Raymond, D. J., Sobel, A. H. 2010 Multiple equilibria in a cloud resolving model using the weak temperature gradient approximation *Journal Of Geophysical Research* 115 D12110.
[26] Vallis, G. K. 1986 El Nino: a chaotic dynamical system? *Science* 232(4747) pp. 243-245.
[27] Garay, B. M. 2015 Indig B. Chaos in Vallis asymmetric Lorenz model for El Nino *Chaos, Solitons and Fractals* 75 pp. 253-262.
[28] Freire, J. G., Bonatto, C., DaCamara, C. C., Gallas, J. A. C. 2008 Multistability, phase diagrams, and intransitivity in the Lorenz-84 low-order atmospheric circulation model *Chaos* 18 033121.
[29] Lu, J., Zhou, T., Zhang, S. 2002 Chaos synchronization between linearly coupled chaotic systems *Chaos, Solitons and Fractals* 14 pp. 529-541.
[30] Palmer, T. N., Doblas-Reyes, F. J., Hagedorn, R., Weisheimer A. 2005 Probabilistic prediction of climate using multi-model ensembles: from basics to applications *Phil. Trans. R. Soc. B* 360 pp. 1991-1998.
[31] Vadivasova, T. E., Sosnovtseva, O. V., Balanov, A. G., Astakhov V. V. 2000 Phase Multistability of Synchronous Chaotic Oscillations Discrete *Dynamics in Nature and Society* 4 pp. 231–243.

IOP Conf. Series: Materials Science and Engineering **173** (2017) 012006 doi:10.1088/1757-899X/173/1/012006

Structural model constructing for optical handwritten character recognition

P A Khaustov, V G Spitsyn and E I Maksimova

Department of computer engineering, Tomsk Polytechnic University, 30, Lenin Avenue, Tomsk, 634050, Russian Federation

E-mail: khaustovpavel@gmail.com

Abstract. The article is devoted to the development of the algorithms for optical handwritten character recognition based on the structural models constructing. The main advantage of these algorithms is the low requirement regarding the number of reference images. The one-pass approach to a thinning of the binary character representation has been proposed. This approach is based on the joint use of Zhang-Suen and Wu-Tsai algorithms. The effectiveness of the proposed approach is confirmed by the results of the experiments. The article includes the detailed description of the structural model constructing algorithm's steps. The proposed algorithm has been implemented in character processing application and has been approved on MNIST handwriting characters database. Algorithms that could be used in case of limited reference images number were used for the comparison.

1. Introduction

The optical handwritten character recognition is a technique that is widely used as a form of information entry from printed paper data records. It is applied in different areas such as medicine, education and sociology for processing questionnaires, task forms, case-records and other types of documents which can be processed automatically [1, 2]. A variety of handwriting makes the problem of the handwritten data recognition singificantly more sophisticated.

The most part of approaches to handwritten characters recognition are based on the features extraction. In these approaches the input image is represented as a feature vector that is used for classification, but this representation is inconvenient to keep track of the classification logic. Such logic monitoring may be required for an assessment of the similarity degree with the reference character images. In turn, this assessment can be used during the preparation for the Unified State Exam (USE). As known, the organizers of the Unified State Exam are often notice the problem of inability of Russian schoolchildren to properly fill the exam forms, according to the offered examples of characters representation. Schoolchildren often neglect the provided rules of answer forms filling and given examples.

That way, one can suppose that an implemention of the optical handwritten character recognition algorithm supporting the writing characters quality assessment in the answer sheets of USE is a relevant task. The main difficulties in the developing process of such algorithm include the ambiguity of writing characters methods, the similar ambiguity of selecting the way for characters similarity estimation, as well as the need for clarifying for the test-taker the difference between his way of

IOP Conf. Series: Materials Science and Engineering **173** (2017) 012006 doi:10.1088/1757-899X/173/1/012006

character representation and the reference one.

A usage of approaches based on features extraction and fuzzy classifiers is not appropriate for the described problem because of the requirement regarding an explanation of the characters similarity. The methods based on a structural components extraction can be used for optical character recognition with classification logic tracking. In these approaches there are no inconvenient for the human perception features, instead, the components of the character image are representated in form of geometric primitives in conjunction with the topological model. Such combination of topological model and structural geometric components is called a structural model. Also, these approaches are handy in case of a small reference images number that is insufficient to train or tune the classifier.

Researchers have made numerous attempts to develop a convenient model for both optical character recognition and classification logic tracking. One of the earliest attempts of Shaw *et al* [3] was to represent a character in form of PDL-expressions. These expressions allow to specify only a small part of the geometrical representations of certain character. Many characters can be defined in many different PDL-expressions.

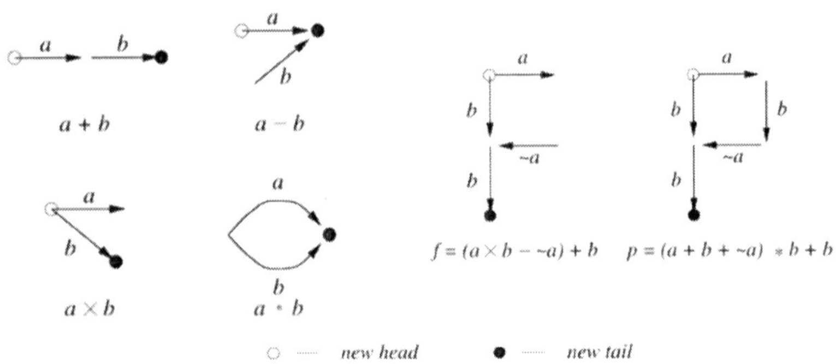

Figure 1. An example of PDL-expressions.

Feder *et al* [4] proposed a usage of PLEX-grammar for character structure representation. PLEX-grammar is based on the use of n-connected nodes (napes) and three types of graphics primitives.

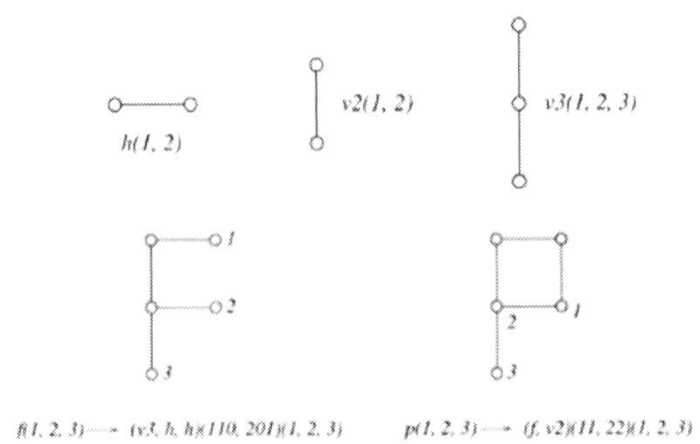

Figure 2. An example of PLEX-model.

There were some other attempts to describe a character representation in a simple way. One of them is Berthod and Maroy's encoding scheme [5]. This scheme is based on the use of the five graphic primitives to describe the character image:

- straight line – T;
- minus (counter-clockwise) curve – M;
- negative (clockwise) curve – P;

- pen-lift – L;
- cusp – R.

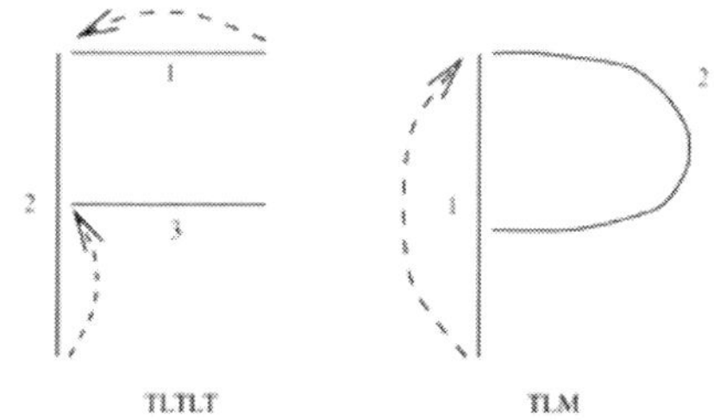

Figure 3. An example of character encoded with Berthod and Maroy's encoding scheme.

The most significant shortcoming of the described model constructing methods is an ability to construct mutliple models for the same character image. Also, multiple character representations may correspond to the same model. The numerous chain codes were proposed to reduce the ambiguity [6, 7].

Due to the ever-increasing power of computers methods based on character model constructing were replaced by the methods based on the artificial neural networks [8], support vector machines [9] and other classifiers [10]. But the same increasing of computing power leads to the new opportunities for models constructing. Now it is possible to obtain more detailed model of character's representation and to use it for the high-performance comparison of two models with respect to both topology and geometry of these models.

In this paper we present a novel approach to optical handwritten character recognition based on constructing a structural model that consists of key points, bends, edges, composite edges and a planar topological model. Also, we propose three different algorithms for structural models comparison.

2. Preprocessing

Firstly, to separate character's pixels from the background a binarization of the original image is performed. To obtain the information about the character's topology it is necessary to perform a skeletonization (thinning) of its graphical representation. The process of skeletonization reduces the variety of graphical representations corresponding to one character.

Since all well-known skeletonization algorithms with high performance have the certain drawbacks, we proposed to use two well-known algorithms one after the other: the Zhang-Suen thinning algorithm [11] and the Wu-Tsai algorithm [12].

During the testing we noted that there are some parts of character representation that are not fully skeletonized after the Zhang-Suen algorithm processing. This is due to the fact that the algorithm aims to preserve the initial topology of the object. During the structural components extracting the cases when there is more than one shortest path between two pixels of not fully skeletonized representation usually happen. Such ambiguity does not allow a proper extracting of the geometric characteristics of the connection between the two key points, and, as a result, the information about the structural components is lost. To avoid of the described problem we suggested to use the Wu-Tsai algorithm in order to eliminate the areas where skeletonization is not fully performed.

It should be noted that the Wu-Tsai algorithm quite often removes small elements belonged to the graphical representation of the character (important in terms of topology inference from a graphical representation). Thus, the usage of a single skeletonization algorithm leads to either a loss of information about the topology of a character graphical representation or to an ambiguity while attempting to retrieve the structural components. In turn, using the Zhang-Suen algorithm, and then the Wu-Tsai algorithm resolves both types of drawbacks.

3. Structural components extraction

To obtain the character's topological model the image obtained after the skeletonization should be processed with multiple runs of Lee algorithm [13], which is, basically, the breadth-first search algorithm for eight-connectivity components. All pixels corresponding to the character graphical representation are considered as vertices.

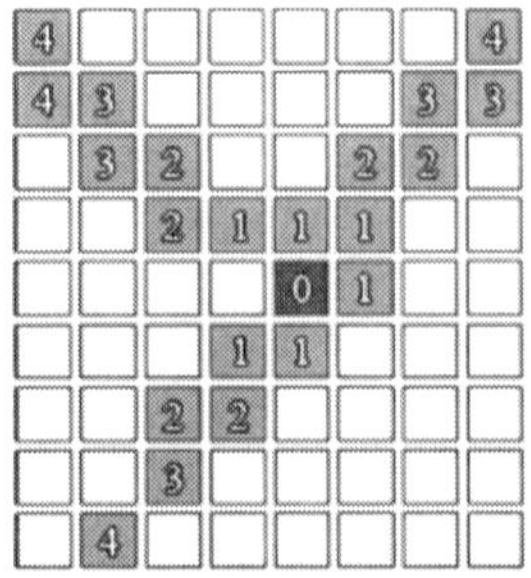

Figure 4. The visualization of the Lee algorithm process.

Then one can select the key points on the skeletonized image as shown in Fig. 4. These points will be considered as the vertices of the graph that corresponds to the final character model. The remaining pixels will be applied for constructing the edges. At first glance, the key points can be attributed only to the pixels, which can be determined based on the positions of black neighboring pixels. However, there are some pixels that cannot be referred to the key points but the connecting edge passing through them significantly changes its direction at such pixels. We will use the term bend for denoting such points. Thus, the most important pixels are divided into two groups:

- pixels, that are the key points of the character's graphical representation (the points where the drawing tool has been lifted, lowered or the direction of its path has been changed);
- pixels, that are related to the bend points, but do not belong to the key points (such points are used to define the shape of connecting edges between the key pixels). Their search can be performed using an iterative algorithm.

Initially we start with an assumption that the key points and bends are indistinguishable. So we need to find the key points candidates. During the experiments we have identified three groups of templates corresponding to the key points candidates: α, β and γ. These templates are shown in Fig. 5. Pixels from groups β and γ have some additional conditions to be considered as the corresponding to the template. Such conditions are shown in the right part of the Fig. 5.

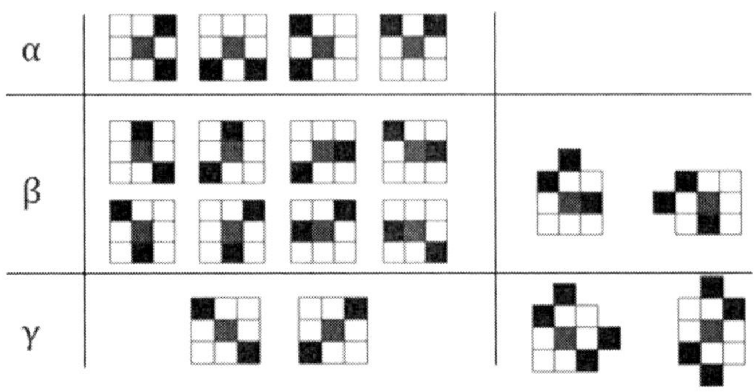

Figure 5. The groups of templates and the corresponding conditions for the key point candidates.

We start iteratively run Lee algorithm to detect the edges connecting the key points of the structural model. We run the bread-first-search waves simultaneously from all the key points which are found at the moment. If two waves from different key points met at some point of the image, we assume that these key points are connected with the edge. One can easily restore the path and, as a result, a direction and a curvature of the corresponding edge. After the current run of the Lee algorithm is finished, all the key points that have only two incident edges with the angle of no less than 120 degrees between them, are considered as bends. If at least one such key point detected, the next run of the Lee algorithm is performed.

The complexity of the algorithm is determined by the complexity of each its iteration. Each iteration includes a separate running of the Lee algorithm that is the breadth-first search algorithm for eight-connectivity components. The complexity of this algorithm is $O(H{\cdot}W)$, where W and H are the image width and height respectively. Moreover, only black pixels after the performed skeletonization will be considered during the image processing. The number of such pixels for the real graphical representation of characters is much less than the total number of image pixels – $W{\cdot}H$, so the described complexity estimation is unreachable. The number of iterations linearly depends on the number of bends B in the final representation of the model. The value of B, in turn, rarely exceeds the value of 15 for handwriting characters of Russian and English alphabets. Thus, the total complexity of the algorithm is $O(B{\cdot}H{\cdot}W)$.

As one can notice, any pixel of the image, except for the key pixels and bends, corresponds to the connecting element between the key points. So, the graphical image of a character can be represented as a planar graph.

We introduce a term of composite edge. A composite edge is a sequence of edges with the bends between them. An edge – is the sequence of pixels forming a path between two pixels, each of which is either a key point or a bend.

4. Structural model constructing

To simplify the graph structure it is necessary to approximate each composition of black pixels between the two bend points by some geometric primitive. The segment or the pair of segments composing the edge is the trivial case of such primitive. However, the cases are possible when the edge cannot be represented as one or two line segments, due to the high curvature of line which connects the considered points. In this case, it is reasonable to approximate the edge with an arc. In such approximation the radius and the position of the circle arc should be computed according to the coordinates of the key points and the curvature of the connecting line. When the curvature degree of the connecting line corresponds to the large circle radius, the larger part of a circle can violate the image bounds, strongly distorting the topological model of the character's graphical representation. So, it is reasonable to approximate such curve by the elliptical arc as it shown in Fig. 6.

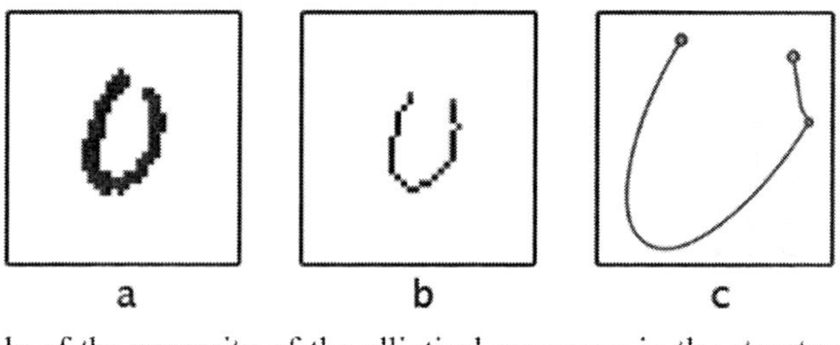

Figure 6. An example of the necessity of the elliptical arc usage in the structural model (c) and the corresponding input image (a) and skeletonized image (b).

The structural model can be represented in the form of a planar graph whose vertices correspond to the key pixels of the character's graphical representation, and the edges – to the sequence of lines and curves connecting these key pixels. Each vertex is described by its position. Each edge, in turn, is described by the identifiers of connected vertices and by its curvature. Moreover, the segments of the connecting edges can be approximated both by line and by circle or ellipse arc. An example of the structural model is shown in Fig. 7.

After being constructed the obtained structural models can be stored on a disk, for instance, in the XML format.

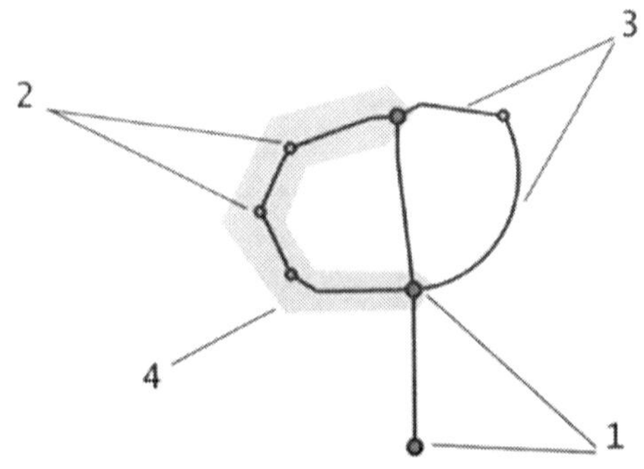

Figure 7. An example of the necessity of the elliptical arc usage in the structural model (c) and the corresponding input image (a) and skeletonized image (b).

5. Structural models comparison

In this paper we propose three algorithms of structural models comparison.

The first one is the algorithm based on the intersection method. This method is a typical intersection method used in assumption that all the graphic primitives of the structural model are considered as a separate planar object.

The second one is the algorithm based on the maximum matching of the minimum weight computation. The example of similarity assessment for two lines is the area of geometric shape between them: the smaller the area – the more similar these lines are. For each pair of edges of the first and the second topological models the numeric value of this area should be computed. Then we can create a bipartite graph. The vertices of the first partite set will correspond to the edges of the first topological model and the vertices of the second partite set – to the edges of second topological model. Each of the edges of obtained graph will correspond to a particular pair of connecting lines of two compared topological models. The weight of an edge in this case will be expressed as the area enclosed between two curves corresponding to this edge. It is required to find the maximum matching of the minimum weight for this graph. Such problem can be solved by an algorithm similar to the of Edmonds-Karp algorithm for searching of maximum flow. The computed weight of matching can be used as a quantitative measure of similarity between two topological models and, as the result, of characters corresponding to these models.

The third algorithm for structural models comparison is based on the probabilistic test to assess the degree of structural models similarity. In this test we represent the structural model as a track for some point. There are two types of actions:

- move a point along the edge to a distance d in some direction v;
- move a point to the nearest key point of the model.

After performing some random set of actions a distance between the resulting point locations is used as measure of similarity.

6. Results

The algorithm was implemented as a console application on C++ programming language using the tools of Open Source Computer Vision Library OpenCV. The training set for an artificial neural network has been generated from a database of CT scans of human lungs provided by the medical establishment. Its images are grouped by the type of disease: cancer, tuberculosis, sarcoidosis and all the others. For each of these four groups the recognition quality has been estimated.

As it was previously mentioned, the proposed method is appropriate in the case when the number of reference images for each of the classes is small.

When the number of reference images is large, the proposed algorithms will have a low performance due to the linear dependence the algorithm's time consumption on the reference images number.

The assessment of the proposed recognition approaches is being performed using the wide-known MNIST dataset of handwritten characters (70000 characters of 10 classes).

Our assessment includes the comparison of the three proposed optical character recognition algorithms based on a structural models constructing:

- method based on the maximum matching of the minimum weight computation (*SMM*);
- intersection method for a structural models (*SMI*);
- probabilistic test to assess the degree of structural models similarity (*SMT*).

Also, to estimate the quality of the optical character recognition, the proposed methods were compared with analogues which can operate with a small number of reference images:

- intersection method for bitmaps [14] (*IM*);
- algorithm based on the support vector machine usage (*SVM*);
- method based on using the probabilistic neural network (*PNN*);
- histogram method (*HM*);
- histogram method with in-out and out-in profiles (*HMP*).

A variable number of reference images E has been used for training or tuning in each of the experiments. Experiments were carried out for odd E values from 3 to 15.

The results of the experiments on MNIST dataset are given in the Table 1.

Table 1. The percentage of correctly recognized characters of MNIST dataset for each of the compared algorithms.

E	SMI	SMM	SMT	IM	SVM	PNN	HM	HMP
3	89.1	**93.2**	78.1	87.4	83.8	70.9	71.5	74.1
5	90.8	**95.1**	79.4	88.7	85.5	73.2	72.2	74.3
7	91.2	**95.1**	80.0	88.9	87.0	74.2	72.6	74.8
9	91.3	**95.2**	80.0	90.4	88.2	74.7	72.8	75.0
11	91.3	**95.2**	80.2	90.4	88.5	75.3	72.9	75.1
13	91.3	**95.3**	80.2	90.6	88.6	75.5	73.0	75.1
15	91.4	**95.3**	80.3	90.7	88.6	75.8	73.1	75.2

As one can see from the results, the algorithm based on the maximum matching of the minimum weight for the given values of E performs better than the others. An advantage of this method is the most significant when the value of E is equal to 5.

Also, it should be noted that the algorithm based on a probabilistic test turned to be insufficiently effective and excelled in quality of recognition only the PNN and methods based on the histograms comparison.

In case of increasing the reference images number to a few thousands the recognition quality of the proposed algorithms improves insignificantly, while their performance becomes much worse. At the same time, when increasing the number of reference images to a few thousand the recognition quality of algorithms based on the support vector machine and PNN significantly improves.

To determine the classes which characters were incorrectly recognized more often than others, and to find the classes to which they have been wrongly classified, we can construct a confusion matrix for the high-efficiency algorithm based on the maximum matching of the minimum weight. The Table 2 contains the experimental data for the reference image number $E = 15$.

The diagonal elements of the matrix are expressed quite clearly. It approved that for each character class the proposed algorithm frequently classifies the corresponding image correctly. Nevertheless, there are pairs of classes, whose structural models are similar enough that, together with the imperfections of the proposed approach, leads to misclassification.

A similar confusion matrix is shown in Table 3 for the proposed algorithm based on the intersections method.

Despite the fact that the number of incorrect classifications is almost doubled (6021 compared to 3288), some pairs of classes became more distinguishable. For instance, the proposed algorithm based on the intersections method significantly better distinguishes classes that correspond to the numbers '*3*' and '*9*'. However some classes became much less distinguishable. Those are, for example, classes corresponding to the digits '*0*', '*6*' and '*9*'.

Table 2. Confusion matrix for the results of the algorithm based the maximum matching computation for the MNIST dataset.

	0	**1**	**2**	**3**	**4**	**5**	**6**	**7**	**8**	**9**
0	**6239**	0	0	0	0	0	379	0	0	285
1	0	**7775**	2	0	2	0	0	98	0	0
2	0	1	**6896**	2	0	0	0	91	0	0
3	0	0	1	**6785**	0	0	0	0	210	145
4	0	11	0	0	**6779**	0	0	12	0	22
5	0	0	0	0	0	**6284**	25	0	0	4
6	317	0	0	0	0	41	**6288**	0	230	0
7	0	115	46	0	3	0	0	**7129**	0	0
8	0	0	0	184	0	0	247	0	**6233**	161
9	285	0	0	154	7	2	0	0	206	**6304**

Table 3. Confusion matrix for the results of the algorithm based on the intersections method for the MNIST dataset.

	0	**1**	**2**	**3**	**4**	**5**	**6**	**7**	**8**	**9**
0	**5628**	0	0	0	0	0	703	0	0	572
1	0	**7730**	1	0	1	0	0	145	0	0
2	0	0	**6912**	2	0	0	0	76	0	0
3	0	0	1	**6916**	0	0	0	0	210	14
4	0	4	0	0	**6790**	0	0	10	0	20
5	0	0	0	0	0	**6242**	15	0	0	56
6	732	0	0	0	0	31	**5740**	0	162	211
7	0	244	37	0	5	0	0	**7007**	0	0
8	0	0	0	184	0	0	456	0	**5462**	723
9	611	0	0	23	6	24	330	0	412	**5552**

7. Conclusion

In this paper the algorithm for optical character recognition has been proposed. This method is based on structural components extraction and has no requirements to the number of reference images.

The proposed approach outperforms its analogues in conditions of a small number of reference images. Another advantage of our method is its high performance as well as low memory consumption.

The further research will be aimed at finding new ways of structural models comparison.

References

[1] Wang T, Wu Dj, Coates A and Ng Ay 2012 End-to-End Text Recognition with Convolutional Neural Networks *Proc. of the Twenty-First Int. Conf. on Pattern Recognition* 3304-3308

[2] Jaderberg M, Simonyan K, Vedaldi A, Zisserman A 2016 Reading Text in the Wild with Convolutional Neural Networks *Int. Journal of Computer Vision* **116(1)** 1-20

[3] Shaw A C 1969 A formal picture description scheme as a basis for picture processing system *Inform. Control* **14** 9-52

[4] Feder J 1971 Plex Languages *Inform Sci.* **3** 225-241

[5] Berthod M, Maroy J P 1979 Learning in syntactic recognition of symbols drawn on a graphic tablet *Comput. Graphics Image Process* **9** 166-182

[6] Freeman H 1961 On the encoding of arbitrary geometric configurations *IRE Transactions on Electronic Computers* **10** 260-268

[7] Bribiesca E 1999 A new chain code *Pattern Recognition* **32** 235-251.

[8] Mehdi L, Solimani A and Dargazany A 2009 Combining wavelet transforms and neural networks for image classification *41st Southeasten Symposium on System Theory, Tullahoma, USA* 44-48

[9] Vapnik V N and Cortes C 1995 Support Vector Networks *Machine Learning* **20(3)** 273-297

[10] Breuel T M, Ul-Hasan A, Azawi M Al and Shafait F 2013 High-Performance OCR for Printed English and Fraktur using LSTM Networks *Proc. of 12th Int. Conf. on Document Analysis and Recognition* 683-687

[11] Zhang T Y, Suen C Y 1984 A Fast, Parallel algorithm for thinning digital patterns *Image Processing and Computer Vision* **3** 236-239

[12] Wu R-Y, Tsai W-H 1992 A new one-pass parallel thinning algorithm for binary images *Pattern Recognition Letters* **13** 715-723

[13] Lee C Y 1961 An Algorithm for Path Connections and Its Applications *IRE Transactions on Electronic Computers* **10(2)** 346-365

[14] Khaustov P A, Spitsyn V G and Maksimova E I 2014 Geneticheskij algoritm poiska mnozhestva krivyh dlja opticheskogo raspoznavanija simvolov s ispol'zovaniem metoda peresechenij [In Russian] *Sovremennye problemy nauki i obrazovanija* **6** 1-7

Applying an instance selection method to an evolutionary neural classifier design

Dmitrii Khritonenko, Vladimir Stanovov, Eugene Semenkin

Siberian State Aerospace University, Krasnoyarsk, Russian Federation, 660037, Krasnoyarsky Rabochy Av., 31

E-mail: HDmitry.91@mail.ru

Abstract. In this paper the application of an instance selection algorithm to the design of a neural classifier is considered. A number of existing instance selection methods are presented. A new wrapper-method, whose main difference compared to other approaches is an iterative procedure for selecting training subsets from the dataset, is described. The approach is based on using training subsample selection probabilities for every instance. The value of these probabilities depends on the classification success for each measurement. An evolutionary algorithm for the design of a neural classifier is presented, which was used to test the efficiency of the presented approach. The described approach has been implemented and tested on a set of classification problems. The testing has shown that the presented algorithm allows the computational complexity to be decreased and the quality of the obtained classifiers to be increased. Compared to analogues found in scientific literature, it was shown that the presented algorithm is an effective tool for classification problem solving.

1. Introduction

Modern developments in information technologies allow huge amounts of data from different areas of human activity to be saved. This can include medical diagnostic data, descriptions of technological processes, the tracking and predicting of economic indexes, text and image recognition, among other types. Using this data allows us to discover unknown and non-trivial dependencies of the real world, and solve classification, clusterisation, prediction and regression problems. The number of measurements in these problems may reach hundreds of thousands and millions, which impose limitations on the set of intelligent data analysis technologies. The instances present in the data could be noisy, may duplicate each other, or contain a lot of useless information. The training instance selection technologies are capable of partially solving these problems. The approach described in this paper refers to wrapper-class and is capable of decreasing the computational complexity of training algorithms, partially eliminating the effect of over-fitting for intelligent data analysis, and increasing the accuracy of the designed classifiers by using an iterative procedure for redistributing instance selection probabilities. As a base model, on which the efficiency of the proposed algorithm was tested, a distributed program system of generating automated neural networks with evolutionary algorithms was chosen [1]. The choice is justified by the fact that the program system has shown its efficiency on a set of test and real problems [2, 3], however, its application in actual problems is not possible in an adequate length of time. The development of a simple and efficient instance selection procedure is a

IOP Conf. Series: Materials Science and Engineering **173** (2017) 012007 doi:10.1088/1757-899X/173/1/012007

topical issue today, and it will allow the effectiveness of using neural technologies in data analysis problems to be increased.

The rest of the paper is organized as follows: Section 2 contains a brief description of the instance selection method used. In Section 3 we describe a neural classier design with the use of evolutionary algorithms. In Section 4 we present the results obtained with the developed methods, compare them to analogues and provide a short discussion. The last section is devoted to conclusions.

2. Instance selection method

Most of the existing instance selection methods are focused not only on data reduction, but also on increasing classification quality [4, 5, 6, 7]. However, these methods usually create a subsample only once, i.e. the goal is to create a subsample, and not to build a high-quality classifier.

The proposed IS method should be used with classifiers which require lots of iterations during the learning process. The main idea that stimulated the development was to create a method, allowing not only the time required to train a classifier to be decreased, but also the classification accuracy to be increased. The method used here has already been used in [8, 9] to train evolutionary fuzzy classifiers, which is why we will give only a short description here.

The method consists of several steps. Firstly, a probability value p_i and a counter U_i are assigned to each instance in the training set $i = 1 \dots N$, where N is the number of instances. The probability values are equal and represent the chance that an instance i would be selected into the training subset, i.e. $p_i = p_j, i, j = 1 \dots N$. The counter values U_i are set to 1. Secondly, the probability values are used to select the instances and start the training process on the subsample. The size of the subsample is fixed and it is calculated with respect to the training sample size (in per cent). Thirdly, the training process continues for a certain number of generations, called the adaptation period. Finally, after the end of the adaptation period, the best individual in the population is used to change the U_i values: if a training example has been classified correctly, then $U_i = U_i + 1$, otherwise $U_i = 1$. U_i values are changed only for instances which are present in the subsample. Before creating a new subsample, the probabilities are updated using the following equation:

$$p_i = \frac{1/U_i}{\sum_{j=1,n} 1/U_j} \tag{1}$$

Increasing the counter leads to a lower probability being chosen, while low counter values force the algorithm to choose a certain instance more often.

This procedure makes the algorithm adaptive, following two main strategies: exploration and exploitation. Instances which have not been selected before, have a counter value of 1, meaning that they have a quite large probability of being chosen, implementing the exploration of new areas of the feature space. Instances which have been chosen several times and have received high counter values get lower probabilities, because they appear to be easier to classify, i.e. they are not so interesting for the training process. However, instances which have been classified incorrectly, get their counters refreshed to $U_i = 1$, making them look like undiscovered instances. This implements the exploitation strategy, i.e. the algorithm uses the information about the classification quality to select the instances of larger interest. At the same time, if one of the instances, which have been classified correctly many times before, gets misclassified with a new model, it gets a higher probability of being chosen in the following adaptation periods. This allows the algorithm to focus not only on instances that are hard to classify, but also to keep the other instances classified correctly as well.

As we are going to use an evolutionary approach with a set of individuals in a population, the best solution for the subsample might not be the best solution for the whole sample. However, the population may still contain a solution with high generalization ability, despite not being the best when checked on the subsample. That is why, after each adaptation period, every individual in the population is checked on the whole training sample, and the best solution for the training sample is remembered, and included in the subsample, as well as the best solution for the subsample.

3. Evolutionary designing of neural classifiers

The problem concerning the automated designing of artificial neural networks is not new, and there is more than one way of solving this problem. One of the approaches developed is the use of evolutionary algorithms [1]. Evolutionary algorithms are an optimization tool in arbitrary search spaces and they have proved their efficiency on a set of real problems. To solve the problem of the evolutionary design of neural classifiers, a genetic programming algorithm (GP) with several modifications is used is this work. The training is performed with different single-criterion unconstrained optimization methods (backpropagation, genetic algorithms, CMA-ES [10] and others).

To start the genetic programming algorithm, it is required to fill the terminal and functional set. As terminal set elements, different types of artificial neurons are used (8 types totally), attributes and sets of attributes of the problem solved, and some constants. The functional set consists of elements joining in one layer operator (plus sign), and operators responsible for layer interconnections (shown in arrows at Fig. 1).

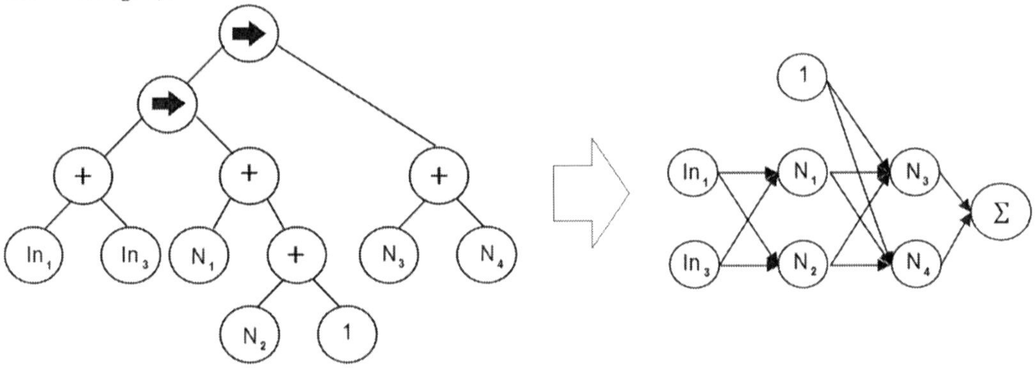

Figure 1. Example of ANN coding in GP.

Where N_k is a neuron with the k-th number, In_i is an attribute of the problem with the number i. It should be mentioned that for this encoding, the insularity condition could be violated. That is why a special procedure for finding and eliminating the violations by changing the operators was used.

After determining the required sets, it becomes possible to automatically design neural networks. However, it is important to remember that the efficiency of all evolutionary algorithms depends on the set of parameters. Because of this, a modification called self-configuration was used [3]. It allows a set of parameters to be refused, the time required for the algorithm to work to be decreased, and the accuracy of the models to be increased. A more general scheme of adjusting the parameters of evolutionary algorithms is presented in [11]. The algorithm configurations are considered, as well as different numerical parameters. The approaches implemented today for single-criterion unconstrained optimization has shown positive results and will soon be transferred to the described algorithm. Together with self-configuration, the uniform crossover scheme has been added to the algorithm [12]. This procedure allows the accuracy of the models obtained to be increased, and it is an additional way of controlling the complexity of the received neural nets.

4. Testing results

To test the efficiency of the described instance selection method, a self-configuring evolutionary algorithm for the automated design of artificial neural networks (ANNEA) has been implemented in a program. A set of computational experiments has been performed on problems taken from the machine learning repositories KEEL [13] and UCI [14].

The following parameters were used for testing the algorithm presented in section 2: number of individuals: 100, number of generations: 10000, subsample size from 5 to 50 percent (with 5% step), adaptation period from 25 to 500 generations (with a 25 generation step). The program system, using the described instance selection algorithm (ANNEA_IS) was compared to its basic version (ANNEA), as well as to analogues found in scientific literature.

Table 1. Numerical characteristics of problems.

Dataset	Number of instances	Number of attributes	Number of classes
Magic	19020	10	2
Page-Blocks	5472	10	5
Australian credit	690	14	2
German credit	1000	20	2
Texture	5500	40	11
Twonorm	7400	20	2
Segment	2310	19	7
Ring	7400	20	2
Penbased	10992	16	10
Satimage	6435	36	6

As an efficiency criterion, the classification error, averaged over all runs, has been used. A run means using stratified 10-fold cross-validation. The total number of runs is 10. In the table below, the testing results for the base algorithm, as well as its modified version, are presented.

Table 2. Testing results.

Dataset	ANNEA			ANNEA_IS		
	Training error, %	Test error, %	Time, min	Training error, %	Test error, %	Time, min
Magic	14.83	15.08	734	13.75	**14.72**	229
Page-Blocks	3.72	**4.23**	180	3.96	**4.23**	51
Australian credit	8.71	10.99	67	10.07	**9.97**	19
German credit	19.56	23.76	129	18.91	**21.70**	47
Texture	6.93	8.05	805	5.90	**7.27**	146
Twonorm	5.21	5.97	481	3.67	**4.24**	131
Segment	5.89	6.14	312	4.69	**5.10**	81
Ring	6.37	6.54	454	4.78	**4.99**	97
Penbased	5.36	7.91	470	3.30	**3.55**	137
Satimage	12.90	14.09	691	12.45	**13.82**	115

It should be noted that applying instance selection for the process of forming neural classifiers with evolutionary algorithms resulted in an increase in the classification accuracy and a decrease in the time required. For the worst case (Page-Blocks dataset), there is no statistical difference for the criteria used (in the sense of the Wilcoxon criterion). On average, the modelling accuracy increases by 14%, and the time required decreases by 1.75-5 times. Let us present the best parameters for the adaptation period and the sizes of the training subsamples for every problem used:

As can be seen from the table, the sizes of the subsamples do not reach the maximal size of 50%. This could be connected to the fact that for efficient learning the algorithm requires around 30% of the total size on average. Further increasing the subsample size definitely increases the model accuracy on the training sample, but leads to over-fitting, and, as a consequence, loss of generalization ability and test sample accuracy. Also it should be noted that for re-training on the new data, the algorithm requires a small adaptation period. A small adaptation period with a fixed number of generations

IOP Conf. Series: Materials Science and Engineering **173** (2017) 012007 doi:10.1088/1757-899X/173/1/012007

allows more training sample rotations to be made and a better description of the problematic areas by the training subset to be provided. This feature allows the algorithm to learn the dataset better.

Table 3. Best algorithm parameters.

Dataset	Subsample size, %	Adaptation period, generations
Magic	35	175
Page-Blocks	25	200
Australian credit	40	50
German credit	35	75
Texture	25	50
Twonorm	25	50
Segment	20	75
Ring	30	125
Penbased	25	75
Satimage	35	50

The presented data is recommended for further usage with the instance selection algorithm when applied to other data analysis technologies and problems.

To compare the implemented program system with analogues, additional experiments were performed. To ensure objectivity in the comparison, the testing conditions were made maximally similar to each other (Table 4). In the case of the conditions being unknown, increased resources were used (Table 5).

Table 4. Comparison to other methods.

Dataset	ANNEA_IS	HEFCA_IS [9]	GP-Coach [15]	Parallel-Fuzzy GBML [16]	FARC-HD [17]
Magic	**14.72**	15.08	20.18	14.89	15.49
Page-Blocks	4.23	**3.25**	8.77	3.62	4.99
Texture	7.27	**4.45**	-	4.77	7.11
Twonorm	4.24	4.81	15.17	**3.39**	4.72
Segment	**5.10**	5.19	24.04	5.90	-
Ring	**4.99**	**5.08**	-	6.73	5.92
Penbased	3.55	3.81	17.80	**3.07**	3.96
Satimage	13.82	12.93	27.50	15.54	**12.68**

Table 5. Comparison to other methods on Australian and German credit.

Algorithm name	Australian credit	German credit	Algorithm name	Australian credit	German credit
SC_GP	9.78	20.5	Bayesian appr.	15.3	32.1
MGP	10.15	21.25	Boosting	24.0	30.
2SGP	**9.73**	**19.85**	Bagging	15.3	31.6
GP	11.11	21.66	RSM	14.8	32.3
Fuzzy classifier	10.9	20.6	CCEL	13.4	25.4
C4.5	10.14	22.27	CART	12.56	24.35
LR	13.04	21.63	MLP	10.14	23.82
k-NN	28.5	28.49	ANNEA_IS	9.97	21.7

Taking the data presented in tables 2, 4 and 5 into consideration, we may conclude the usefulness of applying the instance selection algorithm with the automated generation of neural classifiers. When compared with the base model, the algorithm has shown an improving effect for all ten problems. On

9 of them the accuracy of the resulting model has been improved, and the computational complexity has been decreased on all ten of them. Compared to analogues, the implemented program system with the described algorithm has shown the best result in 30% of the cases. The listed facts allow us to expect further effective usage of the developed algorithms in modern data analysis problems.

5. Conclusions

In this work, an instance selection algorithm has been described, which can be used in intelligent data analysis technologies. The algorithm implements the wrapper strategy, i.e. uses the data received from a certain model when operating. In this work, the approach was considered when using neural classifiers generated by self-configured evolutionary algorithms. As a result of testing and comparing the implemented program system to analogues it has been shown that the proposed approach results in a significant increase in classification quality and a decrease in computational expenses. On average, the time required for the program system to work was decreased by 2.5 times, and the classification quality has increased by 14%. The presented approach could be used not only for the automated design of neural networks, but also for other intelligent data analysis methods, including ensembles.

Acknowledgments

The reported study was funded by RFBR and Government of Krasnoyarsk Territory according to the research project № 16-41-243064.

References

[1] Khritonenko D, Semenkin E 2013 Distributed self-configuring evolutionary algorithms for artificial neural networks design *Vestnik SibGAU* **4 (50)** pp 113–116
[2] Khritonenko D, Semenkin E, Sugak E, Potilitsina E 2015 Solving the problem of city ecology forecasting with neuro-evolutionary algorithms *Vestnik SibGAU* **16 (1)** pp 137–142
[3] Semenkin E, Semenkina M 2013 Classifier Ensembles Integration with Self-Configuring Genetic Programming Algorithm *Lecture Notes in Computer Science: Adaptive and Natural Computing Algorithms* **7824** pp 60-69
[4] Cano J R, Herrera F, Lozano M 2007 Evolutionary Stratified Training Set Selection for Extracting Classification Rules with trade off Precision-Interpretability *Data & Knowledge Engineering archive* **60 (1)** pp 90-108
[5] Cano J R, Herrera F, Lozano M 2003 Using Evolutionary Algorithms as Instance Selection for Data Reduction in KDD: An Experimental Study *IEEE Transactions on Evolutionary Computation* **7 (6)**
[6] Hart P E 1968 The condensed nearest neighbor rule *IEEE Trans Inf Theory* **14** pp 515–516
[7] Olvera-López J A, Ariel Carrasco-Ochoa J, Francisco Martínez-Trinidad J, Kittler J 2010 A review of instance selection methods *Artif Intell Rev* **34** pp 133–143
[8] Stanovov V, Semenkin E, Semenkina O 2015 Self-configuring hybrid evolutionary algorithm for fuzzy classification with active learning *IEEE Congress on evolutionary computation*
[9] Stanovov V, Semenkin E, Semenkina O 2015 Instance Selection Approach for Self-configuring Hybrid Fuzzy Evolutionary Algorithm for Imbalanced Datasets *LNCS: Sixth International Conference on Swarm Intelligence* **9140 (1)** pp 451–459
[10] Hansen N 2010 Variable Metrics in Evolutionary Computation *Habilitation à diriger des recherches, Université Paris-Sud*
[11] Eiben A, Smit S 2011 Parameter tuning for configuring and analyzing evolutionary algorithms *Swarm and Evolutionary Computation* **1(1)** pp 19–31
[12] Semenkin E, Semenkina M 2012. Self-configuring Genetic Algorithm with Modified Uniform Crossover Operator Lecture *Notes in Computer Science: Advances in Swarm Intelligence* **7331** pp 414-421
[13] Alcalá-Fdez J, Sánchez L, Garcia S, Jesus M J, Ventura S, Garrell J M, Otero J, Romero C, Bacardit J, Rivas V M, Fernández J C, Herrera F 2009 KEEL: A 132 software tool to assess

evolutionary algorithms for data mining problems *Soft Comput* **13(3)** pp 307–318

[14] Asuncion A, Newman D 2007 UCI machine learning repository *University of California, Irvine, School of Information and Computer Sciences* [URL]: http://www.ics.uci.edu/~mlearn/MLRepository.html

[15] Berlanga F J, Rivera A J, del Jesus M J, Herrera F 2010 GP-COACH: Genetic programming-based learning of compact and accurate fuzzy rulebased classification systems for high-dimensional problems *Inf. Sci.* **180(8)** pp 1183–1200

[16] Ishibuchi H, Mihara S, Nojima Y 2013 Parallel Distributed Hybrid Fuzzy GBML Models With Rule Set Migration and Training Data Rotation *IEEE Transactions on fuzzy systems* **21(2)**

[17] Alcala-Fdez J, Alcala R, Herrera F 2011 A fuzzy association rulebased classification model for high-dimensional problems with genetic rule selection and lateral tuning *IEEE Trans. Fuzzy Syst.* **19(5)** pp 857–872

Feature Selection for Natural Language Call Routing Based on Self-Adaptive Genetic Algorithm

A Koromyslova[1], **M Semenkina**[1], **R Sergienko**[2]

1Reshetnev Siberian State Aerospace University, Krasnoyarsky rabochy avenue 31, Krasnoyarsk, Russia, 660037
[2]Ulm University, Lise-Meitner-Str. 9, Ulm, Germany, 89081

E-mail: akoromyslova@mail.ru, semenkina88@mail.ru, roman.sergienko@uni-ulm.de

Abstract: The text classification problem for natural language call routing was considered in the paper. Seven different term weighting methods were applied. As dimensionality reduction methods, the feature selection based on self-adaptive GA is considered. k-NN, linear SVM and ANN were used as classification algorithms. The tasks of the research are the following: perform research of text classification for natural language call routing with different term weighting methods and classification algorithms and investigate the feature selection method based on self-adaptive GA. The numerical results showed that the most effective term weighting is TRR. The most effective classification algorithm is ANN. Feature selection with self-adaptive GA provides improvement of classification effectiveness and significant dimensionality reduction with all term weighting methods and with all classification algorithms.

1. Introduction

Natural language call routing is an important problem in the design of modern automatic call services and the solving of this problem could lead to improvement of the call service [21]. Generally natural language call routing can be considered as two different problems. The first one is speech recognition of calls and the second one is topic categorization of users utterances for further routing. Topic categorization of users utterances can be also useful for multidomain spoken dialogue system design [12]. In this work we treat call routing as an example of a text classification application.

In the vector space model [16] text classification is considered as a machine learning problem. The complexity of text categorization with a vector space model is compounded by the need to extract the numerical data from text information before applying machine learning algorithms. Therefore, text classification consists of two parts: text preprocessing and classification algorithm application using the obtained numerical data. Text preprocessing comprises three stages:

- Textual feature extraction.
- Term weighting
- Dimensionality reduction.

The first one is the textual feature extraction based on raw preprocessing of the documents. This process includes deleting punctuation, transforming capital letters to lowercase, and additional procedures such as stop-words filtering [4] and stemming [14]. Stop-words list contains pronouns, prepositions, articles and other words that usually have no importance for the classification. Using stemming it is possible to join different forms of the same word into one textual feature.

The second stage is the numerical feature extraction based on term weighting. For term weighting we use "bag-of-words" model, in which the word order is ignored. There exist different unsupervised and supervised term weighting methods. The most well-known unsupervised term weighting method is TFIDF [15]. The following supervised term weighting methods are also considered in the paper:

Gain Ratio (GR) [3], Confident Weights (CW) [10], Term Second Moment (TM2) [22], Relevance Frequency (RF) [11], Term Relevance Ratio (TRR) [9], and Novel Term Weighting (NTW) [18]; these methods involve information about the classes of the documents.

As a rule, the dimensionality for text classification problems is high even after stop-words filtering and stemming. Due to the high dimensionality, the classification may be inappropriate time-consuming, especially for real-time systems such as natural language call routing. Therefore, the next stage of preprocessing is the dimensionality reduction based on numerical features; it is possible with feature selection or feature transformation. In our research we use a feature selection method based on genetic algorithm or GA-based wrapper.

One of the most complicated problems with GA applications is setting algorithm parameters. A conventional genetic algorithm has at least three methods of selection (proportional, tournament, and rank), three methods of recombination (one-point, two-point, and uniform). Mutation probability requires tuning as well. The amount of various combinations can be estimated at tens. Exhaustive search of combinations requires a lot of time and computational power, especially for such time-consuming problems as GA applications for machine learning. Parameters combination selection at random can be also insufficient as algorithm efficiency on same problem can differ very much for various parameters setting. This problem can be solved with self-configuring GA [17] or co-evolutionary GA [19]. Therefore, we propose a use of self- adaptive GA for the feature selection in the field of natural language call routing.

As classification algorithms we use the k-NN algorithm, the linear SVM, and artificial neural networks (ANN). Some comparative studies of machine learning algorithms in the field of text classification showed high classification effectiveness of k-NN, SVM-based algorithms, and ANN [2, 7, 8, 10, 13].

The tasks of our research are the following:

- Perform research of text classification for natural language call routing with different term weighting methods and classification algorithms.
- Investigate the feature selection method based on self-adaptive GA.

The paper is organized as follows: In Section 2, we describe the considered corpus for natural language call routing. Section 3 describes the considered term weighting methods. The GA-based feature selection is described in Section 4. The self- adaptive GA are described in Section 5. Section 6 contains short description of classification algorithms. The results of numerical experiments are presented in Section 7. Finally, we provide concluding remarks in Section 8.

2. Corpus description

The data for testing and evaluation consists of 292,156 user utterances recorded in English language from caller interactions with commercial automated agents. Utterances are short and contain only one phrase for further routing. The database contains calls in textual format after speech recognition. The database is provided by the company *Speech Cycle* (New York, USA). Utterances from this database are manually labelled by experts and divided into 20 classes (such as appointments, operator, bill, internet, phone and technical support). One of them is a special class TE-NOMATCH which includes utterances that cannot be put into another class or can be put into more than one class.

The database contains 45 unclassified calls and they were removed. The database contains also 23,561 empty calls without any words. These calls were placed in the class TE-NOMATCH automatically and they were also removed from the database. As a rule, the calls are short in the database; many of them contain only one or two words. The average length of an utterance is 4.66 words, the maximal length is 19 words. There are a lot of identical utterances in the database; the corpus contains only 24,458 unique non-empty classified calls. The corpus is unbalanced. The largest class contains 27.05% and the smallest one contains 0.04% of the unique calls.

Due to the very high frequency of a small number of utterances in the corpus, we formulate two different problem definitions.

Problem definition 1. The whole database with 268,550 classified non-empty calls is used for training and test sets forming. Numbers of repetitions of the utterances in training and test sets are used as weights for classification. This problem definition is the closest to the real situation but frequently repeated utterances decrease difference between preprocessing and classification methods. Additionally, there are some identical utterances in training and test sets simultaneously. In this case the over-fitting problem of classification may be hidden. Therefore, this problem definition is not very appropriate for the comparative study.

Problem definition 2. Before training and test samples forming, all utterance duplicates were removed from the database. It means that there is no intersection between training and test sets and frequency of utterances is ignored.

Therefore, the problem definition 1 is suitable for the quality estimation of the real natural language call routing system; the problem definition 2 is the most appropriate for the comparative study of different preprocessing and classification methods.

For statistical analysis we performed 20 different divisions of the database into training and test samples randomly. This procedure was performed for two problem definitions separately. The train samples contain 90% of the calls and the test samples contain 10% of the calls. For each training sample we have designed a dictionary of unique words which appear in the training sample after deleting punctuation and transforming capital letters to lowercase. The size of the dictionary varies from 3,275 to 3,329 words for problem definition 1 and from 3,277 to 3,311 for problem definition 2.

3. Term weighting methods

As a rule, term weighting is a multiplication of two parts: the part based on the term frequency in a document (TF) and the part based on the term frequency in the whole training database. The TF-part is fixed for all considered term weighting methods and is calculated as following:

$$TF_{ij} = \log(tf_{ij} + 1)\ ;\ tf_{ij} = \frac{n_{ij}}{N_j},$$

where n_{ij} is the number of times the ith word occurs in the jth document, N_j is the document size (number of words in the document). The second part of the term weighting is calculated once for each word from the dictionary and does not depend on an utterance for classification. We consider seven different methods for the calculation of the second part of term weighting.

3.1. *Inverse Document Frequency (IDF)*

IDF is a well-known unsupervised term weighting method which was proposed in [15]. There are some modifications of IDF and we use the most popular one:

$$id\ f_{ij} = \log\frac{|D|}{n_i},$$

where $|D|$ is the number of documents in the training set and n_i is the number of documents that have the ith word.

3.2. *Gain Ratio (GR)*

Gain Ratio (GR) is mainly used in term selection [24], but in [3] it was shown that it could also be used for weighting terms. The definition of GR is as follows:

$$GR(t_i,\ c_j)=\frac{\sum_{c\in\{c_j,\ \bar{c}_j\}}\sum_{t\in\{t_j,\ \bar{t}_j\}} M(t,\ c)}{-\sum_{c\in\{c_j,\ \bar{c}_j\}} P(c)\cdot\log P(c)},$$

$$M(t,\ c)=P(t,\ c)\cdot\log\frac{P(t,\ c)}{P(t)\cdot P(c)},$$

where $P(t,\ c)$ is the relative frequency that a document contains the term t and belongs to the category c; $P(t)$is the relative frequency that a document contains the term t and $P(c)$ is the relative frequency that a document belongs to category c. Then, the weight of the term t_i is the max value between all categories as follows:

$$GR(t_i)=\max_{c_j\in C} GR(t_j,\ c_j),$$

where C is a set of all classes.

3.3. Confident Weights (CW)

This supervised term weighting approach has been proposed in [20]. Firstly, the proportion of documents containing term t is defined as the Wilson proportion estimate $p(x,\ n)$ by the following equation:

$$p(x,\ n)=\frac{x+0.5Z^2_{\alpha/2}}{n+Z^2_{\alpha/2}},$$

where x is the number of documents containing the term t in the given corpus, n is the number of documents in the corpus and $\Phi\left(Z_{\alpha/2}\right)=\alpha/2$, where Φ the t-distribution (Students law) when $n < 30$ and the normal distribution when $n \geq 30$.

In this work α=0.95 and $0.5Z^2_{\alpha/2}=1.96$ (as recommended by the authors of the method). For each term t and each class c two functions $p_{pos}(x,\ n)$ and $p_{neg}(x,\ n)$ are calculated. For $p_{pos}(x,\ n)$ x is the number of documents which belong to the class c and have term t; n is the number of documents which belong to the class c. For $p_{neg}(x,\ n)$ x is the number of documents which have the term t but do not belong to the class c; n is the number of documents which do not belong to the class c. The confidence interval $(p^-,\ p^+)$ at 0.95 is calculated using the following equation:

$$M=0.5Z^2_{\alpha/2}\sqrt{\frac{p(1-p)}{n+Z^2_{\alpha/2}}},$$

$$p^-=p-M;\ p^+=p+M.$$

The strength of the term t in the category c is defined as the follows:

$$\mathrm{str}(t,\ c)=\begin{cases}\log_2\dfrac{2p^-_{pos}}{p^-_{pos}+p^+_{neg}}, & \text{if } p^-_{pos}>p^+_{neg},\\ 0, & \text{otherwise.}\end{cases}$$

The maximum strength (*Maxstr*) of the term t_i is calculated as follows:

$$\mathrm{Maxstr}(t_i) = \max_{c_j \in C} \mathrm{str}(t_i, c_j)^2.$$

3.4. Term Second Moment (TM2)
This supervised term weighting method was proposed in [22]. Let $P(c_j|t)$ be the empirical estimation of the probability that a document belongs to the category c_j with the condition that the document contains the term t; $P(c_j)$ is the empirical estimation of the probability that a document belongs to the category c_j without any conditions. The idea is the following: the more $P(c_j|t)$ is different from $P(c_j)$, the more important the term t_i is. Therefore, we can calculate the term weight as the following:

$$\mathrm{TM2}(t_i) = \sum_{j=1}^{|C|} (P(c_j|t) - P(c_j))^2,$$

where C is a set of all classes.

3.5. Relevance Frequency (RF)
The RF term weighting method was proposed in [11] and is calculated as the following:

$$\mathrm{rf}(t_i) = \max_{c_j \in C} \mathrm{rf}(t_i, c_j),$$

$$\mathrm{rf}(t_i, c_j) = \log_2\left(2 + \frac{a_j}{\max\left\{1, \overline{a_j}\right\}}\right),$$

where a_j is the number of documents of the category c_j which contain the term t_i and $\overline{a_j}$ is the number of documents of all the other categories which also contain this term.

3.6. Term Relevance Ratio (TRR)
The TRR method [9] uses *tf* weights and it is calculated as the following:

$$TRR(t_i, c_j) = \log_2\left(2 + \frac{P(t_i|c_j)}{P(t_i|\overline{c_j})}\right),$$

$$P(t_i|c_j) = \frac{\sum_{k=1}^{|T_c|} tf_{ik}}{\sum_{l=1}^{|V|}\sum_{k=1}^{|T_c|} tf_{lk}},$$

$$TRR(t_i) = \max_{c_j \in C} TRR(t_i, c_j),$$

where c_j is a class of the document, $\overline{c_j}$ is all of the other classes of c_j, V is the vocabulary of the training data and T_c is the document set of the class c.

3.7. Novel Term Weighting (NTW)

This method was proposed in [1, 18]. The details of the procedure are the following. Let L be the number of classes; n_i is the number of documents which belong to the ith class; N_{ij} is the number of occurrences of the jth word in all documents from the ith class. $T_{ij} = N_{ij}/n_i$ is the relative frequency of occurrences of the jth word in the ith class; $R_j = \max_i T_{ij}$; $S_j = arg \max_i T_{ij}$ is the class which we assign to the jth word. The term relevance C_j is calculated by the following:

$$C_j = \frac{1}{\sum_{i=1}^{L} T_{ij}} \cdot \left(R_j - \frac{1}{L-1} \cdot \sum_{i=1,\ i \neq S_j}^{L} T_{ij} \right).$$

4. GA-based feature selection

The popular kind of wrappers is feature selection based on genetic algorithms (GA). GA is used as an optimization algorithm for finding the optimal or the sub-optimal subset of the original feature set with the predefined classification algorithm [23]. The details of the GA-based wrapper are the following:

1. Randomly initialize the population of the binary strings with the length equals to dimensionality of the considered classification problem. Set generation counter $g = 0$.
2. Apply the classification algorithm on the validation set for all individuals form the population (with including only features with 1 value in the chromosomes)
2. Set the classification effectiveness measure (i.e. F-score) on the validation set as a fitness function value for all individuals.
3. Save the individual with the best value of fitness function.
4. Check if the generation counter is greater than some predefined value G (maximal number of generations): if $g > G$ than go to the step 7; otherwise go to step 5.
5. Using GA operators: selection, crossover and mutation, form the next population.
6. Increment the generation counter $g = g + 1$. Go to step 2.
7. Put the best individual as a solution of the feature selection procedure; apply the chosen feature subset for the test set.

5. Self-adaptive GA

For solving the problem of GA setting parameters we use the self-configuring algorithm that was proposed in [17]. The scheme of the self-configuring GA is presented in Figure 1.

In the self-configuring GA (SCGA), different types of selection, recombination, and different levels of mutation are performed simultaneously. In the beginning of the algorithm, all types of GA operators have the same probability to be use for a new off-spring generation. After that, the dynamic adaptation of probabilities is performed according "usefulness" of GA operator types in terms of fitness function.

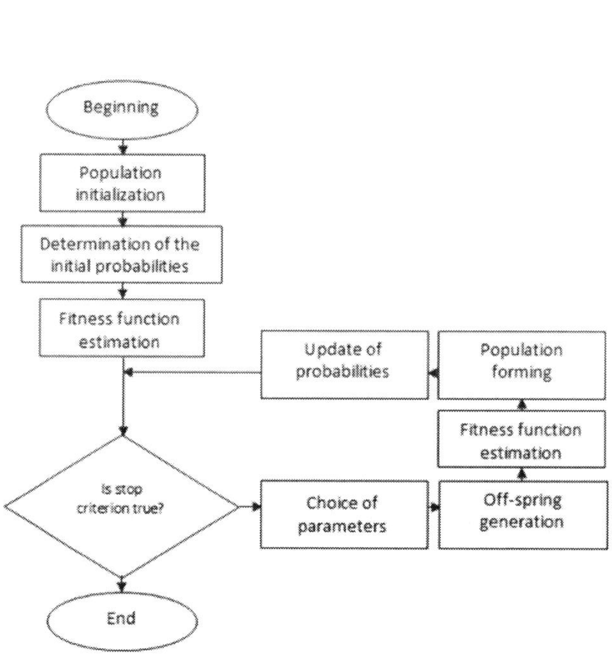

Figure 1. Self-configuring GA (SCGA).

The details of the self-adjusting GA are the following:

1. Put the probabilities of all used types of GA operators: $p_{ij}=1/N_i$, where $j = 1,...,N_i$, N_i is the number of types of the ith operator, $i = 1,...,N$, N is the number of GA operators. In our case N=3: selection, recombination, and mutation.

2. Set threshold probabilities for all used types of GA operators: $\overline{p}_{ij} = 3/(10 \cdot N_i)$.

3. Generate new population. For each off-spring we randomly choose types of selection, recombination, and mutation according probabilities p_{ij}.

4. Recalculate the probabilities with the following:

4.1. For each ith operator do:

4.1.1. Set S_i 0.

4.1.2. For each jth type of the ith operator do:

4.1.2.1. If $p_{ij} < \overline{p}_{ij} + 1/(T \cdot N_i)$AND $p_{ij} > \overline{p}_{ij}$, where T is the number of generations, then:

$$S_i = S_i + (p_{ij} - \overline{p}_{ij});\ p_{ij} = \overline{p}_{ij}.$$

4.1.2.2. If $p_{ij} > \overline{p}_{ij} + 1/(T \cdot N_i)$then: $S_i = S_i + 1/(T \cdot N_i)$; $p_{ij} = p_{ij} - 1/(T \cdot N_i)$.

4.1.2.3. Calculate average fitness function F_{ij} of all off-springs of the current generation that were generated with the jth type of the ith operator.

4.1.3. Find the best type d of the ith operator with the maximal fitness function on the current generation and recalculate its probability: $p_{id} = p_{id} + S_i$

5. Check stop criterion. If TRUE then: END; else: go to the step 3.

Moreover two co-evolutionary algorithms (CEA) [19] have been realized. Both use island model with four islands and cooperative-competitive strategy for resources allocating. First of them (CEA) used four best genetic algorithms (for different types of problems), second one use four independent (on the start and on the probabilities recalculation stage) SCGA (CESCGA).

6. Classification Algorithms

As classification algorithms we use the k-NN algorithm, the linear SVM algorithm, and ANN. The classification criterion is the macro F-score [6] which is appropriate for classification problems with unbalanced classes. For k-NN we performed validation of *k* from 1 to 15 on the validation sample. We used 80% of the train sample for the first level of learning and 20% for the validation. The same scheme of validation is used for feature selection based on self-adjusting GA.

6.1. Genetic Algorithms for Automated Artificial Neural Network Design

Artificial neural network (ANN) is a set of interconnected neurons. Typically, the transfer functions of all the neurons in the neural network are fixed, and the weights are the parameters of the neural network.

In case of GA using for ANN weights tuning: weights are recorded sequentially in the chromosome as a binary code. The number of bits that are used to encode a single weighting coefficient depends on the accuracy and the spread of possible values of weights.

In case of GA using for ANN structure design (GA-ANN):

1. Hidden layers are coded sequentially;
2. Each neuron is encoded in four bits;
3. Each neuron will exist with probability equal to 0.3.
4. If in the network a neuron is not presented, its place in the chromosome is marked with zeros. Otherwise, it is randomly selected as one of the fifteen activation functions, whose number is written in binary code

Since the efficiency of the genetic algorithm for ANN design depends on the dimension of the problem in hand, it is reasonable to avoid the use of uninformative features. The modification of the genetic algorithm for the choice of the most informative features during the automated design of ANN assumes the use of additional bits in the GA chromosomes. These bits determine whether an input is included in the input layer or not.

7. Results of numerical experiments

Tables 1-10 show the results of the numerical experiments for problem definitions 1 and 2 without dimensionality reduction and after feature selection with the self-adaptive GA-based wrapper. The best results in terms of *F*-score are bold (for each case independently). The column "Features" contains number of features after GA-based wrapper performing.

The numerical results showed that the most effective term weighting is TRR. The most effective classification algorithm is ANN. The most effective optimization technique is SCGA. Feature selection with self-configuring GA provides improvement of classification effectiveness and significant dimensionality reduction with all term weighting methods and with all classification algorithms.

Table 1. Results for problem definition 1 with k-NN

Term weighting method	F-score		Features
	All terms	Feature selection	
IDF	0.778	0.800	2781
GR	0.759	0.795	2851
CW	0.763	0.803	2537
RF	0.822	0.863	2509
TM2	0.742	0.778	3054
TRR	**0.851**	**0.887**	2346
NTW	0.847	0.881	2379

Table 2. Results for problem definition 2 with k-NN

Term weighting method	F-score		Features
	All terms	Feature selection	
IDF	0.643	0.667	2678
GR	0.629	0.671	2745

IOP Conf. Series: Materials Science and Engineering **173** (2017) 012008 doi:10.1088/1757-899X/173/1/012008

CW	0.659	0.688	2612
RF	0.655	0.713	2573
TM2	0.613	0.657	2951
TRR	**0.683**	**0.719**	2487
NTW	0.847	0.881	2379

Table 3. Results for problem definition 1 with SVM

Term weighting method	F-score		Features
	All terms	Feature selection	
IDF	0.749	0.762	2617
GR	0.724	0.784	2789
CW	0.776	0.803	2537
RF	0.798	0.851	2386
TM2	0.712	0.765	3012
TRR	**0.812**	**0.874**	2239
NTW	0.803	0.869	2314

Table 4. Results for problem definition 2 with SVM

Term weighting method	F-score		Features
	All terms	Feature selection	
IDF	0.627	0.668	2775
GR	0.619	0.659	2948
CW	0.632	0.672	2896
RF	0.639	0.681	2583
TM2	0.611	0.647	3007
TRR	**0.654**	**0.693**	2469
NTW	0.648	0.685	2541

Table 5. Results for problem definition 1 with SCGA-ANN

Term weighting method	F-score		Features
	All terms	Feature selection	
IDF	0.699	0.796	2679
GR	0.731	0.784	2526
CW	0.766	0.812	2475
RF	0.837	0.850	2337
TM2	0.715	0.771	3145
TRR	**0.844**	**0.891**	2369
NTW	0.832	0.876	2215

Table 6. Results for problem definition 2 with SCGA-ANN

Term weighting method	F-score		Features
	All terms	Feature selection	
IDF	0.679	0.684	2598
GR	0.662	0.675	2657
CW	0.654	0.691	2431
RF	0.682	0.713	2231
TM2	0.621	0.663	2856
TRR	**0.703**	**0.727**	2258
NTW	0.699	0.717	2231

Table 7. Results for problem definition 1 with CEA-ANN

Term weighting method	F-score		Features
	All terms	Feature selection	
IDF	0,673	0,78	2563
GR	0,711	0,758	2408
CW	0,739	0,795	2353
RF	0,82	0,828	2212
TM2	0,69	0,754	3023
TRR	**0,825**	**0,866**	2241
NTW	0,813	0,852	2075

Table 8. Results for problem definition 2 with CEA-ANN

Term weighting method	F-score		Features
	All terms	Feature selection	
IDF	0,655	0,663	2456
GR	0,647	0,657	2529
CW	0,634	0,67	2291
RF	0,655	0,698	2085
TM2	0,599	0,641	2738
TRR	**0,682**	**0,703**	2143
NTW	0,675	0,696	2104

Table 9. Results for problem definition 1 with CESCGA-ANN

Term weighting method	F-score		Features
	All terms	Feature selection	
IDF	0,644	0,653	2384
GR	0,639	0,644	2461
CW	0,621	0,662	2219
RF	0,642	0,686	2021
TM2	0,586	0,629	2681
TRR	**0,674**	**0,691**	2084
NTW	0,667	0,687	2040

Table 10. Results for problem definition 2 with CESCGA-ANN

Term weighting method	F-score		Features
	All terms	Feature selection	
IDF	0,67	0,773	2786
GR	0,697	0,76	2625
CW	0,738	0,785	2560
RF	0,817	0,821	2436
TM2	0,685	0,739	3238
TRR	**0,822**	**0,87**	2452
NTW	0,804	0,853	2314

8. Conclusions

The text classification problem for natural language call routing was considered in the paper. Seven different term weighting methods were applied. As dimensionality reduction methods, the feature selection based on self-configuring GA is considered. k-NN, linear SVM and ANN were used as classification algorithms.

The numerical results showed that the most effective term weighting is TRR. The most effective classification algorithm is ANN. Feature selection with self-configuring GA provides improvement of classification effectiveness and significant dimensionality reduction with all term weighting methods and with all classification algorithms.

Acknowledgements
The reported study was funded by Russian Foundation for Basic Research, Government of Krasnoyarsk Territory, Krasnoyarsk Region Science and Technology Support Fund to the research project №6-41-240822.

References

[1] Akhmedova S, Semenkin E and Sergienko R 2014 Automatically generated classifiers for opinion mining with different term weighting schemes *Informatics in Control, Automation and Robotics (ICINCO'2014)* **2** pp 845–850
[2] Baharudin B, Lee L H and Khan K 2010 A review of machine learning algorithms for text-documents classification *Journal of advances in information technology* **1(1)** pp 4–20
[3] Debole F and Sebastiani F 2004 Supervised term weighting for automated text categorization *In Text mining and its applications* pp 81–97
[4] Fox C 1989 A stop list for general text *In ACM SIGIR Forum* **24** pp 19–21
[5] Gasanova T, Sergienko R, Akhmedova S, Semenkin E and Minker W 2014 Opinion mining and topic categorization with novel term weighting *Proceedings of the 5th Workshop on Computational Approaches to Subjectivity, Sentiment and Social Media Analysis, ACL 2014* pp 84–89
[6] Goutte C and Gaussier E 2005 A probabilistic interpretation of precision, recall and f-score, with implication for evaluation *Advances in information retrieval* pp 345–359
[7] Han E-H S, Karypis G, and Kumar V 2001 Text Categorization Using Weight Adjusted k-Nearest Neighbor Classification
[8] Joachims T 2002 Learning to Classify Text Using Support Vector Machines: Methods, Theory and Algorithms
[9] Ko Y 2012 A study of term weighting schemes using class information for text classification *Proceedings of the 35th international ACM SIGIR conference on Research and development in information retrieval* pp 1029–1030
[10] Kwon O-W and Lee J-H 2003 Text categorization based on k-nearest neighbor approach for web site classification *Information Processing & Management* **39(1)** pp 25–44
[11] Lan M, Tan C L, Su J and Lu Y 2009 Supervised and traditional term weighting methods for automatic text categorization *IEEE Transactions on Pattern Analysis and Machine Intelligence* **31(4)** pp 721–735
[12] Lee C, Jung S, Kim S and Lee G 2009 Example based dialog modeling for practical multi-domain dialog system *Speech Communication* **51(5)** pp 466–484
[13] Morariu D I, Vintan L N and Tresp V 2005 Metaclassification using SVM classifiers for text documents *Intl. Jrnl. of Applied Mathematics and Computer Sciences* **1(1)**
[14] Porter M F 2001 Snowball: A language for stemming algorithms
[15] Salton G and Buckley C 1988 Term-weighting approaches in automatic text retrieval *Information processing & management* **24(5)** pp 513–523
[16] Sebastiani F 2002 Machine learning in automated text categorization *ACM computing surveys (CSUR)* **34(1)** pp 1–47

[17] Semenkin E and Semenkina M 2012 Self-configuring genetic programming algorithm with modified uniform crossover *2012 IEEE Congress on Evolutionary Computation*

[18] Sergienko R, Gasanova T, Semenkin E and Minker W 2014 Text categorization methods application for natural language call routing *11th International Conference on Informatics in Control, Automation and Robotics (ICINCO'2014)* **2** pp 827–831

[19] Sergienko R and Semenkin E 2010 Competitive cooperation for strategy adaptation in coevolutionary genetic algorithm for constrained optimization *IEEE Congress on Evolutionary Computation*

[20] Soucy P and Mineau G W 2005 Beyond tfidf weighting for text categorization in the vector space model *IJCAI* **5** pp 1130–1135

[21] Suhm B, Bers J, McCarthy D, Freeman B, Getty D, Godfrey K and Peterson P 2002 A comparative study of speech in the call center: Natural language call routing vs. touch-tone menus *Proceedings of the SIGCHI conference on Human Factors in Computing Systems* pp 283–290

[22] Xu H and Li C 2007 A novel term weighting scheme for automated text categorization *Seventh International Conference on Intelligent Systems Design and Applications (ISDA'2007)* pp 759–764

[23] Yang J and Honavar V 1998 Feature subset selection using a genetic algorithm *Feature extraction, construction and selection* pp 117–136

[24] Yang Y and Pedersen J O 1997 A comparative study on feature selection in text categorization *ICML* **97** pp 412–420

IOP Conf. Series: Materials Science and Engineering **173** (2017) 012009 doi:10.1088/1757-899X/173/1/012009

Generalization of the mechanisms of cross-correlation analysis in the case of a multivariate time series

O Ja Kravets[1], G V Abramov[2], S Ju Beletskaja[1]

[1] Department of automated and computer systems, Voronezh State Technical University, Voronezh, 394077 Russian Federation
[2] Department of Mathematics and Applied Analysis, Voronezh State University, Voronezh, 394077 Russian Federation

Abstract. The article describes a generalization of the mechanisms of cross-correlation analysis in the case of a multivariate time series and how this allows the optimal lags to be identified for each of the independent variables (IV) using a number of algorithms. The use of generalized mechanisms will allow variables to be analysed and predicted based on the retrospective analysis of multidimensional data. In the available literature, cross-correlation has been defined only for pairs of time series. However, the study of dependent variable (DV) dependencies on multidimensional independent variables that takes into account the vector of specially selected time lags will significantly improve the quality of models based on multiple regression. The idea of multiple cross-correlation lies in the sequential forward shift of each IV row with respect to DV (it transpires that DV is delayed relative to IV) until we obtain a minimum error or the best test of multiple regression. After the completion of all stages of multiple cross-correlation, the synthesis of the model is not a difficult process.

Introduction

The article describes a generalization of the mechanisms of cross-correlation analysis in the case of a multivariate time series and how this allows the optimal lags to be identified for each of the independent variables using a number of algorithms.

The use of the generalized mechanisms will allow variables to be analysed and predicted based on the retrospective analysis of multidimensional data.

Initial data

There are M time series $X_1, X_2, \dots, X_M$ each of N observations. There is a study variable Y - the same time series with parameters, depending, as we believe, on a series of observations.

Multiple regression

The purpose of multiple regression [1, 5] is to provide (1) a calculation of the independent effect of changes in the values of each dependent variable (DV) on an independent variable (IV) and (2) on an empirical basis to predict the values of a dependent variable on the basis of knowledge of the joint effect of IV. The analysis starts by writing an equation that accurately describes the investigated causal relationship. Since this equation can be viewed as a process model, this step is regarded as the developing of the model. The general formula of multiple regression is as follows:

$$Y' = a_0 + b_1X_1 + b_2X_2 \dots + \dots b_MX_M + e. \quad (1)$$

Here the unknown coefficients a_0, b_1, b_2, …, b_M are evaluated in such a way as to minimize the average value of error e. In order to ensure that the application of the method of multiple regression is successful, it is necessary for the model and data to meet five requirements which provide the basis of the use of regression [4, 7].

1. The model must accurately describe the real relationships being studied. This requires that (a) the relation between the variables is linear, (b) that none of the important independent variables are excluded and (c) that no irrelevant variables are included.

2. There should be no errors in the measurement of variables.
3. As for the error - the following conditions are required:
4. Variables must be measured on an interval scale.
a) the mean (the estimated value for each observation) is 0;
b) errors for each observation are not correlated,
c) IVs are not correlated with the error;
d) the deviation of the error remains constant for all values of IV; this condition is called homoscedasticity;
e) the error has a normal distribution.
5. None of the IVs are correlated clearly with any other IV, or with any linear combination of other IVs. If so, then we can say that there is no clear multicollinearity.

If the study adequately satisfies these conditions, we can substitute specific values for Y', X_i and solve the regression equation which describes assumptions about the unknown values of a_0, b_i, using, for example, the least-squares procedure or standard statistical packages.

Interpretation of the results of multiple regression
Consider the coefficient of multiple determination, or R^2 [9]:

$$R^2 = \frac{\sum(Y' - \bar{Y})^2}{\sum(Y - \bar{Y})^2} = \frac{\text{regression\_sum\_of\_squares}}{\text{initial\_sum\_of\_squares}}. \quad (2)$$

This ratio indicates how closely the points that represent data are located to the "straight" line $\bar{Y}$ provided by the model; it is usually called the measure of deviations of DV, which can be explained by the oscillation of all IVs. R^2 varies between 0 and 1; the closer it is to one, the more principal the model is. The R^2 value can always be increased by introducing additional IVs to the model.

Cross-correlation and time lag. The idea of multiple cross-correlation
Sometimes we find that one event affects the other only after some time. The value of this time is called the time lag Δ and it can be determined by using a cross-correlation value [1, 8, 9], defined for two functions Z(t) and V(t) as

$$R_\Delta = \frac{\int Z(t)V(t+\Delta)dt}{\sqrt{\int Z(t)dt \int V(t+\Delta)dt}} \le 1. \quad (3)$$

The closer $|R_\Delta|$ is to 1, the better the optimal time lag Δ adjusts the formula model of the pair regression.

In the available literature, cross-correlation has been defined only for pairs of time series [3, 6, 7, 10]. However, the study of DV dependencies on multidimensional IVs that takes into account the vector of specially selected time lags $\Delta=(\Delta_1, \Delta_2, \ldots, \Delta_M)$, will significantly improve the quality of models based on multiple regression. Fig. 1 shows the idea of multiple cross-correlation. Its essence lies in the sequential forward shift of each IV row with respect to DV (it transpires that DV is delayed relative to IV) until we obtain a minimum error or the best test of multiple regression. The maximum value of the time lag Δ for a specific IV should not exceed N/2, otherwise, there will be difficulties in filling the released elements in the vector of IV.

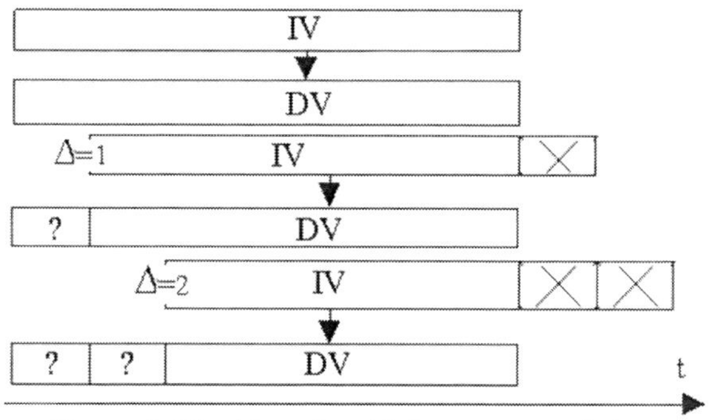

Figure 1. The idea of multiple cross-correlation.

Possible qualitative behaviours of the criterion are shown in Fig. 2-6. Fig. 2 shows the growth of the criterion, which may indicate a big lag, its belonging to a quasi-periodic process with a long period as a fragment of Fig. 3, or the ascending branch of the parabola (Fig. 5-6). In any case, in this situation we are not able to achieve the optimum value Δ.

Fig. 3 corresponds to the quasiperiodic behaviour of the criterion – we can see the minimum value Δ corresponding to the best criteria.

Fig. 4 reflects the situation of the local reduction of the criterion. As with Fig. 2, it is difficult to obtain the optimum value of Δ without additional information about the behaviour of the time lag - we have a situation of negative or zero time lag. Perhaps we are seeing only a part of a parabola or a batch process - but the sample size is too small for definitive conclusions.

Fig. 5 reflects the obvious maximum, Fig. 6 indicates an inefficient sample size or belonging to a quasi-periodic case with a long period. See also the legend to Fig. 2 and 4.

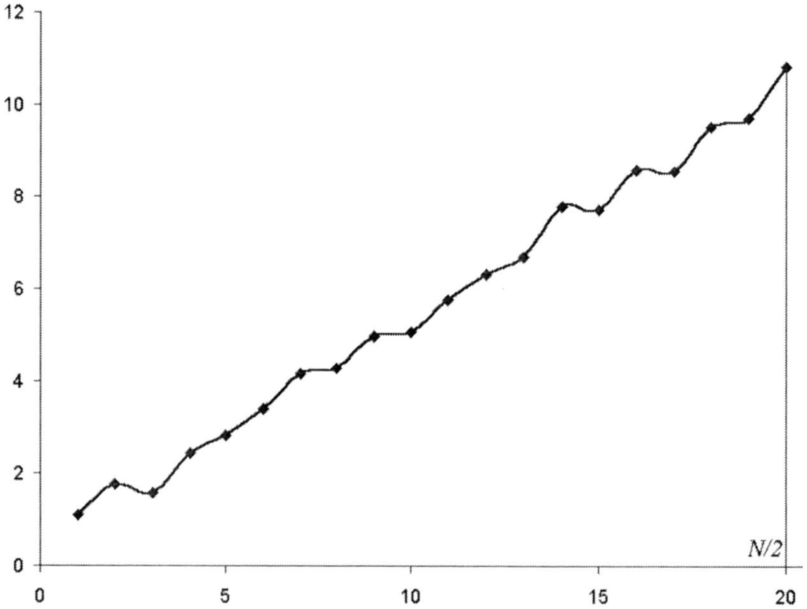

Figure 2. There is no limit to perfection.

IOP Conf. Series: Materials Science and Engineering **173** (2017) 012009 doi:10.1088/1757-899X/173/1/012009

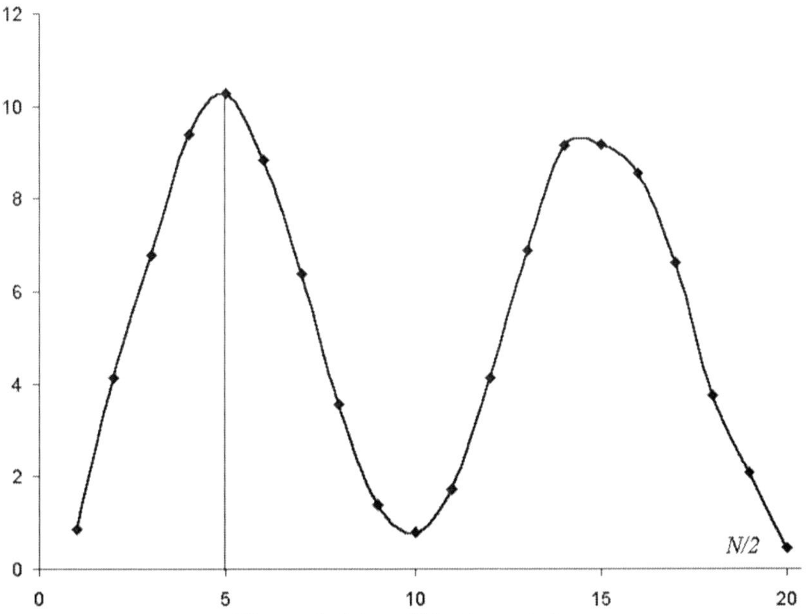

Figure 3. "Found" in the period.

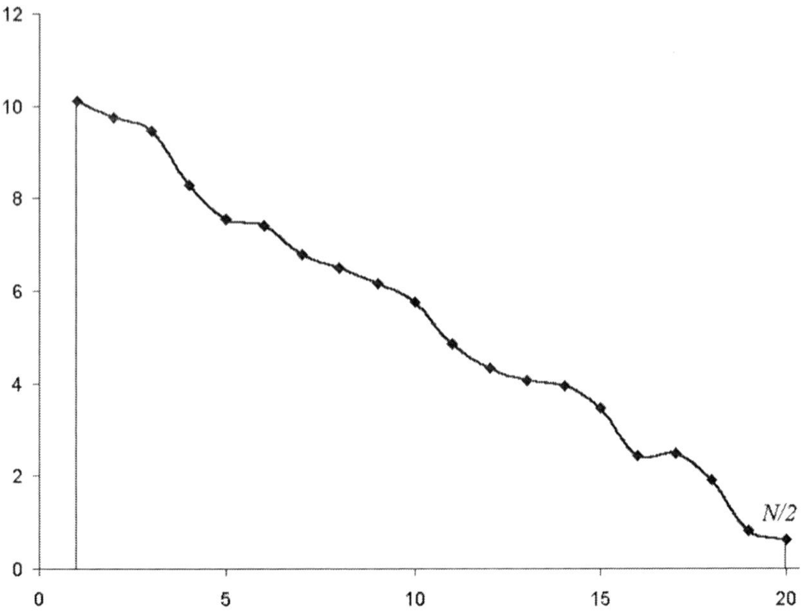

Figure 4. The scale is too small.

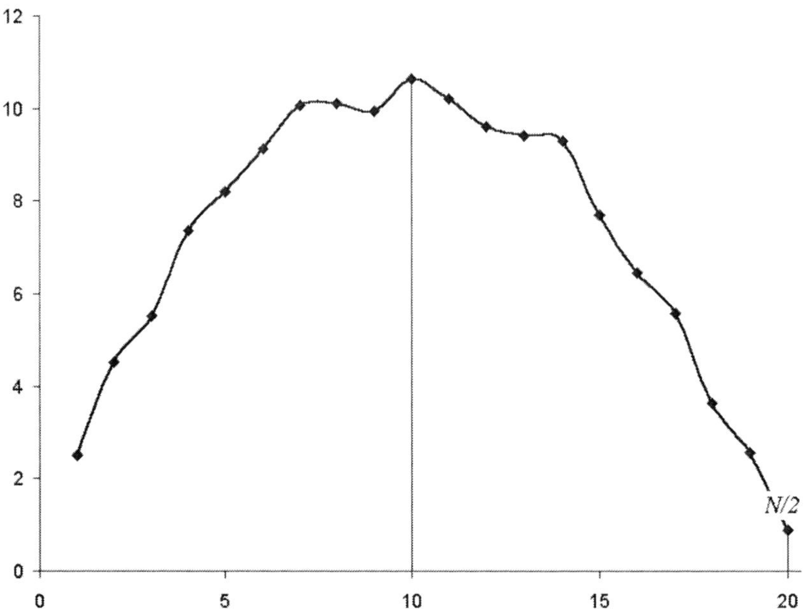

Figure 5. That is what we were looking for.

The situations of criterion behaviour shown in Fig. 2, 3 and 5 are useful for practical purposes.

Tactics for selecting the next (the first) variable

The beginning of the process of determining the optimal time lags of the multivariate model may be different and it is not yet possible to predict which of them is the most optimal one. Here are the main ones:

1. The maximum of the local criterion – we compute the multiple regression parameters and select the IV with the best explanatory properties. Then, for this particular variable, we undertake the procedure of finding the optimal Δ.
2. The minimum of the local criterion (IV with the worst explanatory properties).
3. The random selection of the IV.

After the procedure of determining the optimal Δ for a selected variable, there are two possible tactics to continue the process.

3.1. Reduction of the set of IVs

- searching for paired cross-correlation and remembering $b_i x_i$,
- subtracting $b_i x_i$ and the average explanatory error from the index,
- deleting IV i and repeating the search-cycle (as long as there are variables that make a significant contribution to the model).

3.2. Saving of the processed IV

- adding IV i to the list of variables prohibited for later selection (initially the list is empty);
- repeating the search-cycle (as long as there are variables that make a significant contribution to the model and which are not in the list of variables prohibited for later selection).

An outline flowchart for the process of determining the vector of time lags $\Delta=(\Delta_1, \Delta_2, \ldots, \Delta_M)$ is shown in Fig. 7.

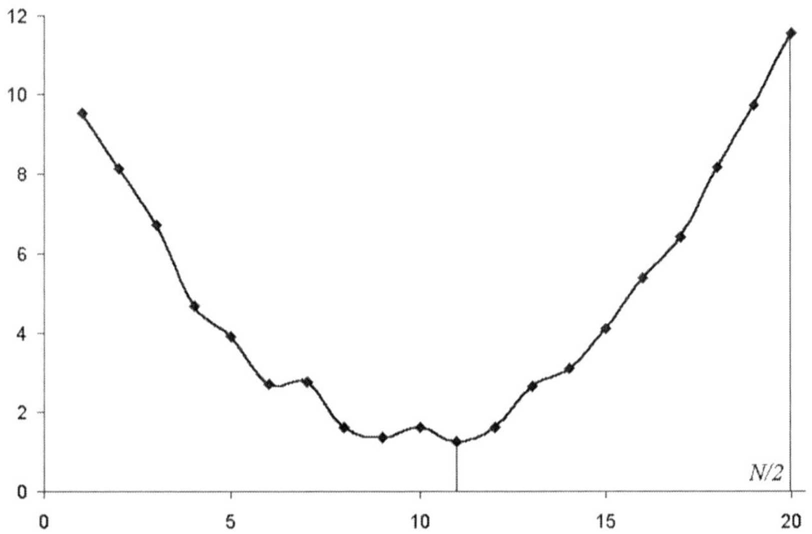

Figure 6. "Where to go?"

Tactics of backward extrapolation

In the case of saving the processed independent variable with an index i (3.2 tactics to continue the process) after the determining of the next element Δ_i of time lags vector $\Delta=(\Delta_1, \Delta_2, \ldots, \Delta_M)$ immediately before the following multiple regression analysis of the resulting model, which will be the next step, it is necessary to fill the elements of the vector X_i for the accuracy of the following analysis - but the filling must be directed "upward" or "leftward" with respect to the vector tail - in the elements with indices 0, -1, -2, ..., $-\Delta_i+1$. We call this process - "backward extrapolation". The complexity of backward extrapolation is not less than the complexity of the traditional method [2], and it is therefore proposed to use one of the following tactics while filling the elements of the vector X_i:

1. zero filling,
2. filling with the mean value of the vector X_i.
3. the use of the "moving average" method,
4. the complete analysis of the time series X_i as a separate object.

Since X_i is already processed, it does not appear necessary to carry out a scrupulous analysis of its autonomous behaviour and in most cases we can use tactics 2 and 3 only.

IOP Conf. Series: Materials Science and Engineering **173** (2017) 012009 doi:10.1088/1757-899X/173/1/012009

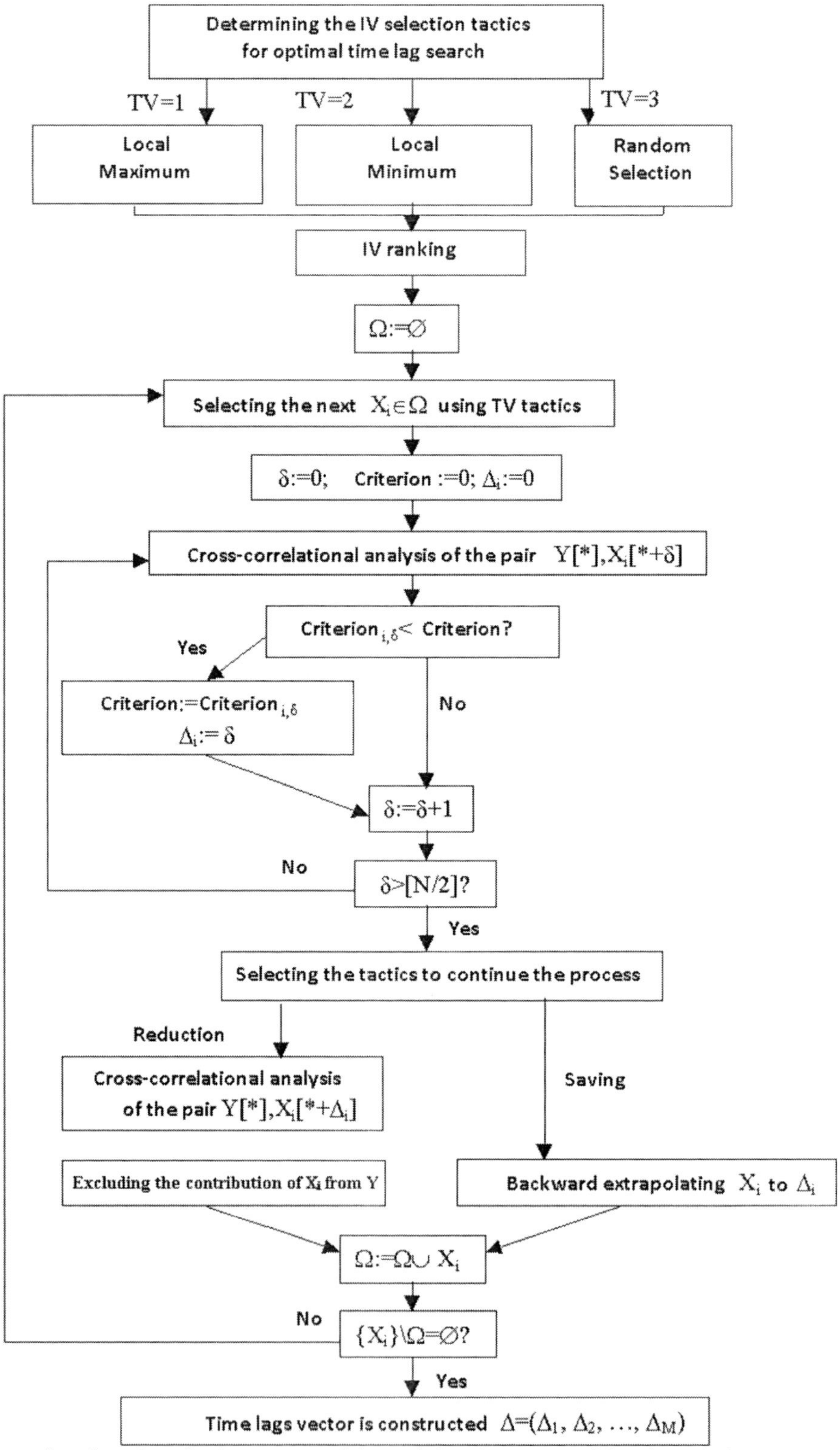

Figure 7. Outline flowchart for the process of determining the vector of time lags $\Delta=(\Delta_1, \Delta_2, \ldots, \Delta_M)$.

Synthesis of the model

After the completion of all the stages of multiple cross-correlation, the synthesis of the model is not a difficult process. We select

$$\Delta_{max}=\max\{\Delta_1, \Delta_2, \ldots, \Delta_M\},$$

then the initial set of X_i is shifted with respect to Y into accordance with Δ_i.

From the resulting set of variables – the independent "shifted" X_i and the dependent Y – we construct a matrix for the final multiple regression analysis in accordance with Fig. 8. The idea of constructing this is to shorten the overall time series by Δ_{max} taking into account the "virtual" filled-in values.

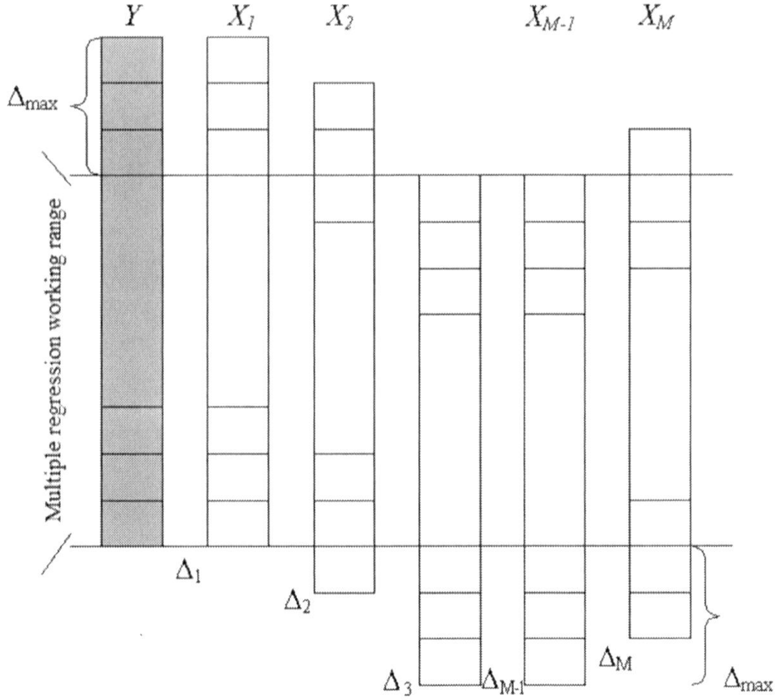

Figure 8. The idea of forming a final matrix of original data: $\Delta_1=0$, $\Delta_2=1$, $\Delta_3=3$, …, $\Delta_{M-1}=3$; $\Delta_M=2$; $\Delta_{max}=3$.

Conclusion

Using the reduced data matrix which is constructed in such a way, multiple regression is carried out and a linear model with a high degree of adequacy is formed.

Thus, the considered idea can be used for detection of difficult multiple-factor dependences of one time series from several ones. At the same time it is possible to set optimum temporal lags. It is important for detection of the postponed response and prediction at least in medicine and ecology.

References

[1] Eliseeva I I, Yuzbashev M M 2005 *General Theory of Statistics* 656 p
[2] Lukashin Y P 2005 *Adaptive methods of short-term time series forecasting* 416 p
[3] Mannheim J B, Rich R K 1997 *Political science. Research Methods* 544 p
[4] Mkhitaryan V S 2005 *Statistics* 671 p
[5] Bohrnstedt G W, Knoke D 1982 *Statistics for Social Data Analysis*
[6] Couch J V 1982 *Fundamentals of Statistics for the Behavioral Sciences*

IOP Conf. Series: Materials Science and Engineering **173** (2017) 012009 doi:10.1088/1757-899X/173/1/012009

[7] Kachigan S K 1982 *Multivariate Statistical Analysis*
[8] Lindeman R, Merenda P P, Gold R Z 1980 *Introduction to Bivariate Multivariate Analysis*
[9] Moore D S 1991 *Statistics*
[10] Tabachnick B G, Fidell L S 1983 *Using Multivariate Statistics*

The optimization of diffraction structures based on the principle selection of the main criterion

O Kravets[1], S Beletskaja[1], Ya Lvovich[2], I Lvovich[2], O Choporov[2], A Preobrazhenskiy[2]

[1]Department of automated and computer systems, Voronezh State Technical University, 14, Moscowsky prospect, Voronezh, 394077, Russia
[2]Voronezh Institute of High Technologies, 73a, Lenina st., Voronezh, 394043, Russia

E-mail: app@vivt.ru

Abstract. The possibilities of optimizing the characteristics of diffractive structures are analysed. A functional block diagram of a subsystem of diffractive structure optimization is shown. Next, a description of the method for the multicriterion optimization of diffractive structures is given. We then consider an algorithm for selecting the main criterion in the process of optimization. The algorithm efficiency is confirmed by an example of optimization of the diffractive structure.

1. Introduction

The lenses of radar signals and the energy converters of electromagnetic waves are, as a rule, characterized by large electrical dimensions, complex geometries, and the presence of absorbing and non-linear elements. The analysis and synthesis of the above-mentioned electrodynamic objects are based on ideas, whereby the physical processes carry the risk of significant and barely controllable error evaluation of their main characteristics, which themselves usually change very quickly when you adjust the frequency, polarization and angle of incidence of electromagnetic waves [1-5]. The measurement of the main characteristics of objects in a wide frequency band and wide angular sector requires a certified or specially equipped polygonal antenna, or a certified anechoic chamber (the cost of which can amount to several million dollars), as well as time-consuming and material resources.

The development of computer-aided design (CAD) for electrodynamic systems allows us to solve completely new problems in the field of antenna-feeder equipment and in the diffraction theory of electromagnetic waves in structures with complex shapes, which requires substantial computing resources, and also to implement research and development, the high degree of difficulty of which has hampered its practical implementation.

In this paper we present a functional diagram showing a subsystem of diffractive structure optimization. Following this, a method of multi-criteria optimization of diffractive structures is described. Then the algorithm for selecting the main criterion is discussed. At the end of the article an example of optimization of a diffractive structure is shown.

2. The functional diagram of the subsystem of optimization of diffractive structures

We have developed a functional diagram showing a subsystem of diffractive structure optimization and this is presented in Figure 1.

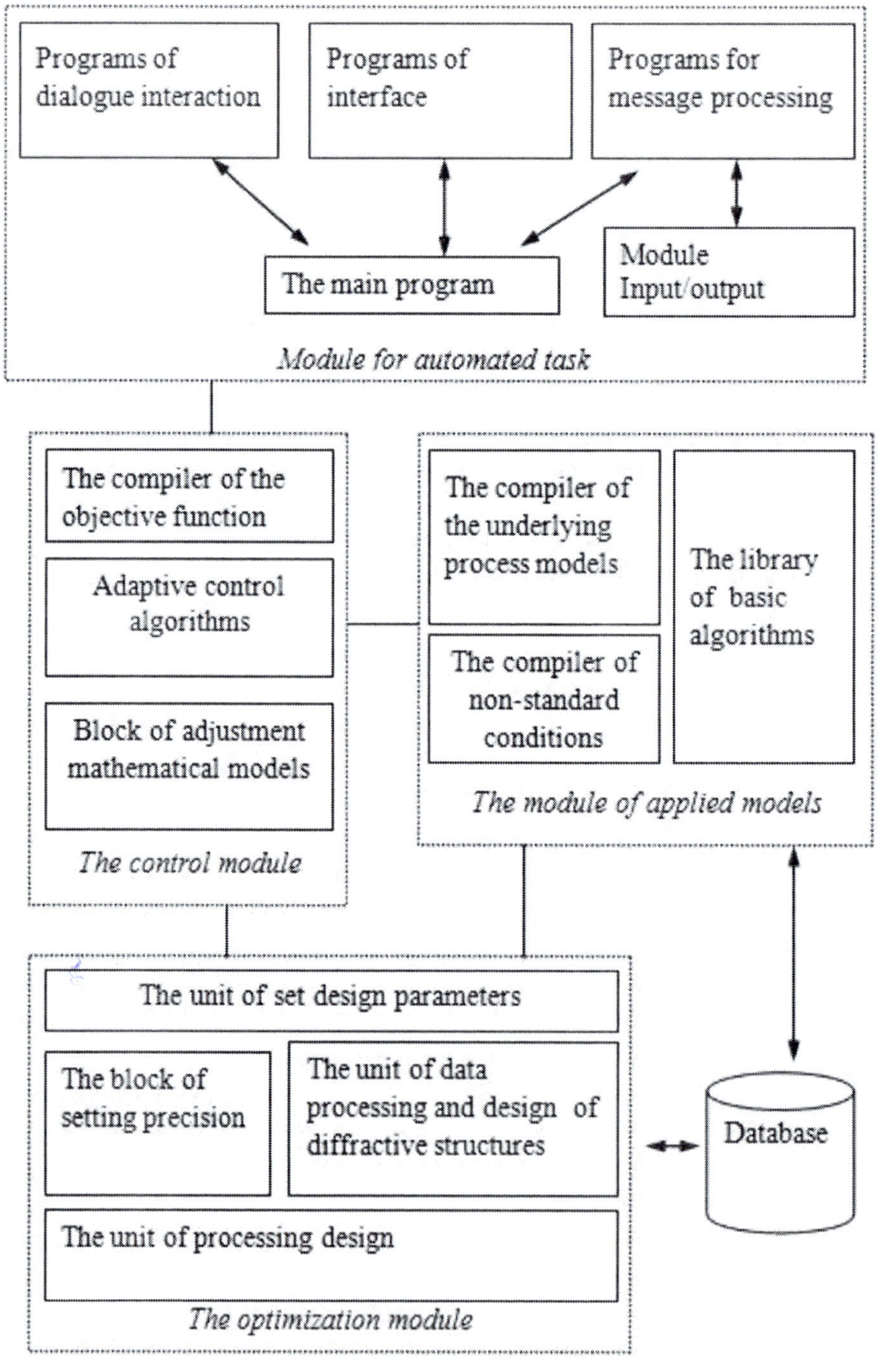

Figure 1. A functional diagram showing a subsystem of optimization of diffractive structure characteristics

The basis of the module of the automated task is the main program transmitting the preferences and settings and creating diffractive structures with a view to adjusting their parameters. To be able to adjust the parameters of the diffraction structures in manual mode, a program processing messages and program dialogue interaction are used. By using the module input and output, the operator can unload from the database the current settings, or set them in manual mode (through manual input or download from the file history settings that were used in the production of diffractive structures).

In the optimization module, a list of operations received from the system top-level management for this particular implementation of the diffractive structure and analysis is given: they set the precision of the optimization and the parameters that characterize the scattering properties of the object. The obtained information is transmitted to the module of applied models characterized by the presence of analytical and numerical models. Input and output parameters for each of the diffraction structures are defined and the target function of optimization is formed. A key feature of the proposed subsystem for the optimization of object parameters is a control module. It uses control algorithms (adjustable model) [6]. This module is responsible for the process of forming the diffraction object, generating and clarifying a mathematical model derived from the applied model module, and transmitting it to the module of the automated tasks: striving for the maximization of the objective function of a number of discrete arguments in imposing functional, topological restrictions. The obtained object parameters and control expressions are logged into a database for the analysis of scattering characteristics of diffractive structures and downloaded in further analyses.

Thus, in the developed subsystem of optimization, the requirements of possible applications for different diffractive structures, the possibility of changing their parameters in the analysis of scattering characteristics of the operator and the loading (restoring) from the database of corresponding parameters in the previous calculations are taken into account.

3. The method of multicriterion optimization of the diffraction structures

Obtainment of the diffractive structures with the desired scattering characteristics is a multi-criteria (multi-factor) challenge [7, 8]. Therefore, optimization of the diffraction structure needs to be multi-criteria –so that it can be carried out in such a way that the operation of the multivariable system [9] will match several criteria, which depend on the output parameters of the diffraction structure.

The method for the main criterion is to be as follows. Depending on the purpose of designing the diffractive structure, as the objective function is selected, the only criterion $f_m(X)$ and the rest of the parameters $f_l(X)$ are taken into account by introducing additional constraints Δl:

$$\begin{gathered} f_m(X) \to \min \\ f_l(X) \le \Delta_l \end{gathered}$$

where $l=1...n$, $l \neq m$.

The main problem in applying this method is the necessity to correctly choose the main criterion and specify the ranges for the remaining criteria. If there are several main criteria, in the case of their mutual contradictions (conflicting criteria), the use of this method is difficult. In addition, an incorrect setting range – for example, placing overly stringent requirements for $f_l(X)$, can lead to an empty set D when manufacturing the diffractive structure with the specified requirements.

On the contrary, with excessively weak restrictions, it is possible to obtain the solution, satisfying some criteria $f_l(X)$ (not meeting the technical specification of the diffraction structure). This can be avoided by using an algorithm, in which each of the quality parameters $f_l(X)$ is designated as the master, and after that the obtained results are compared [10].

In the general case, the main task of optimizing the characteristics of the diffractive structure is by introducing some optimized function $F(X)$, its optimal value:

$$F(X) \to opt \tag{1}$$

For multi-objective optimization [11]:

$$X = \{x_1, x_2 \ldots x_n\}, X \in D, F = \{f_1, f_2 \ldots f_n\}, D \subseteq S \tag{2}$$

where X is the vector of independent variables in some feasible region D (i.e., many possible values of variables), x_i is an unknown, which represents the managed objects (input characteristics) to an optimization problem of the diffraction structures S – space optimization (e.g. – the set of real numbers, R_n).

That is, (1) is the problem of the optimal selection of controlled parameters for a certain class of diffractive structures, which for given n criterion functions f_k, k=1…n there are some restrictions ($X \in D$), (for example, the dimensions of the diffractive structures, etc.).

An additional challenge faced during the optimization of diffraction structures is the interaction of these different criteria among themselves, in addition to independent and (or) neutral criteria, there are conflicting ones (mutually contradictory) [11].

One of the most commonly used methods to solve this problem for multi-objective optimization is the use of some model of trade-off (averaging) on the basis of the construction of the Pareto set in the space of criterion functions):

$$H = D_p = \{(f_1(X), f_1(X), \ldots, f_n(X)) \in R, X \in D_p\} \tag{3}$$

where D_p is the set of all Pareto points. Thus, from the definition of the Pareto set, there is no point outside of this set for which the values for all criteria f_i(X) are not worse than for the points in the Pareto set [10].

The multi-objective optimization problem is the task of finding Pareto optimal sets of decisions, but the decision "in a forehead" would entail significant time costs, and in some cases would be impossible.

It is possible in the design of diffractive structures to make a decision with less effective parameters, but it provides the best resistance to deviations that arise during the design.

Simplifying the construction of a Pareto-optimal solution for a particular class of diffraction structures is one of the most important tasks to reduce time spent on a search and one to a reasonable level [10].

Let's consider the algorithm for selection the main criterion in the design of the specified class of diffraction structures.

4. The algorithm for selection the main criterion

Let the processes of designing diffractive structures be determined by k parameters (that is, there are k criteria (output parameters) for the specified class of diffractive structures), then we write:

$$F(X) = \left[f_1(X), f_2(X) \ldots f_k(X)\right] \tag{4}$$

by virtue of (1) and (2), $F(X)$ must strive to the maximum for X in some space optimization S. In the space S must be a number (and even constrained) allowable solutions – A_d.

However, as noted previously, there are contradictory settings in the design of diffractive structures.

Based on the analysis of optimization methods, it is rational to use a discriminatory method of allocating the main criterion [12].

A mathematical model for this case consists in defining a possible option, which provides a maximum for the target function:

$$K(v_i) = \max\left\{\sum_{k=0}^{n-1} g_k(v_i)\right\}$$
$$g_k(v_i) \geq g_k^{/} \tag{5}$$

where:

$$g_k(v_i) = \sum_{i=0}^{l-1} g_k(p_i) \tag{6}$$

this value specifies parameters, k=0,1,…,m-1; $g_k(p_l)$ – the values of parameters p_l and v_i – options and the option of a diffraction structure (a set of kinds of parameters); $g_k^{/}$ – limits values of the parameters (approximate); $K(v_i)$ is a generalized criterion (the objective function).

Figure 2 presents an algorithm to detect a major parameter that consists of the following:

in step 1 is the ordering of the elements: that is, for each of the n available options of diffractive structures according to the values of the current (reporting) criterion, the items of different types are sorted in ascending (or descending).

It is proposed to introduce the parameter $\sum$ as the sum of values ordered by element types (or, conversely, the sum of the first values is sorted in ascending order).

$\sum$ is a perfect solution for this diffraction structure (that is, the minimum value of the scattered field, the minimum amount of material for forming the diffraction structures, etc.) [13].

In step 2 is the calculation of range parameters for each of the diffraction structures: e is the number of options:

$$\Delta = y_i^e - \Sigma$$

In step 3 the elements e of the second option, which surpass the value of the parameter Δ are discarded.

In the case of there not being at least one element in any embodiment (a feasible solution does not exist), it is necessary to expand the restrictions and return to step 1.

Step 4 proceeds to the next criterion, as has happened until now, until you have considered all of the criteria.

Next, steps 5 and 6 consist in calculating the value of the object function and the job range of its change. The solver uses elements of different options of diffractive structures. The range of change of the object function is as follows:

$$f(v_i) \geq f_i^{\min} = \frac{f_{\max} + f_{i-1}^{\min}}{2}$$

where $f_{i-1}^{\min} = f^{\min}$ and $f^{\min}$ is the sum of the minimum values of the objective function of each option.

In step 7, the final step, the control of elements in each of the options is carried out (a necessary condition for the existence of the decision) and the decision to expand the restrictions on the fits in the absence of elements of at least one option is to be taken.

The result of the algorithm's work is the decision which is the main criterion for the design of diffraction structures obtained in the last iteration.

5. The use a hierarchy analysis method

After making a list of criteria for further use in the algorithm for selecting the main criterion, it is necessary to determine the relative importance of each of them. The most common method is the formation of evaluation in which the relative importance of each criterion is determined by comparing the criteria that can be assessed by comparing the rated points.

It is proposed to use a method of forming the coefficients of criteria importance, based on Thomas Saaty's [14] method of hierarchy analysis. In the method of hierarchy analysis, the pairwise comparison of the criteria will use the following scale for their evaluation: a scale value of 1:9 represents the nine degrees of superiority of one criterion over the other: five values (1, 3, 5, 7, 9) are the main values, and four (2, 4, 6, 8) are the intermediate ones. In this case, if the quality criterion is less important, inverse values are used for its evaluation: 1, 1/2, 1/3, 1/4, 1/5, 1/6, 1/7, 1/8, 1/9.

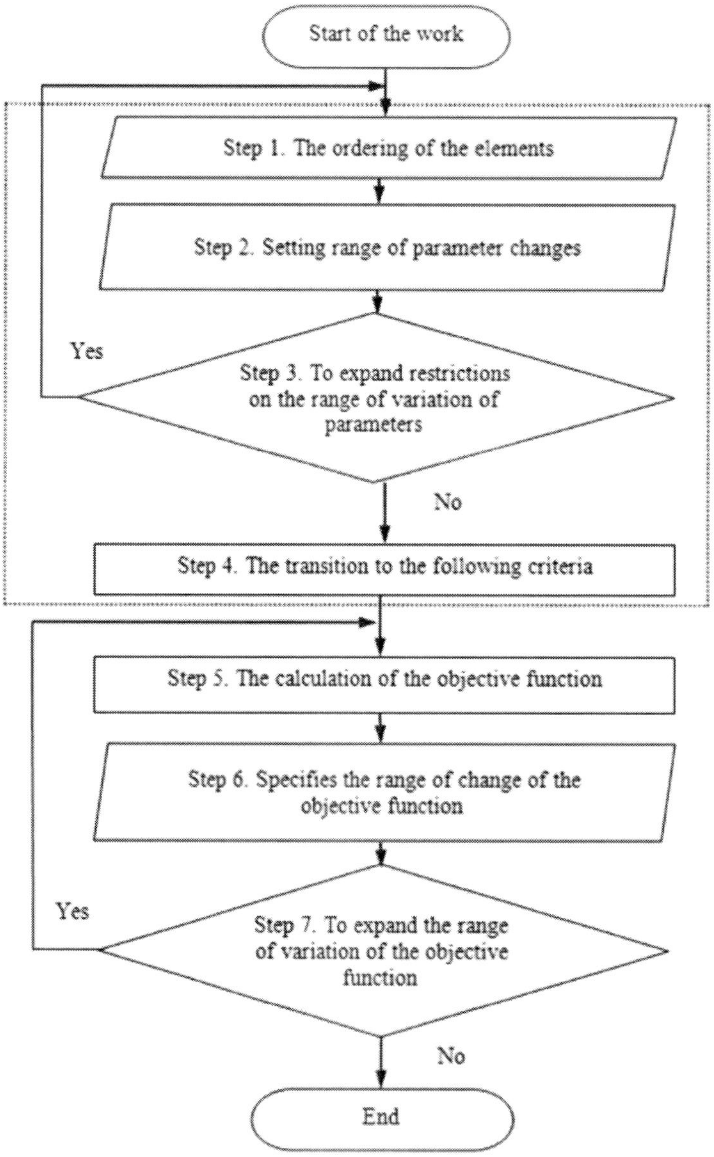

Figure 2. Algorithm for selecting the main criterion for the multi-objective optimization design of diffractive structures.

Below is the matrix of pairwise comparisons of the criteria (Table 1) and the matrix of coordination importance of the criteria (Table 2)

Table 1. The matrix of pairwise comparisons of the criteria.

	Criterion №1	Criterion №2	…	Criterion №n
Criterion №1	1	k_{12}	…	k_{1n}
Criterion №2	k_{21}	1	…	k_{2n}
		…		
Criterion №n	k_{n1}	k_{n2}	…	1

Table 2. The matrix of matching the criteria importance.

	Criterion №1	Criterion №2	…	Criterion №n
Criterion №1	1	k_{12}	…	$k_{2n}\times k_{12}$
Criterion №2	$1/k_{12}$	1	…	$1/k_{2n}$
		…		
Criterion №n	$1/(k_{2n}\times k_{12})$	k_{n2}	…	1

After filling in the matrix, it is necessary to control the index of consistency of a matrix [14]. To check the correctness of its contents, you must define the index of consistency I:

$$I = \frac{K_{\max} - n}{n-1} \tag{7}$$

where n is the number of quality criteria, $K_{\max}$ is the maximum eigenvalue. For a consistent matrix, the index of a match is equal to 0. Using the formula (7) the index of consistency of the matrix is calculated and compared to the average index of consistency for random matrices of the same order.

To apply the proposed algorithms to the optimization of production, the 6 estimated parameters are introduced; prior to filling in the table an expert is exposed to one of the most important quality parameters in the optimization [14] (table 3). We performed a simulation in which the artificial increase (deviation) of the weights (factor of importance) of one of the quality parameters in a fully aligned matrix (for dimensions 3, 4 and 5 quality criteria), the threshold [14] requirement, is set out, and proposed to be used to assess the consistency of the matrix after filling of the coefficients of importance by an expert (table 4).

The method of the main criterion from the point of view of the possibility of using it to develop new types of diffractive structures has advantages related to the fact that when using it, you must not specify the exact values of parameters influencing the characteristics of the object being created, but only the ranges of their change. The scattering characteristics were determined on the basis of the method of integral equations [15].

The main input data is: the possible sector of angles, the ranges of the sizes of the structure and angles of the walls, the required minimum possible level of average radar cross section (RCS) in the given sector of angles. The optimization criterion is defined thus: to define the dimensions of the analysed diffractive structure in which in a given angle sector or several sectors of the viewing angles, the average value of RCS is not lower than a predetermined desired minimum possible level.

The limitations of the method are associated with the size of the required memory, which is required for the elements of the matrix that we get after using the method of moments in solving integral equations. The desirable number of input parameters must not exceed 10 and the number of criteria 3 or 4, as with their subsequent increase, the calculations on modern personal computers require not a few minutes, but several tens of minutes.

Table 3. The rate scale of the quality criteria influence.

Rating	The degree of influence	Comment
1	the equivalence of parameters	
3	some advantage (not essential)	
5	strong superiority (important)	
2,4	intermediate values	2 – the criterion of weak (minimal) advantage 4 – a distinct advantage
6	important criterion	the most important criterion
0	incomparable criteria	Difficult for an Expert to estimate

Table 4. Indices of matrices coherence for the proposed rate scale.

The number of parameters	1	2	3	4	5	6
The value of a random index of consistency	0	0	0,31	0,46	0,49	0,53

6. The results of the optimization of the diffraction structure

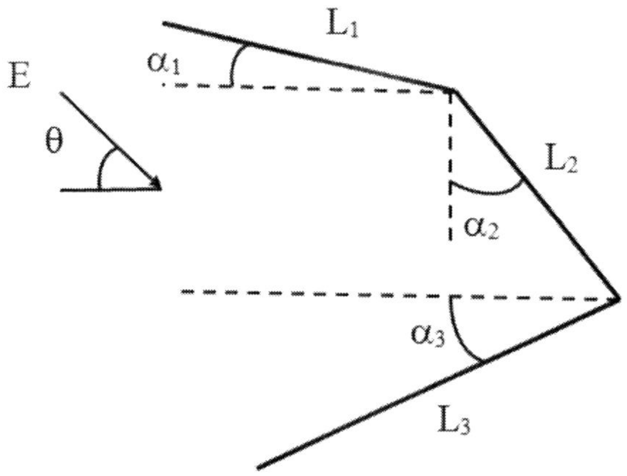

Figure 3. An example of the diffraction structure which is being investigated. Here: L_1, L_2, L_3 are the typical dimensions of the structure , α1, α1, α1 are the angles of inclination of the walls of the hollow structure, E is the vector of the incident electromagnetic wave; θ is the angle of flat electromagnetic wave.

The following limitations for parameters were set: $0°\le\alpha_1\le30°$, $0°\le\alpha_2\le25°$, $0°\le\alpha_3\le31°$, $4.5\lambda\le L_1\le6.4\lambda$, $5.1\lambda\le L_2\le7.2\lambda$, $4.7\lambda\le L_3\le6.8\lambda$, $L= L_1+ L_2+ L_3\le19.5\lambda$, where λ is the length of the incident electromagnetic wave.

It was necessary to design a diffraction structure, which has the value of the average RCS in the range of $10°\le\theta\le26°$ not exceeding 20 dB.

As a result of using this approach the following structure parameters were determined: α_1=5°, α_2=6°, α_3=3°, L_1=5.1λ, L_2=5.7λ, L_1=4.9λ.

7. Conclusion

As can be seen from the above, this paper introduces an efficient method of designing diffractive structures with the desired characteristics, based on the selection of the main criterion. The obtained results can be used to solve problems related to the management of the scattered electromagnetic field.

References

[1] Preobrazhenskiy A P, Yaroslavtsev N P 2005 Modern Radar Systems for Measuring the Radar Characteristics *Vestnik Voronezhskogo gosudarstvennogo tekhnicheskogo universiteta* **1(8)** 29-32

[2] Lvovich I, Preobrazhenskiy A 2015 The Possibilities of Improvement of Wireless Coverage Inside Buildings *Information Technology Applications* **1** 124-130.

[3] Lvovich Ya Ye, Lvovich I Y, Preobrazhenskiy A P, Choporov O N 2014 The Use of "Ant" Algorithm in Constructing Models of Objects That Have Maximum Average Values of the Scattering Characteristics *Life Science Journal* **11(12)** 463-466

[4] Golovinov S O, Preobrazhenskii A P, Lvovich I Y 2013 Modeling the Millimeter Wave Propagation in Urbanized Areas Based on a Combined Algorithm *Telecommunications and Radio Engineering* **72(2)** 139-145

[5] Lvovich I Ya, Preobrazhensky A P, Choporov O N 2013 Analysis of Potential of Error-Correcting Capabilities of Codes *Life Science Journal* **10(4)** 830-834

[6] Tsypkin J Z 1984 *Fundamentals of information theory of identification* 320 p

[7] Lvovich I Ya, Preobrazhensky A P, Savchenko V E 2015 The Possibility of Developing the CAD Subsystem for the Analysis of Scattering Characteristics of Hollow Structures *Global Scientific Potential* **9(54)** 66-68

[8] Golovinov S O, Preobrazhenskii A P, Lvovich I Y 2013 Modeling the Millimeter Wave Propagation in Urbanized Areas Based on a Combined Algorithm *Telecommunications and Radio Engineering* **72(2)** 139-145

[9] Granichin O N 2003 *Introduction to stochastic optimization methods and assessment: textbook* 131 p

[10] Vasilyev F P 2002 *Optimization Methods* 196 p

[11] Lotov V A, Pospelova I I *Multi-criteria decision-making tasks: textbook* 197 p

[12] Horn J, Nafpliotis N, Goldberg D E 1994 A niched Pareto genetic algorithm for multiobjective optimization *Proceedings of the First IEEE Conference on Evolutionary Computation* pp 82-87

[13] Klykov Y I, Gorkov L N 1980 *Data Banks for decision-making* 208 p

[14] Saaty T 1993 *Decision Making. Analytic hierarchy process* 278 p

[15] Preobrazhenskiy A P 2005 Estimation of possibilities of combined procedure for calculation of scattering cross section of two-dimensional perfectly conductive cavities *Telecommunications and Radio Engineering* **63(3)** 269-264

IOP Conf. Series: Materials Science and Engineering **173** (2017) 012011 doi:10.1088/1757-899X/173/1/012011

Modeling of dialogue regimes of distance robot control

E V Larkin[1], **A N Privalov**[2]

[1] Tula State University, 92 "Lenina" prospect, Tula, 300012, Russia,
[2] Tula State Pedagogical University named after L.N. Tolstoy, 125 "Lenina" prospect, Tula, 300026, Russia.

E-mail: elarkin@mail.ru

Abstract. Process of distance control of mobile robots is investigated. Petri-Markov net for modeling of dialogue regime is worked out. It is shown, that sequence of operations of next subjects: a human operator, a dialogue computer and an onboard computer may be simulated with use the theory of semi-Markov processes. From the semi-Markov process of the general form Markov process was obtained, which includes only states of transaction generation. It is shown, that a real transaction flow is the result of «concurrency» in states of Markov process. Iteration procedure for evaluation of transaction flow parameters, which takes into account effect of «concurrency», is proposed.

1. Introduction

At present mobile robots (MR) are rather of widely use in such domains of human activity as environment monitoring, anti-terror systems, intelligence activity etc. [1, 2]. The main feature of the current state of mobile robotics is a restriction of functions which MR can do absolutely independently. This feature deprives the robot of a real autonomy, and that is why, from existing natural levels of hierarchy of control systems (strategic, tactic and functional logic) only tasks of two lowest levels are realized in practice today [3], such as:

- tactic level, at which MR receives commands for solving tasks, modifies a genetic algorithms of their solving with taking into account of current states of onboard robot equipment and environment, and generates execution commands to onboard units;
- functional logic level, at which control commands are realized, feedbacks are closed and coordination of functioning of onboard equipment is executed.

Main features of tactic and functional logical levels tasks are the strong restrictions both for time intervals of sensors and actuator poll, and the lag time between sensor information reading and actuator command generation. Besides, there is the problem of synchronization of operation of control loops with the flow of commands from an outer human operator. That is why, the definition of mobile robot control time characteristics at present is rather actual, and not a properly solved task [4, 5].

2. Common model of distance control process

The principle of MR control [3] is shown on fig. 1 a.

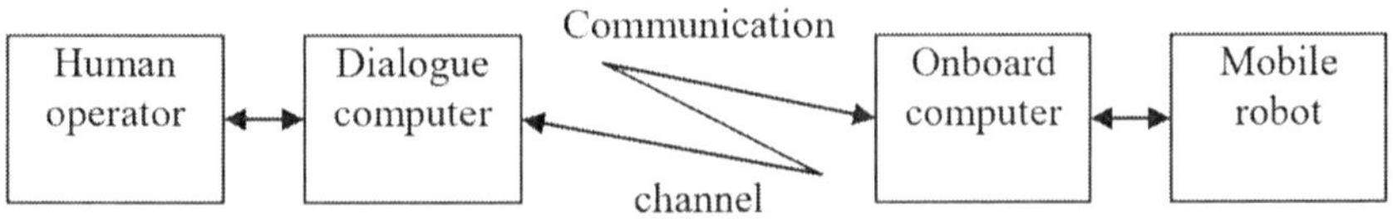

a)

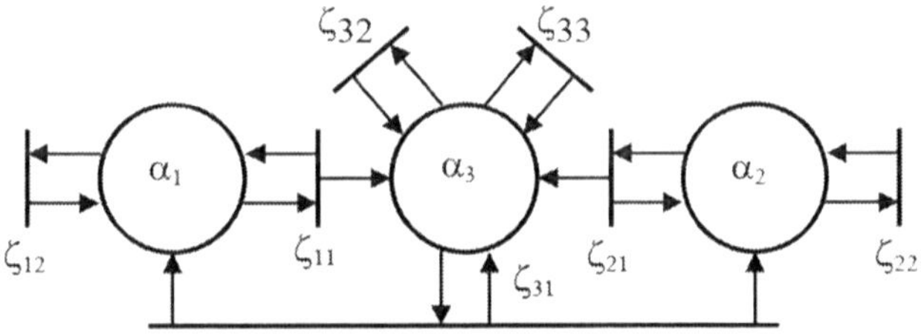

b)

Figure 1. Functional diagram of MR control (a), and Petri-Markov net describing the process of control (b).

MR is directed by a human operator, which is situated at the remote control point. An operator interacts with the dialogue computer and generates a flow of commands, which are transmitted through communication channel onto the onboard computer. The onboard computer decodes commands and activates the proper algorithm to execute the commands. The execution of commands leads to switching of MR state. The information about the current state of the robot and its environment is transmitted backward through the onboard computer, communication channel and dialogue computer to a human operator.

Due to the fact, that three subjects: a human operator, the dialogue computer and the onboard computer are in uninterrupted interaction, the primary model, which explains the control principle, was obtained with use of Petri-Markov nets apparatus [6]. Petri-Markov nets (fig. 1 b.), describing interactions in the system under investigation is as follows:

$$\sigma = \{\mathrm{A}, \mathrm{Z}, \rho_{\alpha\zeta}, \rho_{\zeta\alpha}, \Lambda\}, \tag{1}$$

where $\mathrm{A} = \{\alpha_1, \alpha_2, \alpha_3\}$ - is the set of places, which describes algorithms of interacted subjects functioning; $\mathrm{Z} = \{\zeta_{11}, \zeta_{12}, \zeta_{21}, \zeta_{22}, \zeta_{31}, \zeta_{32}, \zeta_{33}\}$ - is the set of transitions, which describes the logics of subjects interaction; $\rho_{\alpha\zeta}$ - is the adjacency matrix of size 3×7, which reflects the set of places to the set of transitions; $\rho_{\zeta\alpha}$ - is the adjacency matrix of size 7×3, which reflects the set of transitions to the set of places; Λ - matrix of logical conditions of semi-steps execution from transitions to places;

$$\rho_{\alpha\zeta} = \begin{pmatrix} 1 & 1 & 0 & 0 & 0 & 0 & 0 \\ 0 & 0 & 0 & 1 & 0 & 1 & 1 \\ 0 & 0 & 1 & 1 & 1 & 0 & 0 \end{pmatrix}; \tag{2}$$
$$\begin{matrix} 11 & 12 & 31 & 32 & 33 & 21 & 22 \end{matrix}$$

$$\rho_{\zeta\alpha} = \begin{pmatrix} 1 & 0 & 1 \\ 1 & 0 & 0 \\ 1 & 1 & 1 \\ 0 & 0 & 1 \\ 0 & 0 & 1 \\ 0 & 1 & 1 \\ 0 & 1 & 0 \end{pmatrix}; \quad (3)$$

$$\Lambda = \begin{pmatrix} [\alpha_1, \zeta_{11}] & 0 & [\alpha_1, \zeta_{11}] \\ [\alpha_1, \zeta_{12}] & 0 & 0 \\ [\alpha_3, \zeta_{31}] & [\alpha_3, \zeta_{31}] & [\alpha_3, \zeta_{31}] \\ 0 & 0 & [\alpha_3, \zeta_{32}] \\ 0 & 0 & [\alpha_3, \zeta_{33}] \\ 0 & [\alpha_2, \zeta_{21}] & [\alpha_2, \zeta_{21}] \\ 0 & [\alpha_2, \zeta_{22}] & 0 \end{pmatrix}, \quad (4)$$

where vector $[\alpha_m, \zeta_n]$ describes the procedure of semi-steps execution from the place α_m to the transition ζ_n, $m \in \{1, 2, 3\}$, $n \in \{11, 12, 31, 32, 33, 21, 22\}$.

3. Models of algorithms

Places $\alpha_1, \alpha_2, \alpha_3$ describe sequences of operations of the next subjects: α_1 - a human operator; α_2 - the dialogue computer; α_3 - the onboard computer. A human operator realizes the sequence of operations as a decision-making creative process, based on qualification, intuition, instructions, description of hardware/software, etc. The dialogue computer and the onboard computer realize sequences of their operations as interpretation of some algorithms. Anyway sequences of operations develop in time. After completion of current operation algorithm switches to one of possible next operation in the stochastic way, from the point of view of an external observer. Due to random time of execution of operation and stochastic way of switching, the natural model of sequences of operations in a system under investigation is the semi-Markov processes [7, 8]

$$\alpha_i = \{A_i, \boldsymbol{r}_i, \boldsymbol{h}_i(t)\}, i \in \{1, 2, 3\}, \quad (5)$$

where $A_i = \{a_{1_i}, ..., a_{j_i}, ..., a_{J_i}\}$ - is the set of states; $\boldsymbol{r}_i = (r_{j_i, m_i})$ and $\boldsymbol{h}_i(t) = [h_{j_i, m_i}(t)]$ - are the adjacency and the semi-Markov matrices correspondingly, both matrices of size $J_i \times J_i$.

In most common case $\forall r_{j_i, m_i} = 1$ (full graph with loops). The elements of semi-Markov matrix satisfy the following conditions:

$$\sum_{m_i=1}^{J_i} \int_0^{\infty} h_{j_i, m_i}(t) dt = 1; \quad \frac{h_{j_i, m_i}(t)}{\int_0^{\infty} h_{j_i, m_i}(t) dt} \neq \lim_{\tau \to \infty} \delta(t - \tau), \quad (6)$$

where $\delta(t - \tau)$ - is shifted Dirac δ-function.

Restrictions (6) point that processes $\alpha_1, \alpha_2, \alpha_3$ are the ergodic ones.

In accordance with Petri-Markov net (1), in every semi-Markov process $\alpha_1, \alpha_2, \alpha_3$ with sets of states A_1, A_2, A_3, there are subsets of states simulating generation of transactions to the adjacent semi-Markov process. Processes α_1, α_2 have one such subset. Without loss of generality one can assume, that states under consideration have numbers from the first till S_i-th, $S_i \leq J_i$, $i \in \{1, 2\}$. So, sets A_1, A_2 are as follows:

$$A_i = \{a_{1_i}, \ldots, a_{s_i}, \ldots, a_{S_i}, a_{S_i+1}, \ldots, a_{j_i}, \ldots, a_{J_i}\}, i \in \{1, 2\}. \tag{7}$$

In semi-Markov process α_3 there are two subsets of states, which simulate generation of transactions. States of first subset generate transactions to α_1. States of second subset generate transactions to α_2. The named states have numbers from first till S_3-th, and from (S_3+1)-th till U_3-th. Thus set A_3 is as follows:

$$A_3 = \{a_{1_3}, \ldots, a_{s_3}, \ldots, a_{S_3}, a_{S_3+1}, \ldots, a_{u_3}, \ldots, a_{U_3}, a_{U_3+1}, \ldots, a_{j_3}, \ldots, a_{J_3}\}. \tag{8}$$

For evaluation of parameters of the transactions flow one should simplify semi-Markov processes $\alpha_1, \alpha_2, \alpha_3$ till processes which contain states of generation of transactions only. For this purpose in processes α_1, α_2 one should split every state a_{s_i} onto two states: starting one ${}^b a_{s_i}$ and absorbing one ${}^e a_{s_i}$, $1_i \le s_i \le S_i$, $i \in \{1, 2\}$. In semi-Markov process α_3 states a_{s_3}, a_{u_3}, $1_3 \le s_3 \le S_3$, $1_3 \le u_3 \le U_3$ are split onto starting states ${}^b a_{s_3}$, ${}^b a_{u_3}$ and absorbing states ${}^e a_{s_3}$, ${}^e a_{u_3}$.

When splitting semi-Markov processes α_1, α_2 are converted to processes

$$\alpha_i' = \{A_i', \boldsymbol{r}_i', \boldsymbol{h}_i'(t)\}, i \in \{1, 2, 3\}, \tag{9}$$

where

$$A_i' = \{{}^b a_{1_i}, \ldots, {}^b a_{s_i}, \ldots, {}^b a_{S_i}, a_{S_i+1}, \ldots, a_{j_i}, \ldots, a_{J_i}, {}^e a_{1_i}, \ldots, {}^e a_{s_i}, \ldots, {}^e a_{S_i}\},$$
$$i \in \{1, 2\};$$
$$A_3' = \{{}^b a_{1_3}, \ldots, {}^b a_{s_3}, \ldots, {}^b a_{S_3}, {}^b a_{S_3+1}, \ldots, {}^b a_{u_3}, \ldots, {}^b a_{U_3}, a_{U_3+1}, \ldots,$$
$$a_{j_3}, \ldots, a_{J_3}, {}^e a_{1_3}, \ldots, {}^e a_{s_3}, \ldots, {}^e a_{S_3}, {}^e a_{S_3+1}, \ldots, {}^e a_{u_3}, \ldots, {}^e a_{U_3}\} \tag{10}$$

Both the adjacency matrix $\boldsymbol{r}'$ and semi-Markov matrix $\boldsymbol{h}_i'(t)$ in (9) are of size $(J_i + S_i) \times (J_i + S_i)$, when $i \in \{1, 2\}$ and $(J_3 + U_3) \times (J_3 + U_3)$, when $i = 3$. Matrices are formed as follows:

columns with numbers from 1_i till S_i, if $i \in \{1, 2\}$, and from 1_3 till U_3, if $i = 3$, should be transferred to columns with numbers from $J_i + 1_i$ till $J_i + S_i$, if $i \in \{1, 2\}$, and from $J_3 + 1_3$ till $J_3 + U_3$, if $i = 3$;

columns with numbers from 1_i till S_i, if $i \in \{1, 2\}$, and from 1_3 till U_3, if $i = 3$, should be fulfilled with zeros;

rows with numbers from $J_i + 1_i$ till $J_i + S_i$, if $i \in \{1, 2\}$, and from $J_3 + 1_3$ till $J_3 + U_3$, if $i = 3$, should be fulfilled with zeros.

Semi-Markov processes (9) with divided states cease to be ergodic ones. For them probabilities and densities of time of wandering from state ${}^b a_{m_i}$ to state ${}^e a_{n_i}$ may be found, where $1_i \le m_i \le S_i$, if $i \in \{1, 2\}$, and $1_3 \le m_i \le U_3$, if $i = 3$; $J_i + 1 \le n_i \le J_i + S_i$, if $i \in \{1, 2\}$, and $J_3 + 1 \le m_i \le J_3 + U_3$, if $i = 3$. Probabilities, densities and expectations of time of wandering are as follows:

$$h''_{m_i, n_i}(t) = L^{-1}\left[{}^r\boldsymbol{I}_{n_i} \cdot \sum_{k=1}^{\infty} \{L[\boldsymbol{h}_i'(t)]\}^k \cdot {}^c\boldsymbol{I}_{m_i}\right]; \tag{11}$$

$$p_{m_i, n_i} = \int_0^\infty h''_{m_i, n_i}(t)dt; \tag{12}$$

$$f_{m_i, n_i}(t) = \frac{h''_{m_i, n_i}(t)}{p_{m_i, n_i}}; \tag{13}$$

$$T_{m_i, n_i} = \int_0^\infty t f_{m_i, n_i}(t)dt, \tag{14}$$

where L and L^{-1} - are direct and inverse Laplace transforms correspondingly; ${}^{r}\boldsymbol{I}_{n_i}$ - is the row vector of size $J_i + S_i$, if $i \in \{1, 2\}$, and of size $J_3 + U_3$, if $i = 3$, n_i-th element of which is equal to one, and all other elements are equal to zero; ${}^{c}\boldsymbol{I}_{m_i}$ - is the column vector of size $J_i + S_i$, if $i \in \{1, 2\}$, and of size $J_3 + U_3$, if $i = 3$, m_i-th element of which is equal to one, and all other elements are equal to zero.

Formulae (11) ÷ (14) should be applied for all possible combinations of indexes of pairs (m_i, n_i). As a result semi-Markov processes $\alpha_1'', \alpha_2'', \alpha_3''$, which include states of transaction generation only, are formed. Every switching of processes formed born one transaction. Processes are as follows:

$$\alpha_i'' = \{A_i'', \boldsymbol{r}_i'', \boldsymbol{h}_i''(t)\}, i \in \{1, 2, 3\}, \tag{15}$$

where $A_i'' = \{a_{1_i}'', \dots, a_{m_i}'', \dots, a_{S_i}''\}$, if $i \in \{1, 2\}$, and $A_3'' = \{a_{1_3}'', \dots, a_{m_3}'', \dots, a_{U_3}''\}$, if $i = 3$; $\boldsymbol{r}_i'' = (r_{m_i, n_i})$ - is the adjacency matrix of size $S_i \times S_i$, if $i \in \{1, 2\}$ and $U_3 \times U_3$, if $i = 3$; $\boldsymbol{h}_i''(t) = [h_{m_i, n_i}''(t)]$ - is the semi-Markov matrix of the same size, as adjacency matrix.

It should be noted, that semi-Markov processes (15) were obtained from ergodic processes (5) by means of execution of equivalent transforms, so processes (15) are ergodic too. For the ergodic semi-Markov processes rightly are the next restrictions

$$\sum_{n_i=1}^{S_i} p_{m_i, n_i} = 1, \text{ if } i \in \{1, 2\}, \text{ and } \sum_{n_i=1}^{U_i} p_{m_i, n_i} = 1, \text{ if } i = 3. \tag{16}$$

For evaluation of parameters of transactions flow it is necessary to transform semi-Markov processes (15) to more convenient abstract models. Due to the fact, that every switching in (15) born a transaction, structures of such models are as it is shown on fig. 2 a, b with solid lines.

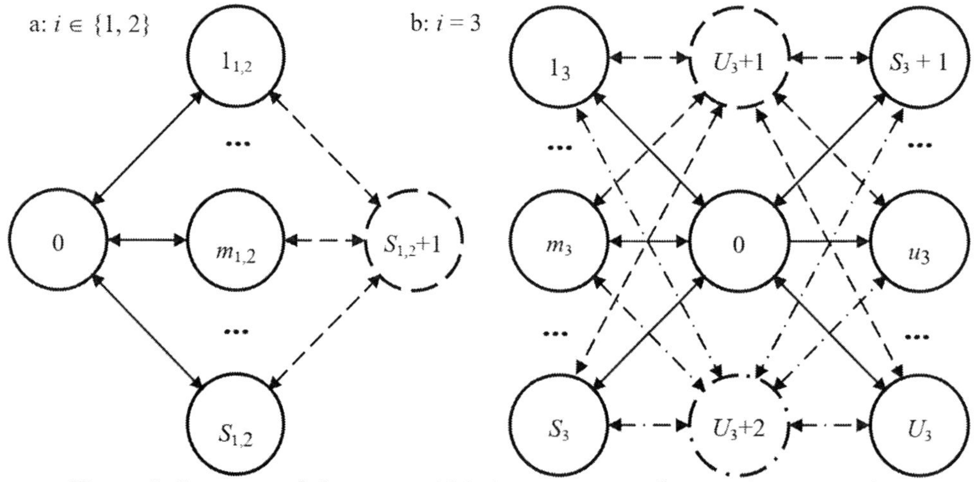

Figure 2. Structures of abstract semi-Markov processes of transactions generation.

States a_{0_i}'' into abstract models were inserted artificially. Density of time of residence in this state is described by Dirac δ-function. Insertion into the structures the state a_{0_i}'' is caused by the necessity of re-switching the process in cycle to obtain a flow of transactions.

Abstract semi-Markov process is as follows:

$$\alpha_i''' = \{A_i''', \boldsymbol{r}_i''', \boldsymbol{h}_i''(t)\}, i \in \{1, 2, 3\}, \tag{17}$$

where $A_i'''=\{a''_{0_i}, a''_{1_i}, ..., a''_{m_i}, ..., a''_{S_i}\}$, if $i \in \{1, 2\}$, and $A_3'''=\{a''_{0_3}, a''_{1_3}, ..., a''_{m_3}, ..., a''_{U_3}\}$, if $i = 3$, are the sets of states; $r_i'''=(r_{m_i,n_i})$ - is the adjacency matrix of size $(S_i+1)\times(S_i+1)$, if $i \in \{1, 2\}$ and $(U_3+1)\times(U_3+1)$, if $i = 3$; $h_i''(t)=[h'''_{m_i,n_i}(t)]$ - is the semi-Markov matrix of the same size.

$$r_i'''=\begin{pmatrix} 0 & 1 & \dots & 1 & \dots & 1 \\ 1 & 0 & \dots & 0 & \dots & 0 \\ & & \dots & & & \\ 1 & 0 & \dots & 0 & \dots & 0 \\ & & \dots & & & \\ 1 & 0 & \dots & 0 & \dots & 0 \end{pmatrix};$$

$$h_i''(t)=\begin{pmatrix} 0 & \pi_{1_i}\delta(t) & \dots & \pi_{m_i}\delta(t) & \dots & \pi_{M_i}\delta(t) \\ f_{1_i}(t) & 0 & \dots & 0 & \dots & 0 \\ & & \dots & & & \\ f_{m_i}(t) & 0 & \dots & 0 & \dots & 0 \\ & & \dots & & & \\ f_{M_i}(t) & 0 & \dots & 0 & \dots & 0 \end{pmatrix};$$

$$M_i = \begin{cases} S_i, \text{ when } i \in \{1, 2\}; \\ U_3, \text{ when } i = 3. \end{cases} \tag{18}$$

Probabilities of switching from state 0 into other states of semi-Markov process are as follows

$$\pi_{m_i} = \frac{T_{m_i}}{\tau_{m_i}}, \tag{19}$$

where T_{m_i} - is the expectation of time of residence of ergodic semi-Markov process (15) in the state a''_{m_i}; τ_{m_i} - is the expectation of time of returning to state a''_{m_i}.

Density and expectation of time of residence of process (15) in the state a''_{m_i} are as follows:

$$f_{m_i}(t) = \sum_{m_i=1}^{M_i} h''_{m_i}(t); \tag{20}$$

$$T_{m_i} = \int_0^\infty t \cdot f_{m_i}(t) dt, \tag{21}$$

$$M_i = \begin{cases} S_i, \text{ when } i \in \{1, 2\}; \\ U_3, \text{ when } i = 3. \end{cases}$$

For evaluation of the returning time expectation to the state a''_{m_i} one should to split the state a''_{m_i} to ${}^b a''_{m_i}$ and ${}^e a''_{m_i}$. This may be done by means of transferring of column of matrix $h_i''(t)$ with number m_i to column with number S_i, if $i \in \{1, 2\}$, and to column with number U_3, if $i = 3$. The column with number m_i, and the row with number S_i, if $i \in \{1, 2\}$, and U_3, if $i = 3$, should be fulfilled with zeros. As a result matrix $\tilde{h}_i''(t)$ of size $(S_i+1)\times(S_i+1)$, if $i \in \{1, 2\}$, and of size $(U_3+1)\times(U_3+1)$, if $i = 3$, is formed. Expectation of time of returning is as follows:

$$\tau_{m_i} = \begin{cases} \int_0^\infty t \cdot L^{-1}\left[{}^r I_{S_i+1} \cdot \sum_{k=1}^{\infty} \{L[\tilde{h}_i''(t)]\}^k \cdot {}^c I_{m_i} \right] dt \text{ when } i \in \{1,2\}; \\ \int_0^\infty t \cdot L^{-1}\left[{}^r I_{U_i+1} \cdot \sum_{k=1}^{\infty} \{L[\tilde{h}_i''(t)]\}^k \cdot {}^c I_{m_i} \right] dt \text{ when } i = 3; \end{cases} \tag{22}$$

where ${}^{c}\boldsymbol{I}_{m_i}$ - is the column vector of size $S_i + 1$ $(U_i + 1)$, m_i-th element is equal to one, and other elements are zeros; ${}^{r}\boldsymbol{I}_{S_i+1}$ (${}^{r}\boldsymbol{I}_{U_i+1}$) - is the row vector of size $S_i + 1$ $(U_i + 1)$, $(S_i + 1)$-th ($(U_i + 1)$-th) element of which is equal to one, and all other elements are equal to zeros.

As the transactions are generated by means of wandering through different trajectories, train of transactions, formed when wandering on separate trajectory may be considered as a separate flow. Generation of transactions when wandering on the set of trajectories may be considered as a stochastic combination of flows. In accordance with results obtained in [9] which are verified with Monte-Carlo method [10] such flow may be considered as Poisson one. Thus densities of time between transactions are as follows:

$$f_{m_i}(t) = \lambda_{m_i} \exp(-\lambda_{m_i} t), \tag{23}$$

where $\lambda_{m_i} = \dfrac{1}{T_{m_i}}$ - is the parameter of density of flow.

So the process (17) may be considered as strongly Markov one.

4. Models of interaction as «concurrency» of subjects

Let us return back to Petri-Markov net (1). Interactions on the transitions $\zeta_{11}, \zeta_{21}, \zeta_{31}$ are as follows. The execution of semi-steps $[\alpha_1, \zeta_{11}]$, $[\alpha_2, \zeta_{21}]$ and $[\alpha_3, \zeta_{31}]$ means, that Markov process (17) after natural completion of residence in states a''_{m_i} return to the state a''_{0_i} and then follow the switching into states $a''_{1_i}, ..., a''_{m_i}, ..., a''_{S_i}$ with probabilities $\pi_{1_i}, ..., \pi_{m_i}, ..., \pi_{S_i}$, if $i \in \{1, 2\}$, and into states $a''_{1_3}, ..., a''_{m_3}, ..., a''_{U_3}$ with probabilities $\pi_{1_i}, ..., \pi_{m_i}, ..., \pi_{U_i}$, if $i = 3$. Besides, switching of any process into state a''_{0_i}, leads to a generation of transaction to the adjacent Markov process, herewith the process, which receives transaction, restarts. For modeling of such situation onto structures of processes additional states a''_{S_i+1}, if $i \in \{1, 2\}$, or a''_{U_3+1} and a''_{U_3+2}, if $i = 3$ are inserted. Switching from states inserted to states $a''_{1_i}, ..., a''_{m_i}, ..., a''_{S_i}$ is done in time, which is defined by δ-function with probabilities $q_{1_i}, ..., q_{m_i}, ..., q_{S_i}$, if $i \in \{1, 2\}$, and to states $a''_{1_3}, ..., a''_{m_3}, ..., a''_{U_3}$ with probabilities $q_{1_i,1}, ..., q_{m_i,1}, ..., q_{U_i,1}$, if $i = 3$, and transaction was received from α'''_1, and probabilities $q_{1_i,2}, ..., q_{m_i,2}, ..., q_{U_i,2}$, if it is α'''_2 transaction (fig. 2 a, b, dash and dash-dot lines, correspondingly).

Thus in every state of processes α'''_i, $i \in \{1,2\}$ develops «concurrency» [11] between natural switching and switching through interruption. In every state of α'''_3 develops «concurrency» between natural switching, interruption from first process and interruption from second process. «Concurrency» of two and three subjects is modeled with Petri-Markov nets, shown on fig. 3 a and b correspondingly.

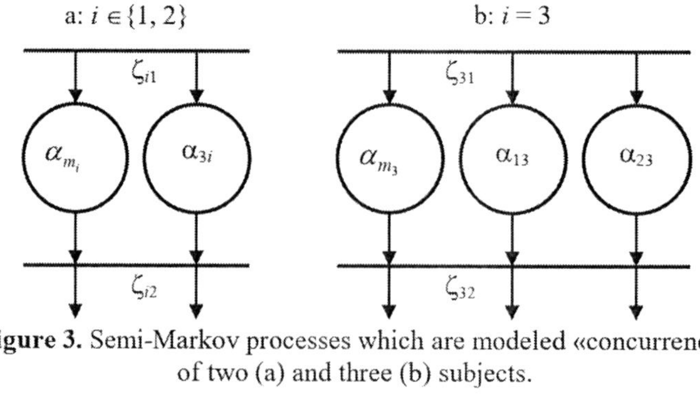

Figure 3. Semi-Markov processes which are modeled «concurrency» of two (a) and three (b) subjects.

IOP Conf. Series: Materials Science and Engineering **173** (2017) 012011 doi:10.1088/1757-899X/173/1/012011

Petri-Markov net which describes «concurrency» of two subjects, includes two places α_{m_i}, α_{3i}, and transitions $\bar{\zeta}_{i2}$ $i \in \{1, 2\}$. Places α_{m_i} are the models of states a''_{m_i} of Markov process. Places α_{31}, α_{32} are the models of transactions received.. Semi-steps $(\bar{\zeta}_{i1}, \alpha_{m_i})$ and $(\bar{\zeta}_{i1}, \alpha_{3i})$ are executed simultaneously. «Winner» is that place, from which semi-step will be executed the first, $(\alpha_{m_i}, \bar{\zeta}_{i2})$ or $(\alpha_{31}, \bar{\zeta}_{i2})$. Petri-Markov net which describes «concurrency» of three subjects, includes three places α_{m_3}, α_{13}, α_{23}, and transitions $\bar{\zeta}_{31}$, $\bar{\zeta}_{32}$. Place α_{m_3} is the model of states a''_{m_3} of Markov process. States α_{13}, α_{23} are the models of transactions received from the first and the second process, correspondingly. Semi-steps $(\bar{\zeta}_{31}, \alpha_{m_3})$, $(\bar{\zeta}_{31}, \alpha_{13})$ and $(\bar{\zeta}_{31}, \alpha_{23})$ are executed simultaneously. «Winner» is that place, from which semi-step will be executed first, $(\alpha_{m_3}, \bar{\zeta}_{32})$, $(\alpha_{13}, \bar{\zeta}_{32})$ or $(\alpha_{23}, \bar{\zeta}_{32})$.

Let us denote the density of flow of transactions from α_3 to α_1 as λ_{31}, and from α_3 to α_2 as λ_{32}. Then density of time of execution of semi-step to $\bar{\zeta}_{i2}$ the first, is defined as

$$f_{\zeta 2}(t) = \lambda_{m_i} \exp[-t(\lambda_{m_i} + \lambda_{3i})] + \lambda_{3i} \exp[-t(\lambda_{m_i} + \lambda_{3i})], \; i \in \{1, 2\}, \tag{24}$$

where λ_{m_i} - are the parameters of densities of flows of transactions from states a''_{m_i} to the process α_3; λ_{3i} - are the parameters of densities of flows of transactions from the process α_3 to the process α_i, $i \in \{1, 2\}$.

Weighted conditional density of the time of execution of semi-steps $(\alpha_{m_i}, \bar{\zeta}_{i2})$ and $(\alpha_{31}, \bar{\zeta}_{i2})$, the first, are as follows

$$\begin{bmatrix} \overline{h'''_{m_i,S_i+1}(t)} \\ \overline{h'''_{m_i,0_i}(t)} \end{bmatrix} = \begin{pmatrix} \dfrac{\lambda_{3i}}{\lambda_{3i} + \lambda_{m_i}} \exp[-t(\lambda_{m_i} + \lambda_{3i})] \\ \dfrac{\lambda_{m_i}}{\lambda_{3i} + \lambda_{m_i}} \exp[-t(\lambda_{m_i} + \lambda_{3i})] \end{pmatrix}, \; i \in \{1, 2\}. \tag{25}$$

Let us denote density of flow of transactions from α_1 to α_3 as λ_{13}, from α_2 to α_3 as λ_{23}. Then density of time of execution of semi-step to $\bar{\zeta}_{i2}$ the first, is as follows

$$\begin{aligned} f_{\zeta 2}(t) = \lambda_{13} \exp[-t(\lambda_{m_3} + \lambda_{13} + \lambda_{23})] + \lambda_{23} \exp[-t(\lambda_{m_3} + \lambda_{13} + \lambda_{23})] + \\ + \lambda_{m_3} \exp[-t(\lambda_{m_3} + \lambda_{13} + \lambda_{23})] \end{aligned} . \tag{26}$$

Weighted conditional density of the time of execution of semi-steps $(\alpha_{m_3}, \bar{\zeta}_{32})$, and $(\alpha_{13}, \bar{\zeta}_{32})$, $(\alpha_{23}, \bar{\zeta}_{32})$ are as follows:

$$\begin{bmatrix} \overline{h'''_{m_3,S_3+1}(t)} \\ \overline{h'''_{m_3,S_3+2}(t)} \\ \overline{h'''_{m_3,0_3}(t)} \end{bmatrix} = \begin{pmatrix} \dfrac{\lambda_{13}}{\lambda_{13} + \lambda_{23} + \lambda_{m_3}} \exp[-t(\lambda_{m_3} + \lambda_{13} + \lambda_{23})] \\ \dfrac{\lambda_{23}}{\lambda_{13} + \lambda_{23} + \lambda_{m_3}} \exp[-t(\lambda_{m_3} + \lambda_{13} + \lambda_{23})] \\ \dfrac{\lambda_{m_3}}{\lambda_{13} + \lambda_{23} + \lambda_{m_3}} \exp[-t(\lambda_{m_3} + \lambda_{13} + \lambda_{23})] \end{pmatrix}. \tag{27}$$

Parameters (25) and (27) present parameters of Markov process with the structure, which is shown on fig 2 a, b with solid, dash and dash-dot lines. Process is as follows

$$\overline{a_i'''} = \{\overline{A_i'''}, \overline{r_i'''}, \overline{h_i'''(t)}\}, \; i \in \{1, 2, 3\}, \tag{28}$$

where $\overline{\overline{A_i'''}} = \{a''_{0_i}, a''_{1_i}, \ldots, a''_{m_i}, \ldots, a''_{S_i}, a''_{S_i}\}$, if $i \in \{1, 2\}$, and $\overline{\overline{A_3'''}} = \{a''_{0_3}, a''_{1_3}, \ldots, a''_{m_3}, \ldots, a''_{U_3}, a''_{U_3+1}, a''_{U_3+2}\}$, if $i = 3$; $\overline{\overline{\boldsymbol{r}_i'''}} = (r_{m_i,n_i})$ - is the adjacency matrix of size $(S_i+1)\times(S_i+1)$, if $i \in \{1, 2\}$ and $(U_3+2)\times(U_3+2)$, if $i = 3$; $\overline{\overline{\boldsymbol{h}_i'''}}(t) = \left[\overline{\overline{h'''_{m_i,n_i}}}(t)\right]$ Markov matrix of the same size, as adjacency matrix;

$$\overline{\overline{\boldsymbol{r}_i'''}} = \begin{pmatrix} 0 & 1 & \ldots & 1 & \ldots & 1 & 0 \\ 1 & 0 & \ldots & 0 & \ldots & 0 & 1 \\ & & & \ldots & & & \\ 1 & 0 & \ldots & 0 & \ldots & 0 & 1 \\ & & & \ldots & & & \\ 1 & 0 & \ldots & 0 & \ldots & 0 & 1 \\ 0 & 1 & \ldots & 1 & \ldots & 1 & 0 \end{pmatrix}, i \in \{1, 2\}; \quad (28)$$

$$\overline{\overline{\boldsymbol{h}_i'''}}(t) = \begin{pmatrix} 0 & \pi_{1_i}\delta(t) & \ldots & \pi_{m_i}\delta(t) & \ldots & \pi_{S_i}\delta(t) & 0 \\ \overline{\overline{h'''_{1_i,0_i}}}(t) & 0 & \ldots & 0 & \ldots & 0 & \overline{\overline{h'''_{1_i,0_i}}}(t) \\ & & & \ldots & & & \\ \overline{\overline{h'''_{m_i,0_i}}}(t) & 0 & \ldots & 0 & \ldots & 0 & \overline{\overline{h'''_{m_i,S_i+1}}}(t) \\ & & & \ldots & & & \\ \overline{\overline{h'''_{S_i,0_i}}}(t)1 & 0 & \ldots & 0 & \ldots & 0 & \overline{\overline{h'''_{m_i,S_i+1}}}(t) \\ 0 & q_{1_i}\delta(t) & \ldots & q_{m_i}\delta(t) & \ldots & q_{S_i}\delta(t) & 0 \end{pmatrix}, i \in \{1, 2\}; \quad (29)$$

$$\overline{\overline{\boldsymbol{r}_3'''}} = \begin{pmatrix} 0 & 1 & \ldots & 1 & \ldots & 1 & 0 & 0 \\ 1 & 0 & \ldots & 0 & \ldots & 0 & 1 & 1 \\ & & & \ldots & & & & \\ 1 & 0 & \ldots & 0 & \ldots & 0 & 1 & 1 \\ & & & \ldots & & & & \\ 1 & 0 & \ldots & 0 & \ldots & 0 & 1 & 1 \\ 0 & 1 & \ldots & 1 & \ldots & 1 & 0 & 0 \\ 0 & 1 & \ldots & 1 & \ldots & 1 & 0 & 0 \end{pmatrix}, i = 3; \quad (30)$$

$$\overline{\overline{\boldsymbol{r}_3'''}} = \begin{pmatrix} 0 & \pi_{1_i}\delta(t) & \ldots & \pi_{m_i}\delta(t) & \ldots & \pi_{U_i}\delta(t) & 0 & 0 \\ \overline{\overline{h'''_{1_3,0_3}}} & 0 & \ldots & 0 & \ldots & 0 & \overline{\overline{h'''_{1_3,U_3+1}}} & \overline{\overline{h'''_{1_3,U_3+2}}} \\ & & \ldots & & & & & \\ \overline{\overline{h'''_{m_3,0_3}}} & 0 & \ldots & 0 & \ldots & 0 & \overline{\overline{h'''_{m_3,U_3+1}}} & \overline{\overline{h'''_{m_3,U_3+2}}} \\ & & \ldots & & & & & \\ \overline{\overline{h'''_{U_3,0_3}}} & 0 & \ldots & 0 & \ldots & 0 & \overline{\overline{h'''_{U_3,U_3+1}}} & \overline{\overline{h'''_{U_3,U_3+1}}} \\ 0 & q_{1_3,1}\delta(t) & \ldots & q_{m_3,1}\delta(t) & \ldots & q_{U_3,1}\delta(t) & 0 & 0 \\ 0 & q_{1_3,2}\delta(t) & \ldots & q_{m_i,2}\delta(t) & \ldots & q_{U_3,1}\delta(t) & 0 & 0 \end{pmatrix}, i = 3. \quad (31)$$

Let us denote probabilities and densities, which are obtained when preliminary evaluation as ${}^0\pi_{m_i} = \pi_{m_i}$, ${}^0\overline{\overline{\boldsymbol{h}_i'''}}(t) = \overline{\overline{\boldsymbol{h}_i'''}}(t)$, on the l-th stage of evaluation as ${}^l\pi_{m_i}$, ${}^l\overline{\overline{\boldsymbol{h}_i'''}}(t)$. Then parameters of flows of transactions may be obtained with the use of the following iteration procedure.

1. Parameters ${}^0\pi_{m_i}$, ${}^0\overline{\overline{\boldsymbol{h}_i'''}}(t)$ are calculated excluding interaction, as it was described above.

2. Recalculation of parameters ${}^{l+1}\pi_{m_i} = \varphi_\pi\left[{}^{l}\pi_{m_i},\ {}^{l}\overline{\boldsymbol{h}_i'''(t)}\right]$, ${}^{l+1}\overline{\boldsymbol{h}_i'''(t)} = \varphi_h\left[{}^{l}\pi_{m_i},\ {}^{l}\overline{\boldsymbol{h}_i'''(t)}\right]$, where φ_π and φ_h - are the functions, which are defined by dependencies, obtained above.

3. Iterations should be repeated until satisfactory difference between previous and current meanings of parameters will be obtained.

5. Conclusion

The analytical model of mobile robot dialogue regimes has been developed. The model describes the interaction of three subjects: a human operator, the dialogue computer and the onboard computer. The operation of every subject is dissected into elementary operations for which time of execution and probabilities of transition to the next possible operation may be simply evaluated. This permit to tune a mobile robot control system to solving of the concrete tasks with concrete conditions of their execution, and concrete condition of environment. The result obtained may be used for working out other dialogue systems, for example for industry robots.

Further continuation of investigations in this domain may be directed to an improvement of iteration procedure, to optimization of dialogue algorithms and its adaptation to characteristics of both a human operator and a mobile robot, to optimization the transaction flows in concrete systems etc.

References

[1] Kahar S, Sulaiman R, Prbuwono S, Ahmad N and Abu Hassan M 2012 A review of wireless technology usage for mobile robot controller *International conference on system engineering and modeling (ICSEM 2012)* **34** pp 7 - 12

[2] Ivutin A, Larkin E and Kotov V 2015 Established routine of swarm monitoring systems functioning *Advances in Swarm and Computational Intelligence* pp 415–422

[3] Tzafestas S 2014 *Introduction to Mobile Robot Control* 750 p

[4] Larkin E and Ivutin A 2014 Estimation of Latency in Embedded Real-Time Systems *3-rd Mediterranean Conference on Embedded Computing (MECO-2014)* pp 236 – 239

[5] Li Y, Malik S and Wolf A 2016 Efficient microarchitecture modeling and path analysis for real time software *Proceedings of 16-th Real time systems simposium IEEE* pp 298 - 307

[6] Ivutin A, Larkin E, Lutskov Y and Novikov A 2014 Simulation of concurrent process with Petri-Markov nets *Life Science Journal* **11(11)** pp 506 – 511

[7] Korolyuk V and Swishchuk A 1995 Semi-Markov Random Evolutions pp 59 – 91

[8] Larkin E, Ivutin A, Kotov V and Privalov A 2016 Semi-Markov modeling of commands execution by mobile robot *Lecture notes in artificial intelligence: Proceedings of first International conference «Interactive collaborated robotics ICR 2016»* pp 189 – 198

[9] Grigelionis B 1963 On the convergence of sums of random step processes to a Poisson process *Theory of Probability & Its Applications* **8** pp 177 - 182

[10] Ivutin A and Larkin E 2015 Simulation of Concurrent Games *Bulletin of the South Ural State University. Series: Mathematical Modelling* **8** pp 43 – 54

[11] Cleaveland R and Smolka S 1996 Strategic directions in concurrency research *CSUR* **28** pp 607 – 625

The application of the multi-alternative approach in active neural network models

S Podvalny, E Vasiljev

Voronezh State Technical University 14 Moscow Av., Voronezh, 394026, Russia

E-mail:spodvalny@yandex.ru

Abstract. The article refers to the construction of intelligent systems based artificial neuron networks are used. We discuss the basic properties of the non-compliance of artificial neuron networks and their biological prototypes. It is shown here that the main reason for these discrepancies is the structural immutability of the neuron network models in the learning process, that is, their passivity. Based on the modern understanding of the biological nervous system as a structured ensemble of nerve cells, it is proposed to abandon the attempts to simulate its work at the level of the elementary neurons functioning processes and proceed to the reproduction of the information structure of data storage and processing on the basis of the general enough evolutionary principles of multialternativity, i.e. the multi-level structural model, diversity and modularity. The implementation method of these principles is offered, using the faceted memory organization in the neuron network with the rearranging active structure. An example of the implementation of the active facet-type neuron network in the intellectual decision-making system in the conditions of critical events development in the electrical distribution system.

1. Introduction

Process of a circuitry development for the intelligent systems based on neuron networks is based on a direct analogy between the tasks solvable by these systems and the motives of higher nervous activity of living organisms.

Currently, the implementation of said reduced analogy is limited primarily to the attempts to reproduce electrochemical processes in the artificial biological neuron systems by way of the development of different neuron excitation and inhibition patterns and the elementary organization of their interrelation.

In its classic form a neuron network is seemingly the functional:

$$y_{j_N}(x) = F\left(\sum_{i_N} b_{i_N j_N} \cdot \ldots \cdot F\left(\sum_{i_2 j_2} b_{i_2 j_2} \times F\left(\sum_{i_1 j_1} b_{i_1 j_1} x_{i_1 j_1} + c_{j_1}\right) + c_{j_2}\right) + \ldots + c_{j_N}\right), \qquad (1)$$

where b, c are the vectors of configurable parameters (weight coefficients); r is the number of the network layer; j_r is a neuron number r in the layer; i_r is the entry number in the neuron; N is the number of layers of the network; x, y are the vectors of the input and output variables of the network;

$x_{ir,jr}$ is the i element of the vector x as an input to neuron j in the r layer; $F(b,c,x)$ is the neuron activation function, e.g., the sigmoidal form.

The formal justification for the use of neuron networks in the of decision-making problems is the theorem of completeness, which says that: every continuous function on a closed bounded set may be uniformly approximated by the functions computed by the neuron-type network (1), if the activation function $F(b,c,x)$ of a neuron is continuously twice differentiable.

However, in practice these formal grounds for the implementation of artificial properties of biological networks were not sufficient, and the artificial neuron networks of type (1) have inherent fundamental inconsistency with their biological prototype.

These inconsistencies are associated primarily with the attempt to reproduce biological processes using some functional with a predetermined constant structure, which alteration is made only at the lowest level, namely, the parametric one: the network is unable to alter its structure in the process of its training, i.e. it is passive.

A failure of passive neuron network to alternate its structure results in the shortcomings of these network as such, that are not peculiar to their biological prototypes [1,2].

The most important of these are:

the retraining problem which consists in the growth of a network error in case of presentation of an *a priori* unknown excessive amount of training vectors. *In vivo* neuron networks the memory elements demonstrate their high selectivity, whereas the nature of memory is cumulative, making it possible to keep previously stored data without the distortion on the network in amount that is practically unlimited;

low generalizing properties, establishing the relationship of "private-general" between the recognizable situations, for which the network should have developed a multi-level, hierarchical structure;

the lack of functional autonomy of the elements of artificial neuron networks resulting in a non-linear increase of the number of adjustable parameters alongside with the increase of the tasks number, i.e., the manifestation of the «curse of dimensionality» in the process of their training. In biological neuron networks there are no such limitations of a recurrently stored data, they were not observed.

In order to eliminate these shortcomings in this work a new approach is developed to the construction of a neuron network model of the intelligent control system based on the evolutionary principles of the organization of its functioning that are comparatively general, that is, the principles of multialternativity [3-5]:

the multilevel (hierarchy) principle which is implemented by the organization in the biological systems of at least two-level control one of which functions in the domain of tolerance variations of the system state, and the other responds to the critical deviations only from this condition;

the principle of diversity of the control algorithms and the separation of functions (multi-mode and switch structures), which is closely associated with the multilevel principle and reflects the flexibility of the control algorithms in the changing conditions of the system operation;

the principle of modularity of the structure which provides a variety of levels and modes of control based on the combinations of a limited set of elementary modules of a system.

As shown below, a set of these principles described above can solve the problem of constructing the neuron network intelligent systems which are not a result of the qualitative complexity of their algorithm, but by means of increasing gradually the number of simple ingredients in the system that are gradually embedded into the structure of the neuron network.

2. Active neuron network based on the multialternativity principles

Modern understanding of the nervous system as a structured ensemble of nerve cells suggests that biological neural networks are characterized by:

- independent storage of events of neuroactivity in the form of ensemble of associated neurons that are excited differentially by the particular sensory signals only. This independence eliminates the retraining;

- neuron network activity ensuring the restructuring of relations between the structure of ensembles after each training event (the formation of new stable ensemble and its embedding into the total structure of the network).

To implement these properties in an artificial neuron network the authors propose to abandon the attempts to simulate a biological nervous system at the level of the elementary functioning of neuronal processes and proceed to the reproduction of the data store and processing structure.

This structure may implement the faceted principle (from the French *facette* – a face) of objects classification can be used which is characterized by the fact that for every event, the ensemble (set) $\{f,s\}$ is formed from the signs-facets f, whose set of values s defines itself the specific object $a(f,s)$:

$$a \in A,\ |A| = z,\ f = \{f_1, f_2, ..., f_n\},\ s = \{s^{f_1}, s^{f_2}, ..., s^{f_n}\},\ s^{f_i} = \{s_1^{f_i}, s_2^{f_i}, ..., s_t^{f_i}\},\ i = \overline{1,n}. \qquad (2)$$

Faceted principle of data storage allows us to combine different objects in a network structure $A^i = a_1^i \cap a_2^i \cap ... \cap a_z^i$ separately for each feature f_i, feature where the introduction of this additional feature or object does not require any adjustment of a previously existing structure of relations but complements it.

Graphic illustration of the relations (2) describing the memory faceted organization in the neuron network is shown in Fig. 1.

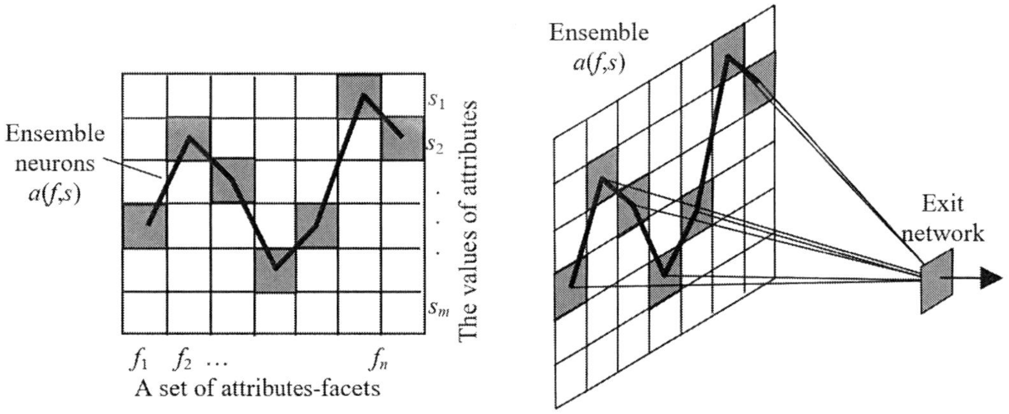

Figure 1. Faceted organization of memory in neural networks.

Fig. 2 shows a possible structure of the faceted type neuron network with three hierarchical levels of ensembles (sets) of $a(f,s)$ neurons which can be excited or inhibited by specific receptor-facets f and from other ensembles.

In particular, the excitation of f_4, f_5, f_6 receptors activates the $a_{1,2}$ ensemble and further, in the absence of inhibition by the $a_{1,1}$ ensemble, the $a_{2,1}$ ensemble is excited which is activated indirectly, i.e., not by the primary receptor f, but in case of a previous activation of $a_{1,2}$.

Thus, a controlled sequence of events occurs in the system, which corresponds to the $a_{1,2}$ and $a_{2,1}$ ensembles. At the same time the excitation of the $a_{2,1}$ ensemble in the absence of inhibition by other ensembles, such as $a_{2,2}$, $a_{2,3}$, lead to the activation of an $a_{3,1}$ event.

As a result, under the external influence of $\{f_4, f_5, f_6\}$ the sequence of events with the specified excitation has been formed.

3. Example of the active faceted type neuron network

As an example, the electric power distribution network is shown, consisting of an input high voltage substation and thirty low voltage transformer substations shown in Fig. 3 [6].

We assume that the power transmission cables indicated by dashed lines are reserved and shall be connected to the system in a critical event: an overload or failure of the main power cables.

Now we set the problem of the development of the decision making neuron network model of an enterprise providing in case of a critical event the command to connect the corresponding spare line.

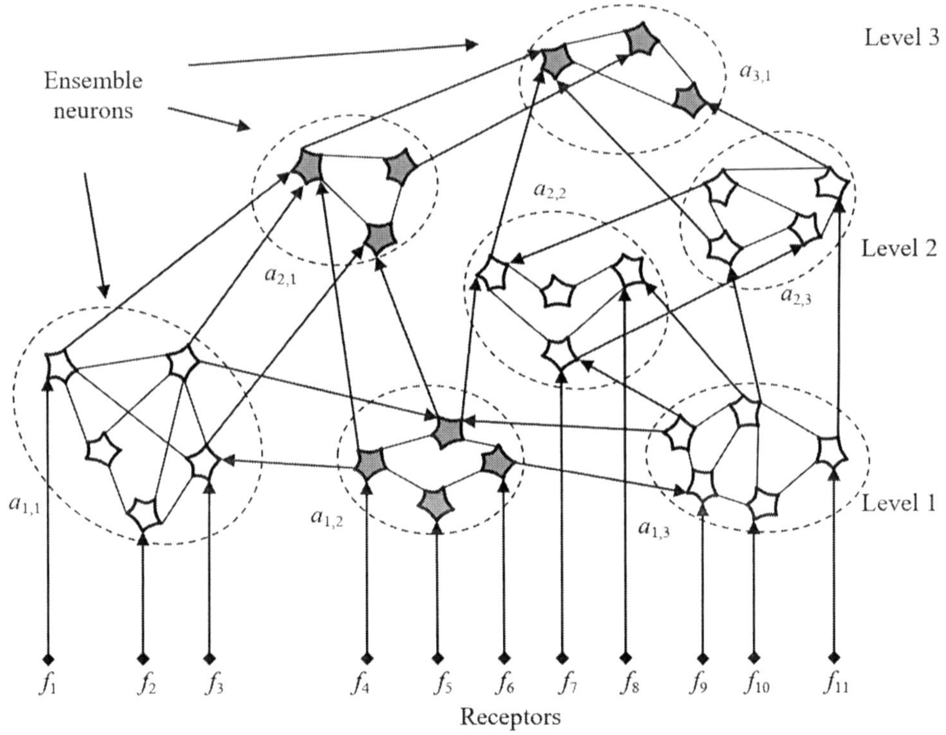

Figure 2. The structure of the neural network type facet.

The numbers of these events with the substation numbers to be connected to the failed line are shown in Table 1 below.

Table 1. The set of critical events.

Number events	1	2	3	4	5	6	7	8	9	10	11	12	13	14	15	16	17	18	19	20	21	22	23	24	25	26	27	28	29	30
Number of substations	0-1	1-2	2-3	3-4	3-8	8-9	8-7	7-6	6-5	0-26	26-27	27-22	23-28	28-29	29-30	23-22	22-21	21-20	20-19	0-10	10-16	16-18	18-17	18-15	15-11	11-12	12-13	13-14	27-24	24-25

For this task a prototype two-layer passive network with thirty inputs and ten neurons in the first layer and eleven outputs was chosen.

The numbering of the reserve line is shown in Table 2.

A list of decisions taken (the matrix of decision rules) is shown in Table 3.

The training of a neuron network was carried out by error back propagation and consisted of the presentation of thirty critical situation vectors x and their corresponding solutions y, that were required to restore the network operability. Furthermore, in the experiments 2 ... 13 (Table 4) in order to evaluate the properties of generalizing neuron network, the training was conducted on incomplete samples, whereas in the process of confirming the situations were presented for recognition that were not participating in the training.

Table 2. The numbering of the reserve line.

Number of the line	1	2	3	4	5	6	7	8	9	10	11
Number of substations	4-9	7-11	0-5	5-17	11-14	17-19	19-26	0-30	30-26	9-12	20-25

Table 3. The matrix of decision rules.

Number event	Number of the reserve line										
	1	2	3	4	5	6	7	8	9	10	11
1	0	0	1	0	0	0	0	0	0	0	0
2	0	0	1	0	0	0	0	0	0	0	0
3	0	0	1	0	0	0	0	0	0	0	0
4	1	0	0	0	0	0	0	0	0	0	0
5	0	0	1	0	0	0	0	0	0	0	0
...	...	...	...	...	...	...	...	...	...	...	...
29	0	0	0	0	0	0	0	0	0	0	1
30	0	0	0	0	0	0	0	0	0	0	1

Table 4. Single failure study results.

Test number	Situations that are not included in the training set	Numbers of unrecognized situations in the network testing
1	Not	Not
2	9	9,20
3	8,9	7,8,9,10,24,25,26
4	7,8,9	8,9,11,26
5	5,7,8,9	5,6,7,8,9,10,11
6	3,5,7,8,9	5,6,7,8,9,10,11
7	2,3,5,7,8,9	1,2,3,5,7,8,9,11
8	22	9,11,22,26
9	21,22	20,21,22
10	15	11
11	14,15	10,11,15
12	13,14,15	11,15
13	12,13,14,15	4,12,15,29,30

With the same purpose simultaneous failures of two main lines were simulated in the network training previously on a complete set of single failures (Table 5).

The test results are summarized in Tables 4, 5 and testify to the following:

- from thirteen options of trained networks only one network (test 1, Table 4) fully recognized all emergencies charged on it that were included in the training set;
- the majority of networks do not recognize events that are not included in the training set (Table. 4), i.e., passive neuron networks demonstrate the absence of extrapolating properties and a complete mismatch of their structure to the structure of critical events. In particular, in the test 2, Table 4 the situation 9 was not recognized which is a special case of the six (*sic*!) situations numbered as 1, 2, 3, 5, 7, 8 (see. Fig. 1, tabl.1,2), each of those are normally recognized by the network. Similar

results are shown by tests 3 ... 7; 8, 9; 11 ... 13, shown in Table 1. These facts are an evidence of the absence of generalizing ability of the obtained networks;

- in case of an occurrence of two events, where each event is recognized individually by the network in all experiments (Table 5), no fully correct decision has been obtained indicating that the interpolating and selectivity properties of a network are low.

Table 5. Double failure study results.

Test number	Situations that are created in the process of the network testing	Numbers of unrecognized situations in the network testing
1	1 и 20	1,20
2	1 и 10	1,10
3	10 и 20	10,20
4	24 и 20	24
5	10 и 11	11

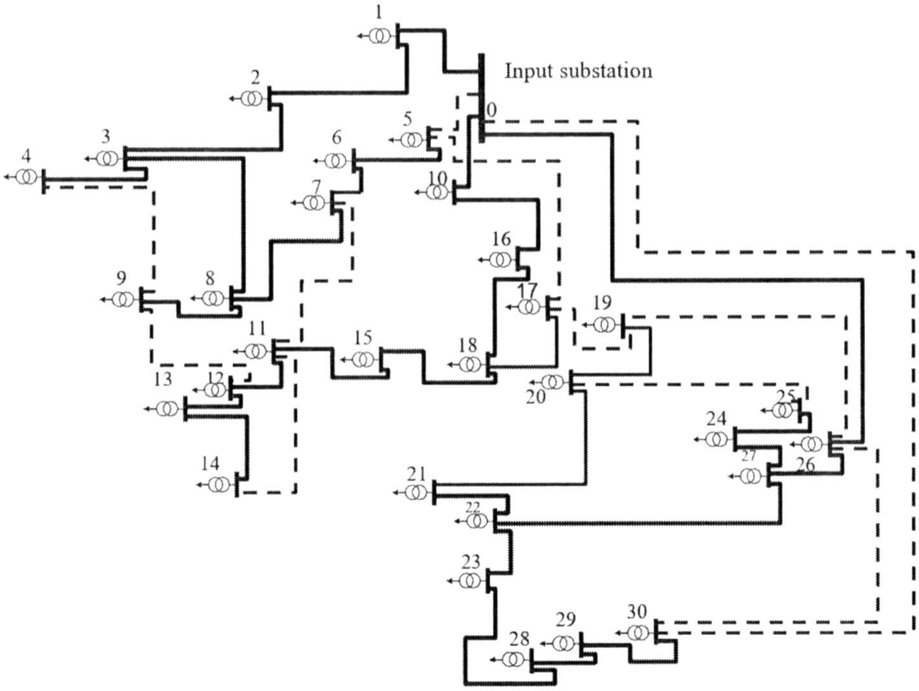

Figure 3. The configuration of the electrical network under consideration.

In general, the low functionality properties revealed in the constructed passive neuron network does not permit to use it for the responsible control tasks of critical situations. Any attempts to change the network configuration, the number of layers, the number of neurons and the training method do not lead to a substantial improvement in the network because of its general fundamental limitations: the network is not able to change its structure in the process of training, i.e., it is passive.

Elimination of this limitation is possible by way of the transition to the concept of a network design based on the evolutionary principles of multialternativity: the modularity, multi-level structure and division of functions that which keads to the creation of active neuron networks.

Let's proceed to the construction of a neuron network of the active faceted type, the general principles of which were described above.

To obtain the perfect solution, the list of facets $f = \{f_1, f_2, ..., f_n\}$, n=30, is a set of numbers of de-energized electric power substations; the number of events-objects z = 30; the number (quantity) of each parameter values t = 2: $s^{f_i} = \{0;1\}$.

Graphic representation of the faceted network fragment is shown in Fig. 4, where the ensembles (sets) of neurons F realize the AND logical operation. Numbering of the F ensembles corresponds to the event numbers.

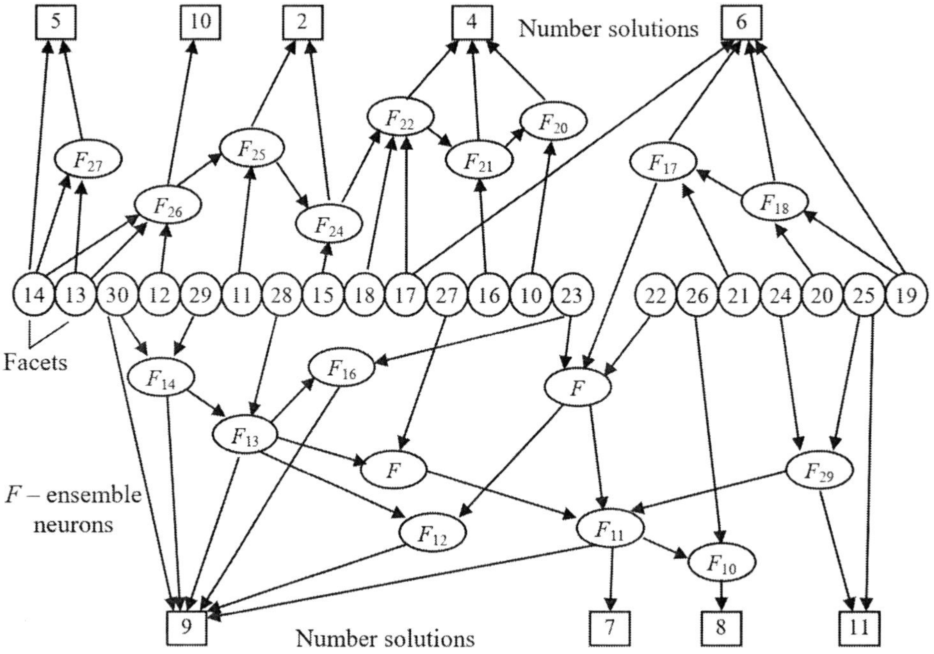

Figure 4. Faceted active neural network structure.

An important feature of the resulting model is its structural similarity of the object model, i.e., the structure of this model displays an object environment.

For example, if the electric substations 12, 13 and 14 are de-energized, it corresponds to the event 26 (F_{26} in Fig. 4) and to the decision to activate the spare reserve line 10 (see. Fig. 3) in case of failure of the main line 11-12. At the same time, from Fig. 3 it follows that the special case of the situation are the de-energized substations 13 and 14. This case corresponds to event 27 and the adoption of solutions by the model for the activation of a reserve line 5.

Thus, the use of the model will lead to the activation of two redundant lines Nos. 10 and 5. Analysis of the electrical circuit in Fig. 3 shows that in case of a failure of the primary line 11-12 both solutions are equivalent. It follows that the model has generalizing properties to the extent which is allowed by the object structure. Indeed, if the network in the process of training is charged with the situation 20 (F_{20} in Fig. 4) which is mapped in turn to the solutions 2, 4, 5 and 10, then it would be sufficient to correct parry of the critical situations 21, 22, 24, 25, 26, 27, 28, that were not included in the training whereas all of them are the special, particular cases of the general situation 20.

It should be noted that the emergence of generalizing properties of the faceted neuron network is explained by the presence in its inherent hierarchical relationships, i.e., its multi-level structure.

Repeated experiments reported in Tables 3 and 5, for a constructed faceted neuron network has fully confirmed its advantages over the passive network: in all cases of single and double failures the active neuron network correctly recognized the situation and made an appropriate decision.

The learning procedure in this neuron network includes:

adding a new object in the form of an ensemble of characteristic values $a(f,s)_{z+1}$;

the inclusion of each feature-facet of an object in the network structure of this feature.

Here the pre-formed relationships in the network are not destroyed, thus eliminating the possibility of the network (partial or total) retraining phenomenon. Since the block nature of the training is reduced to a simple addition of new situations data and does not require a multi-parameter optimization, the network preserves its highly selective properties: for each emergency situation a single solution is generated with a superposition of these solutions, in case of multiple failures. For example, while the occurrence of the events 10 and 11 the model shall work out the decisions 7, 8 and 9, fully restoring the power supply (see. Figure 3).

4. Conclusion

Using passive models of neuron networks in the decision making and control systems faces significant difficulties for their implementation due to the propensity of these models to retraining and their low extrapolating options.

Designing the neuron systems is based on the evolutionary principles of multialternativity provides the creation of multiple-action neuron networks with reconfigurable structure, the properties of which resemble to a greater extent their biological prototypes. The comparative analysis of examples of the implementation of passive and active neuron networks in this work demonstrates the following:

a multi-level hierarchical scheme of internal connections in the active network provides high generalizing ability of the system in the process of the decisions making in the situations that were not encountered in the process of their training;

modular structure allows to build-in the new ensembles implementing the initial structure of the active neuron system, without facing the particular restrictions of the «curse of dimensionality» and the retraining effect;

faceted memory organization following the «one event - one group» rule provides an opportunity to build-up an unlimited number of selected events in the active system and the practical implementation of the principle of information variety.

References

[1] Jandel M 2014 Biologically relevant neural network architectures for support vector machines *Neural Networks* **49** pp 39-50

[2] Weng J, Luwang T, Lu H and Xue X 2008 Multilayer in-place learning networks for modeling functional layers in the laminar cortex *Neural Networks* **21** pp 150-159

[3] Podvalny S L and Vasiljev E M 2015 A multi-alternative approach to control in open systems: origins, current state, and future prospects *Automation and Remote Control* **76(8)** pp 1471-1499

[4] Podvalny S L and Vasiljev E M 2015 Evolutionary principles for construction of intellectual systems of multi-alternative control *Automation and Remote Control* **76(2)** pp 311-317

[5] Podvalny S L, Vasiljev E M and Barabanov V F 2014 Models of Multi-Alternative Control and Decision-Making in Complex System *Automation and Remote Control* **75(10)** pp 1886-1891

[6] Vasiljev E M and Govorov R A 2015 Active neural network control model of critical objects *Bull. Voronezh State Tech.* **11(3)** pp 31-36

IOP Conf. Series: Materials Science and Engineering **173** (2017) 012013 doi:10.1088/1757-899X/173/1/012013

A study of fuzzy logic ensemble system performance on face recognition problem

A Polyakova[1], **L Lipinskiy**[1]

[1] Reshetnev Siberian State Aerospace University
31 Krasnoyarskiy Rabochiy Ave., 660037, Krasnoyarsk, Russia

E-mail: polyakova_nasty@mail.ru

Abstract. Some problems are difficult to solve by using a single intelligent information technology (IIT). The ensemble of the various data mining (DM) techniques is a set of models which are able to solve the problem by itself, but the combination of which allows increasing the efficiency of the system as a whole. Using the IIT ensembles can improve the reliability and efficiency of the final decision, since it emphasizes on the diversity of its components. The new method of the intellectual informational technology ensemble design is considered in this paper. It is based on the fuzzy logic and is designed to solve the classification and regression problems. The ensemble consists of several data mining algorithms: artificial neural network, support vector machine and decision trees. These algorithms and their ensemble have been tested by solving the face recognition problems. Principal components analysis (PCA) is used for feature selection.

1. Introduction

Nowadays informational technologies (IT) find their application in different aspects of society. The power of IT and modern computational technologies is an important tool for solving complex application problems.

Problems for which the intellectual intelligent technologies are used have the following characteristic features: the incompleteness and inaccuracy of initial data, the high computational complexity of the results and their formalization.

Such data analysis problems are difficult to solve using a separate technology. For example, when it is impossible to obtain a formalized mathematical description for the data mining problem, one of the good solutions could be to solve the problem with an artificial neural network (ANN) [1, 2]. When you need to make a decision based on the experience of experts in the subject area of the problem, it is better to use a fuzzy logic system (FLS) to extract the expert experience [3].

The advantage of using support vectors machines is the lack of necessity of the pre-study behavior data [4]. The method of support vectors machine is based on solving a quadratic programming problem, which always has a unique solution.

Ultimately, the use of ensembles of IIT can significantly improve the efficiency of solving applied problems in many areas of human activity.

The ensemble may consist of different methods or of the same one but with the special settings each. For example, if the ensemble consists of several ANN, as described in [5], the separate neural network has a number of layers, number of neurons and other parameters may also vary.

The ensemble can be formed with various methods. In [6], the ensemble is made by means of genetic programming (GP), in [7, 8] the three-stage evolutionary approach is used and in [9] the ensemble based on a C4.5 algorithm.

The new scheme for the ensemble decision design based on fuzzy logic is proposed in this paper.

2. Methodology

The authors propose a ensemble approach based on fuzzy logic system to solve the classification and regression problems (Fig. 1.).

Suppose we have a sample $(\bar{x}_i, y_i{}^j), i = 1,2,\dots,n$, where n - sample size; $\bar{x}_i = (x_{i,1}, x_{i,2}, \dots, x_{i,m})^T$ - set of independent variables, m - the number of variables in the problem; $y_i{}^j = f(\bar{x}_i)$ - corresponding to $\bar{x}_i$ dependent variable, $j = \overline{1,v}$ - a finite number of classes (names, labels). It is required to effectively design the fuzzy controller capable of forming algorithms ensembles.

Fuzzy controller makes a decision of the algorithms choice depending on the proximity of the test object to the objects in the training sample and the accuracy of the classifier on the nearest subject.

The fuzzy controller has to provide the minimal solution error by the efficient distribution of computational resources between algorithms.

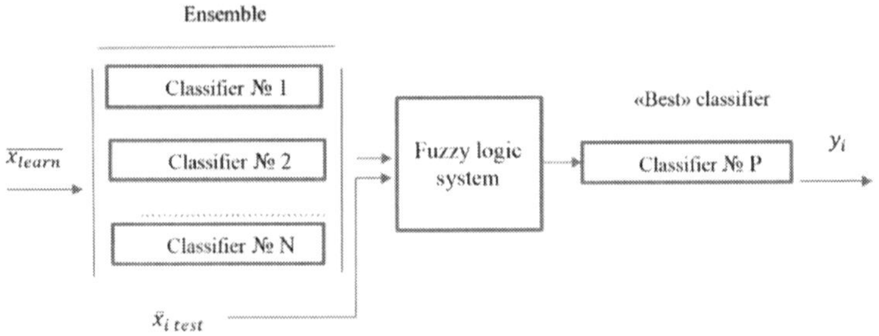

Figure 1. The ensemble approach based on fuzzy logic system to solve classification problems.

The sample $(\bar{x}_i, y_i)$ should be divided into 2 parts: training and testing. The task is to synthesize the critical rule based on the train sample that distinguish the classes with the least possible error, i.e. to recognize the new, not encountered earlier objects. The test sample is needed to assess the quality of the decision rule.

The scheme of the proposed approach is presented below:

1. All members of the ensemble are trained using the train sample X_{learn}.
2. Formed an ordered list of algorithms as follows:

2.1. Determine t=1, and Ensemble $= \emptyset$, $\varepsilon > 0$ - accuracy (threshold), N - the maximum number of algorithms in the ensemble, X= X_{learn}.

2.2. $Algorithm_t$ is the algorithms with the greatest proportion of correctly classified objects on X. For the regression it is the algorithm with the least error on X. Ensemble = Ensemble $\cup$ $Algorithm_t$.

2.3. A new set is formed $X = X \setminus \{x_i: Algorithm_t(x_i) = y_i\}$. Determine t=t+1.

2.4. The error is calculated: $Error\ (Ensemble) = \frac{|X|}{|X_{learn}|} * 100\%$.

For classification problems $|X|$ - a set of incorrectly classified objects. For regression problems $|X|$ - a set of objects for which an error value is less than the threshold, where the threshold set by the user.

Repeat 2.2. - 2.4. as long as the $X \neq \emptyset$, or $Error > \varepsilon$, or $|Ensemble| < N$.

In this scheme, the learning algorithm is performed on the entire training set X_{learn}. The ensemble is constructed of the methods that have shown the greatest efficiency in solving the classification problem (regression). Decision making algorithm is based on fuzzy controller.

The rule base and the following linguistic variables are formed to construct the fuzzy controller:

1. Linguistic variable №1 (Fig. 2.): The proximity of the object in training sample to the test sample. Term-set = {low, medium, far}. Basic variable: the distance in the Euclidean metric.

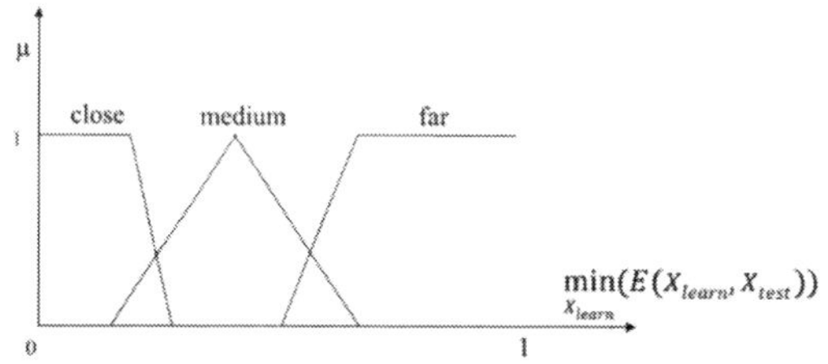

Figure 2. Schedule of underlying variable for linguistic variable №1.

2. The linguistic variable №2 (Fig. 3.): classifier error (regression error) on a sample object. Term - set = {low, medium, high}. Basic variable: the distance in the Euclidean metric.

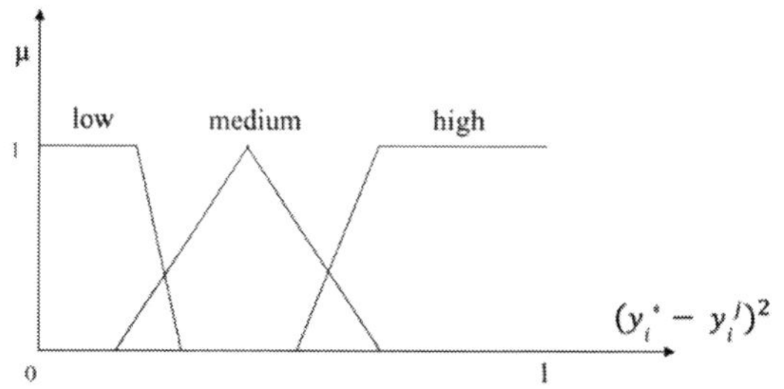

Figure 3. Schedule a base variable for linguistic variable №2.

3. Linguistic variable №3 (Fig. 4.): the credibility of the classifier (regression algorithm). Term-set = {low, medium, high}. Basic variable: a dimensionless quantity in the range from 0 to 1, where 0 - complete lack of confidence in the algorithm, and 1 - absolute trust.

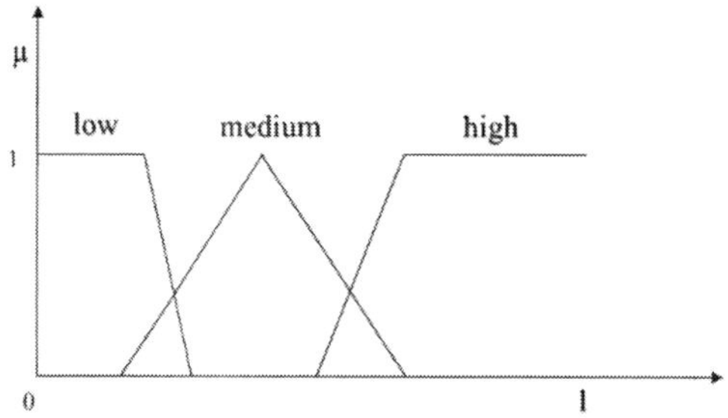

Figure 4. Schedule a base variable for linguistic variable №3.

The output of a fuzzy controller is the degree of confidence to the classifier (regression algorithm). The qualifier (the regression algorithm) with the higher credibility is selected.

Suppose there are N classification algorithms (regression), $i = \overline{1, N}$.

The rule base is as follows:

- IF the Algorithm has a *low* error AND the closest object from its training sample is *close* to the test object, the degree of confidence to the Algorithm is *high*
- IF the Algorithm has a *medium* error AND the closest object from its training sample is *medium* to the test object, the degree of confidence to the Algorithm is *average*

....

- IF the Algorithm has a *high* error AND the *closest* object from its train sample is *far* to a test object, the degree of confidence to the Algorithm *lower*

Fuzzy credibility output is calculated for each algorithm. The membership functions for linguistic variables have a triangular form.

The developed decision making algorithm using fuzzy controller is the following:

1. The object $\bar{x}_{i\ test}$ from the test sample is an input to the ensemble.
2. Calculates the value of linguistic variables:
   a. Calculate $\rho(\bar{x}_{i\ test}, x_j), \forall x_j \in X_{learn}$. Object x_j, for which $min_{x_j \in X_{learn}} \rho(\bar{x}_{i\ test}, x_j)$ is said to be x^*. Calculate the value of the function membership $\mu_1\left(min_{x_j \in X_{learn}} \rho(\bar{x}_{i\ test}, x_j)\right)$ for the linguistic variable "The proximity of the object of training sample to the test sample".
   b. Calculate $(y_i^*(x^*) - y_i^j)^2\ \forall j\ (j = 1,\ v$ - a finite number of labels). Object y_i^j, for which $min_{y_i^j}(y_i^*(x^*) - y_i^j)^2$ is called $y_i^{j^*}$. Calculate the value of the function membership $\mu_2\left(min_{y_i^j}(y_i^*(x^*) - y_i^{j*})^2\right)$ for linguistic variable " The success of the classifier (regression algorithm) on a sample object".
3. On the basis of linguistic variables $\mu_1{}^k, \mu_2{}^k$ the k (where $k = 1, N$ - number of algorithms in the ensemble) is calculated as the degree of confidence (degree of membership) to each algorithm with center of gravity method $\mu_k = F(\mu_1{}^k, \mu_2{}^k)$.

On the basis of the values determined by the membership the kth algorithm in rule with the maximum degree of confidence is determined: $k^* = arg(\max_k(\mu_k))$.

The traditional classification methods work with objects, the parameters of which are given exclusively in the clear form, making them difficult to work with fuzzy objects. Currently for the classification of such objects the methods based on fuzzy logic are being actively developed. These methods form a classes with blurred boundaries and the object can simultaneously belong to several of them with varying degrees of membership.

When an object belongs to the different clusters equally or does not belong to one, the algorithm may be wrong. This problem could be successfully solved using fuzzy classification. In this case, the fuzzy inference is performed by using the classifiers or regression algorithms combination. Instead of a clear answer to the question "what class does the object has?", it determines the level of a membership to a particular class. Thus, the statement "object A belongs to the class 1 with a probability of 90%, Class 2 - 10%" is true and more convenient.

With the help of a software system that implements the algorithm described above on the basis of the formation of the ensemble fuzzy controller, the practical face recognition problems for their images has been resolved.

3. Face recognition problem

The automated human-machine system, working in the dialog mode is called a dialogue system. In the dialog mode, such system responds to all commands from the user, and also appeals to him for confirmation of the execution of specific actions or for obtaining additional information. [10]

Currently, dialog systems are applied in the most different fields of human activity. As a result of human-machine communication many parameters can be determined, such as age, sex, emotions, and other characteristics of the person. [11]

Pattern recognition problem is the main task of data mining. The human face has a central role in social interaction, it is not surprising that the automatic processing of information for the face is an important and a very active area of research. Face detection is a special case of pattern recognition problem [12].

Solving face recognition problem performed in many computer vision systems. The most common systems are the biometric verification and identification systems. Application areas of such systems are biometric scanners and passport, system of technical vision of robots, etc.

Face recognition based on the image is done in two stages:

1. Face positioning on the image;
2. Personality identification based on face images.

The complexity of this problem is determined by some factors:

1. Variability of the image from lighting (directivity, the type and number of light sources);
2. Different spatial positions (relative to the camera) and the scales of faces to their images;
3. The high variability of individuals because of the anatomical features of human variability and type of appearance (for a period of time).

Regarding this, in order to obtain reliable results for face recognition algorithms to their image, it is necessary to perform several requirements for images [13]:

1. All images must be converted from an RGB (color model) to grayscale;
2. Pictures should be centered;
3. Images must be the same size.

The classical scheme of the face recognition algorithm is shown in figure 5:

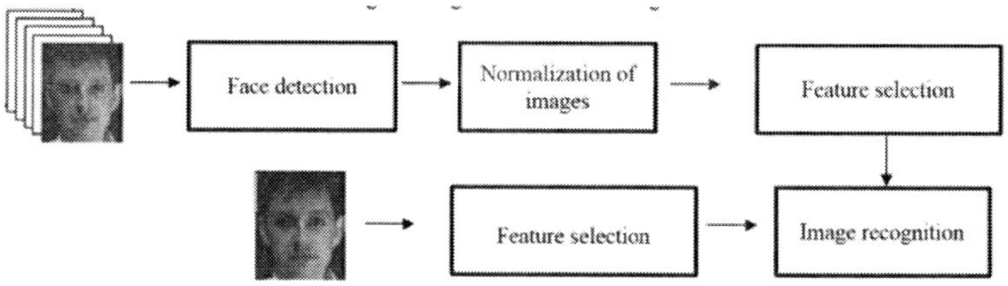

Figure 5. Stages of the classical scheme of the face recognition algorithm.

Extracting feature vectors is based on the image. Each pixel of the analyzed image becomes a component of the vector, transforming the black and white image into a vector space [14] (Fig. 6.).

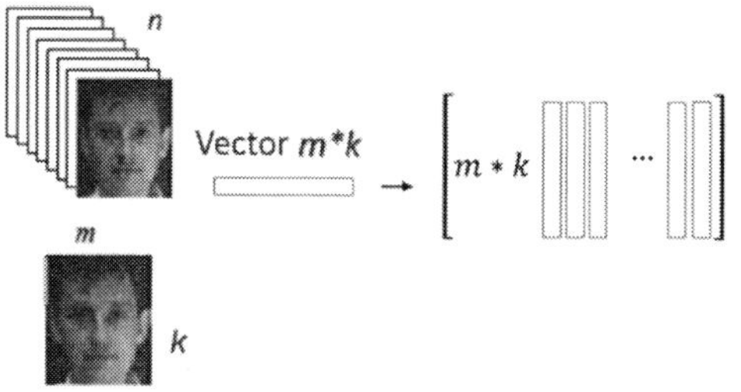

Figure 6. Extraction the feature vector of the image dimension m × k.

Using of all set of numerical attributes in the process of solving the classification problem can significantly slow down the algorithm and reduce the accuracy of the solution. Therefore it is important to reduce the dimension in the process of problems solving with the feature extraction techniques. In order to select the most suitable characteristics, statistical methods such as factor analysis, as well as more complex, for example, based on genetic algorithms (GA) may be used.

In [15, 16] the procedure of extraction of informative features based on adaptive multicriteria genetic algorithm is considered, its efficiency in combination with different classification models is researched.

In this paper, the principal component analysis (PCA) is used for feature selection. Principal component analysis is based on determining the minimum number of factors which make the greatest contribution to the dispersion of the data.

4. Experimental results

Three image databases was used for the experimental research:

1. AT & T Facedatabase [17];
2. Faces95 [18].

AT & T Facedatabase contains images of 40 individual speakers of 10 images each. Images of persons are represented in the resolution of 92 × 112 pixels.

For some subjects, the images were taken at different times, varying the lighting, facial expressions (open / closed eyes, smiling / not smiling) and facial details (glasses / no glasses). All the images were taken against a dark homogeneous background with the subjects in an upright, frontal position (with tolerance for some side movement).

Database of Faces95 contains images of 74 individual speakers of 20 images each. Images of persons are represented in the resolution of 180 × 200 pixels.

As the methods for the ensemble the following algorithms presented in the software system RapidMiner Studio were selected [19]:

1. Artificial neural network (ANN);
2. The method of decision trees (WJ48);
3. Support Vector Machines (SVM).

The principal component analysis is used to reduce dimensionality. Splitting into the test and training samples was made in a random way in the ratio of 70/30. The training of all the algorithms included in the ensemble was done on all training set. All tasks algorithms were run 10 times.

All the results were averaged. Also was the T-Test operator was applied (significance level - 0.05) by Student's criterion. A comparative analysis of the effectiveness of each method individually with the ensemble, which includes all 3 algorithm, was done.

The criterion of effectiveness is classification accuracy - the proportion of correctly recognized items in the sample.

Components number decreased from 10304 to 399 after using of the PCA method for the first problem. The results are shown in Table 1.

Table 1. Comparative analysis of efficiency for task №1 (all components).

T-test		77.42%	61.83%	77.42%
		ANN	**WJ48**	**Ensemble**
72.25%	**SVM**	0.00493818	0.000159513	0.004938181
77.42%	**ANN**		3.02454E-05	-
61.83%	**WJ48**			3.02454E-05

According to Table 1 the ensemble and artificial neural network have shown the best result. The analysis by Student's criterion shows that the artificial neural network and ensemble have statistically significant distinctions in comparison with all other algorithms, but statistically not significant among themselves.

A comparative analysis at 10, 50 and 100 components was improved. The results are shown in Tables 2 - 4.

Table 2. Comparative analysis of efficiency for task №1 (number of components – 10).

T-test		91.501%	62.748%	91.833%
		ANN	**WJ48**	**Ensemble**
82.498%	**SVM**	0.000351169	0.00022555	0.000282261
91.501%	**ANN**		3.49373E-07	0.105112517
62.748%	**WJ48**			2.41443E-07

Table 3. Comparative analysis of efficiency for task №1 (number of components – 50).

T-test		95.916%	63.418%	95.916%
		ANN	**WJ48**	**Ensemble**
93.333%	**SVM**	0.122178256	6.85314E-06	0.122151877
95.916%	**ANN**		1.5052E-07	-
63.418%	**WJ48**			1.50376E-07

Table 4. Comparative analysis of efficiency for task №1 (number of components – 100).

T-test		95.966%	62.167%	95.666%
		ANN	**WJ48**	**Ensemble**
93.667%	**SVM**	0.000486134	1.06871E-08	0.000486754
95.666%	**ANN**		2.30975E-08	-
62.167%	**WJ48**			2.31149E-08

According to the results which were presented in the tables 2-4, we can conclude that with 10 components the ensemble has the best mean precision value, while the difference from the other algorithms are statistically significant.

The ensemble has the same mean precision value as neural network. Comparison of the results obtained by the ensemble and the neural network demonstrated that differences are not statistically significant.

In paper [20] the recognition results of the methods Eigenfaces and N-PCA for database Faces95 are presented (with various train and test percentage of database are presented).

Table 5. Comparative analysis of efficiency for task №1 for some algorithms.

	Accuracy
SVM	93.667%
ANN	**95.966%**
WJ48	62.167%
Ensemble	**95.666%**
Eigenfaces	92.50%
N-PCA	93.75%

According to the results which were presented in the table 5, we can conclude that the ensemble and artificial neural network has the best mean precision value.

Components number decreased from 36000 to 432 after using of the PCA method for the second problem. The results are shown in Table 6.

Table 6. Comparative analysis of efficiency for task №2 (all components).

T-test		83.262%	71.203%	83.2638%
		ANN	**WJ48**	**Ensemble**
76.065%	**SVM**	3.60889E-08	0.001098475	3.5693E-08
83.262%	**ANN**		6.53248E-07	0.022045801
71.203%	**WJ48**			6.5058E-07

According to Table65 the ensemble have shown the best result. The analysis by Student's criterion shows that the ensemble have statistically significant distinctions in comparison with all other algorithms.

Also let's conduct comparative analysis for 10, 50 and 100 components. The results are shown in Tables 7 - 9.

Table 7. Comparative analysis of efficiency for task №2 (number of components – 10).

T-test		87.708%	73.611%	89.398%
		ANN	**WJ48**	**Ensemble**
63.008%	**SVM**	3.56777E-10	3.22703E-06	8.83219E-11
87.708%	**ANN**		8.0621E-09	0.000747201
73.611%	**WJ48**			5.39682E-10

Table 8. Comparative analysis of efficiency for task №2 (number of components – 50).

T-test		91.921%	72.917%	92.269%
		ANN	**WJ48**	**Ensemble**
87.222%	**SVM**	3.30767E-06	7.5287E-09	8.23075E-07
91.921%	**ANN**		2.48827E-09	0.004601526
72.917%	**WJ48**			2.16802E-09

Table 9. Comparative analysis of efficiency for task №2 (number of components – 100).

T-test		92.386%	71.992%	92.454%
		ANN	**WJ48**	**Ensemble**
89.027%	**SVM**	2.51327E-05	7.9544E-10	2.13644E-05
92.386%	**ANN**		3.34761E-10	0.356727509
71.992%	**WJ48**			2.07942E-10

According to the results which were presented in the tables 7-9, we can conclude that the ensemble has the best mean precision value with any number of the components, while differences of the results obtained by the ensemble and alternative methods are statistically significant.

In paper [21] the recognition results of the methods Eigenfaces, LDA and Gabor SVM for database AT&T are presented (with various train and test percentage of database are presented).

Table 10. Comparative analysis of efficiency for task №2 for some algorithms.

	Accuracy
SVM	89.027%
ANN	92.386%
WJ48	71.992%
Ensemble	**92.454%**
Eigenfaces	82.7%
LDA	92.6%
Gabor SVM	90.2%

According to the results which were presented in the table 10, we can conclude that the ensemble has the best mean precision value. According to the results presented above, one may conclude that collective decision making allows to increase the accuracy of the solution.

Acknowledgements
This research is supported by the Russian Foundation for Basic Research, within project № 16-01-00767.

References

[1] Freeman J A, Skapura D M 1991 *Neural Networks Algorithms Applications and Programming Techniques*
[2] Ossowski S 2002 *Neural networks for information processing* 344 p
[3] Takagi T, Sugeno M 1985 Fuzzy identification of systems and its applications to modeling and control *IEEE Trans. on Systems, Man, and Cybernetics*
[4] Quinlan J R 1993 *C4.5: Programs for Machine learning*
[5] Hansen L K 1990 Neural network ensembles *IEEE Trans. Pattern Analysis and Machine Intelligence* **12 (10)** pp 993-1001
[6] Semenkin E S, Shabalov A A, Efimov S N 2011 Automated design of intelligent information technology ensembles by means of genetic programming techniqie *Bulletin of the Siberian State Aerospace University named after academician M.F. Reshetnev* **3** pp 77-81
[7] Bukhtoyarov V V, Semenkin E S 2013 Evolutionary three-stage approach for designing of neural networks ensembles for classification problems *Lecture notes in computer science* **1** pp 467-477
[8] Bukhtoyarov V V, Semenkina O E 2010 Comprehensive evolutionary approach for neural network ensemble automatic design *2010 IEEE World Congress on Computational Intelligence - 2010 IEEE Congress on Evolutionary Computation*
[9] Zhou Z-H, Jiang Y 2004 NeC4.5: Neural Ensemble Based C4.5 *IEEE Trans. Knowl. Data Eng.* **16(6)** pp 770-773
[10] Special Interest Group on Discourse and Dialogue - URL: http://www.sigdial.org/.
[11] Pantic M, Rothkrantz L J M 2003 Toward an Affect-Sensitive Multimodal Human-Computer Interaction *Proceedings of the IEEE, Spec. Issue on Human-Computer Multimodal Interface* **91(9)** pp 1370-1390
[12] Zhao W, Chellappa R, Rosenfeld A, Phillips P J 2003 Face Recognition: A Literature Survey *ACM Computing Surveys* pp 399-458
[13] Shervin E. 2012 Face Recognition using Eigenfaces or Fisherfaces *Mastering OpenCV with Practical Computer Vision Projects*

[14] Yang M H, Kriegman D J, Ahuja N 2002 Detecting faces in images *IEEE Transactions on Pattern Analysis and Machine Intelligence* **24** pp 34-58
[15] Brester C, Semenkin E, Sidorov M, Minker W 2014 Self-adaptive multi-objective genetic algorithms for feature selection *Proceedings of International Conference on Engineering and Applied Sciences Optimization (OPT-i'14)* pp 1838–1846
[16] Brester Ch, Sidorov M, Semenkin E 2014 Acoustic Emotion Recognition: Two Ways of Features Selection Based on Self-Adaptive Multi-Objective Genetic Algorithm *Proceedings of the International Conference on Informatics in Control, Automation and Robotics (ICINCO)* pp 851–855
[17] AT&T Facedatabase - URL: http://www.cl.cam.ac.uk/research/dtg/attarchive/facedatabase.html
[18] Collection of Facial Images: Faces95 - URL: http://cswww.essex.ac.uk/mv/allfaces/faces95.html
[19] RapidMiner Studio URL: https://rapidminer.com/
[20] Bansal A K, Chawla P 2013 Performance evaluation of face recognition using PCA and N-PCA *International Journal of Computer Applications* **76(8)** pp 14–20.
[21] Daoud E 2015 Face Recognition Using Fuzzy Clustering and Kernel Least Square *Journal of Computer and Communications* **3** pp 1-7

IOP Conf. Series: Materials Science and Engineering **173** (2017) 012014 doi:10.1088/1757-899X/173/1/012014

LTI system order reduction approach based on asymptotical equivalence and the Co-operation of biology-related algorithms

I S Ryzhikov, E S Semenkin, Sh A Akhmedova

Siberian State Aerospace University, 31 Krasnoyarskiy Rabochiy Ave., 660037, Krasnoyarsk, Russia

E-mail: ryzhikov-88@yandex.ru

Abstract. A novel order reduction method for linear time invariant systems is described. The method is based on reducing the initial problem to an optimization one, using the proposed model representation, and solving the problem with an efficient optimization algorithm. The proposed method of determining the model allows all the parameters of the model with lower order to be identified and by definition, provides the model with the required steady-state. As a powerful optimization tool, the meta-heuristic Co-Operation of Biology-Related Algorithms was used. Experimental results proved that the proposed approach outperforms other approaches and that the reduced order model achieves a high level of accuracy.

1. Introduction

In this study optimization problems were solved by a meta-heuristic called Co-Operation of Biology-Related Algorithms (COBRA) [1], which is based on the cooperation of five similar nature-inspired algorithms. The workability of the COBRA approach and its usefulness were shown in [1] on a set of benchmark functions.

The proposed algorithm was applied to the order reduction of linear time-invariant (LTI) system models. The described problem is important in system theory and can be applied to single input single output (SISO) systems and for multiple input multiple output (MIMO) systems. Various approaches to solving the order reduction problem exist, and we suggest a novel one that in some cases outperforms other approaches. Many of these approaches are based on combining the stability equation [2] or a similar technique and a reduced extremum problem, which normally involves the minimization of the cumulative error. The minimization problem can be efficiently solved using evolutionary optimization algorithms, nature inspired optimization algorithms or other mimetic algorithms, since these algorithms are a powerful tool for solving complex black-box optimization problems. Since applying these algorithms, the described problem solving methods have started to be developed again.

In the following studies, the same order reduction problem was considered. The current investigation is related to the research below and solves the same problems, so the results can be compared. In the studies [3] and [4] a genetic algorithm is used for solving the extremum problem, in the study [5] the big bang big crunch algorithm is used and in [6] the cuckoo search algorithm is used.

The last three approaches use the stability equation method. Thus the parameters of the denominator can be calculated and these parameters provide the stability of the LTI model.

In this paper all the parameters of the denominator are to be identified. The stability of the LTI system is ensured by using a criterion that is based on the transient matrix eigenvalues and their implementation into the objective function. The equality of model steady-states is carried out by the nominator coefficient determination, which is based on the asymptote value and requires the denominator coefficients.

The approach performance is estimated on the basis of specific criteria values and comparing them with criteria values for different approaches.

This paper contains three main sections, as well as an introduction and a conclusion. The first main section states the main principles of the meta-heuristics we applied to reduced optimization problem solving. The method of forming the objective function and of determining the solution is described in the second section. In the third section the experimental results are described: we compare the result with the initial model of higher order and with the results found in other work.

2. Co-Operation of Biology-Related Algorithms

The meta-heuristic approach called Co-Operation of Biology-Related Algorithms (COBRA) was originally developed for solving real-valued optimization problems [1]. The approach outlined above is based on the collective work of five well-known nature-inspired algorithms, namely Particle Swarm Optimization (PSO) [7], Wolf Pack Search (WPS) [8], the Firefly Algorithm (FFA) [9], the Cuckoo Search Algorithm (CSA) [10] and the Bat Algorithm (BA) [11].

The algorithm COBRA involves generating five populations which are then executed in parallel cooperating with each other, so each algorithm corresponds to its own population. The key advantage of the optimization method COBRA consists of its self-tuning work. Thus there is no need to choose the population size for components of COBRA. The approach settings were taken as they were recommended by the authors of the papers mentioned above.

The number of individuals in the population of each component-algorithm can increase or decrease depending on whether the fitness value improves: if the fitness value does not improve over a given number of generations, then the size of all populations increases and vice versa. Besides, on every generation the component algorithms compete with each other, so a "winner algorithm" can be determined. Namely the algorithm with the best population's average fitness value is considered as a "winner". The population of the "winner" algorithm "grows" by accepting individuals removed from other populations. The migration operator of the given approach consists in replacement of the worst individuals of each population by the best individuals of others.

Experimental results showed that the meta-heuristic approach COBRA is reliable and works efficiently. Moreover, it was established that the meta-heuristic COBRA outperforms its component-algorithms and can be used instead of the described bio-inspired optimization techniques.

While solving the optimization problems described in this study, populations were initialised uniformly on the closed interval [-10; 10]. The minimal population size for each component was equal to 6 individuals and calculations were terminated when the number of function evaluations was equal to 2500. This number for the criterion evaluation was taken from paper [6].

In this paper, the decision was taken to use the transformation of the main criterion to a fitness function as follows:

$$fitness(x) = \left(1 + C(x)\right)^{-1} \quad (1)$$

where $x \in R^X$ is an alternative, R^X is the search space and $C(\cdot): R^X \to [0, +\infty)$ is a criterion. According to this definition, $fitness(\cdot): R^X \to (0, 1]$.

3. The order reduction problem and a proposed approach

Let the SISO LTI system model be determined by the following differential equation

$$\sum_{i=0}^{n} a_i \cdot x^{(i)}(t) = \sum_{i=0}^{m} b_i \cdot u^{(i)}(t) \tag{2}$$

where $a_i \in R, i = \overline{1,n}$ and $b_i \in R, i = \overline{1,m}$ are the coefficients, $n : n > m$ is the order, $t \in R^+ \cup \{0\}$ is the time variable, $x^{(i)}(t)$ is the i-th derivative of the output, $u^{(i)}$ is the i-th derivative of the control input and $x^{(0)}(t) = x(t)$.

For the initial value $x^{(i)}(0) = 0$ equation (2) after using the Laplace transformation can be represented with a transfer function

$$G(s) = \sum_{j=0}^{m} b_j \cdot s^j \cdot \left(\sum_{i=0}^{n} a_i \cdot s^i \right)^{-1} . \tag{3}$$

The MIMO LTI system can be represented similarly,

$$\begin{aligned} &\frac{d}{dt} X(t) = A \cdot X(t) + B \cdot U(t), \\ &Y(t) = C \cdot X(t) + D \cdot U(t), \end{aligned} \tag{4}$$

where $Y(t): R^+ \cup \{0\} \to R^{N_o}$ is the output function, N_o is the number of outputs, $U(t): R^+ \cup \{0\} \to R^{N_c}$ is the input function, N_c is the number of inputs, $X(t): R^+ \cup \{0\} \to R^{N_z}$ is the space variable, the system matrix is $A \in R^{N_z \times N_z}$, the control matrix is $B \in R^{N_z \times N_c}$, the output matrix is $C \in R^{N_o \times N_z}$ and the feed-forward matrix is $D \in R^{N_o \times N_c}$. The transfer function for the MIMO system is the following:

$$W_s(s) = C \cdot (s \cdot I - A)^{-1} \cdot B + D. \tag{5}$$

In this paper we consider the MIMO system with two inputs and two outputs, thus its transient function (5) can be determined with the expression

$$W_s(s) = \begin{pmatrix} W^{1,1}(s) & W^{1,2}(s) \\ W^{2,1}(s) & W^{2,2}(s) \end{pmatrix}, \tag{6}$$

where $W^{i,j}(s) = \dfrac{D^{i,j}(s)}{N^{i,j}(s)}$, $D^{i,j}, N^{i,j}$, $i,j = \overline{1,2}$, are the denominator and nominator, respectively. Factoring out the denominator gives

$$W_s(s) = \frac{1}{D_s(s)} \cdot \begin{pmatrix} N_s^{1,1}(s) & N_s^{1,2}(s) \\ N_s^{2,1}(s) & N_s^{2,2}(s) \end{pmatrix},$$
$$D_s(s) = \prod_{i,j} D^{i,j}(s),\ N_s^{p,q}(s) = N^{p,q}(s) \cdot \prod_{\substack{i,j \\ i \neq p \wedge j \neq q}} D^{i,j}(s). \tag{7}$$

It is known that linear system (1) is in a steady-state in the case of a stable system and input $u(t) = H(t)$, $H(t)$, the unit-step function, is a constant

$$a^s = \lim_{t \to +\infty} x(t) = \frac{a_0}{b_0} \tag{8}$$

the coefficients a_0 and b_0 are given in (3). This means that if the model output asymptote is a^s, the parameters of the reduced model are to be determined by fraction (8).

We need to make a model of the 2nd order which is determined by the following transfer function

$$G_m(s,p) = \frac{p_2 \cdot s + a^s \cdot p_1}{s^2 + p_0 \cdot s + p_1}, \tag{9}$$

for the SISO systems and

$$G_m(s,p) = \frac{1}{D_m} \cdot \begin{pmatrix} N_m^{1,1} & N_m^{1,2} \\ N_m^{2,1} & N_m^{2,2} \end{pmatrix},$$
$$D_m(s,p) = s^2 + p_0 \cdot s + p_1,\ N_m^{i,j}(s,p) = p_{1+2(i-1)+j} \cdot s + a^s_{2(i-1)+j} \cdot p_1, \tag{10}$$

for the MIMO systems. Hence, the parameters p are to be identified. As can be seen, the absolute term of the nominator is defined by (8) to ensure the asymptotic equality.

We want the model with reduced order to be a good estimation of the LTI system, so its response is close to the response of the model with higher order on the same control input $u(t)$. The response is a function on a time domain and for both models it can be found by solving the Cauchy problem for (2) and (4). Since we consider the unit-step function as an input and its first derivative is the Dirac delta function, the invariant Laplace transformation is useful:

$$x(t,p) = L^{-1}\left(G_m(s,p) \cdot L(u(t))\right),\ \hat{x}(t) = L^{-1}\left(G(s) \cdot L(u(t))\right), \tag{11}$$

for the SISO LTI systems and

$$x^{i,j}(t,p) = L^{-1}\left(\frac{N_m^{i,j}(s,p) \cdot L(u(t))}{D_m(s,p)}\right),\ \hat{x}^{i,j}(t) = L^{-1}\left(\frac{N_s^{i,j}(s) \cdot L(u(t))}{D_s(s)}\right), \tag{12}$$

for the MIMO LTI systems. Let $u(t) = H(t)$, so the parameters are the solution of the extremum problem

IOP Conf. Series: Materials Science and Engineering **173** (2017) 012014 doi:10.1088/1757-899X/173/1/012014

$$C_{siso}(p)=\sum_{i=0}^{N}\left(\hat{x}(t_i)-x(t_i,p)\right)^2, \quad C_{siso}(p)\to\min_{p\in R^3}, \tag{13}$$

or

$$C_{mimo}(p)=\frac{\sum_{i,j}\sum_{k=0}^{N}\left(\hat{x}^{i,j}(t_k)-x^{i,j}(t_k,p)\right)^2}{2+N_s}, \quad C_{mimo}(p)\to\min_{p\in R^{2+N_s}}, \tag{14}$$

where $t_i=\frac{T\cdot i}{N}, i=\overline{1,N}$, T is the final time, N is the number of points, and all the responses are calculated due to (11) or (12) for the SISO problems and MIMO problems, respectively.

To ensure that the reduced order model is stable, the criteria (13) and (14) were modified: the stability condition based on the eigenvalues was implemented. A LTI system of the 2nd order can be described with an equation from the denominator of (3), $a_0^*\cdot s^2+a_1^*\cdot s+a_2^*$, whose parameters are related to the coefficients of the system matrix, so the system is asymptotically stable when

$$-\frac{a_1^*}{a_0^*}<0 \tag{15}$$

because inequality (14) means that all the eigenvalues are less than zero. However, the 2nd order LTI system can always be determined by the equation (9), so $a_0^*=1$ and thus

$$-a_1^*<0\Rightarrow a_1^*>0\Rightarrow p_0>0 \tag{16}$$

In this work a penalty function was used to provide the implementation of the stability condition into criteria (13) and (14). Modified criteria are as follows,

$$\tilde{C}_{siso}(p)=C_{siso}+c\cdot P(p_0)\to\min_{p\in R^3} \tag{17}$$

$$\tilde{C}_{mimo}(p)=C_{mimo}(p)+c\cdot P(p_0)\to\min_{p\in R^{2+N_s}}, \tag{18}$$

where $P(\cdot):R\to R^+\cup\{0\}$ is a static penalty function and $c>0$ is a coefficient. The penalty function activates as the parameter becomes less than zero:

$$P(x)=\begin{cases}0, & x>0\\ \|x\|, & x\le 0\end{cases} \tag{19}$$

To analyse solution accuracy three more criteria are used. These criteria are involved in comparing the efficiency of approaches. Let $x(t)=x(t,p^*), \hat{x}(t)$ be the solutions of (11) or (12), the input is $u(t)=H(t)$ and $p^*=\arg\min C_{siso}(p)$ or $p^*=\arg\min C_{mimo}(p)$, depending on the problem. The first criterion we want to calculate is the integral square error

$$I_1 = \int_0^{+\infty} \left(x(t) - \hat{x}(t) \right)^2 dt \quad (20)$$

Its estimation was used to identify the parameters by solving problems (13) or (14). The integral (20) is divergent if $\lim_{t \to +\infty} x(t) \neq \lim_{t \to +\infty} \hat{x}(t)$, and for this reason expression (8) was implemented and the stability condition is required.

The next criteria are the relative integral square error and the criteria are given in [3] which are proposed to check the accuracy of the model. Both criteria are expressed by the fraction:

$$I_2 = \frac{\int_0^{+\infty} \left(x(t) - \hat{x}(t) \right)^2 dt}{\int_0^{+\infty} \left(x(t) - x(+\infty) \right)^2 dt}, \quad (21)$$

and for $u(t) = \delta(t)$,

$$I_3 = \frac{\int_0^{+\infty} \left(x(t) - \hat{x}(t) \right)^2 dt}{\int_0^{+\infty} \left(x(t) \right)^2 dt} \quad (22)$$

The result of the inverse Laplace transformation (11) and (12) was found symbolically for the current problems, where the initial and reduced order models are linear. Due to the decomposition of fraction (3) each LTI 2nd order model belongs to one of the following solution classes:

- $b_0^2 - 4 \cdot b_1 = 0, b_0 \neq 0$;
- $b_0^2 - 4 \cdot b_1 = 0, b_0 = 0$;
- $b_0^2 - 4 \cdot b_1 < 0,\ a_0 = a_s \cdot b_0 \neq 0$;
- $b_0^2 - 4 \cdot b_1 < 0,\ a_0 = a_s \cdot b_0 = 0$;
- $b_0^2 - 4 \cdot b_1 > 0,\ b_0 \neq 0$;
- $b_0^2 - 4 \cdot b_1 > 0,\ b_0 = 0$,

and there is a symbolic expression of the LTI model output, which is the unit-step function $H(t)$ or impulse function $\delta(t)$ response, for each solution class. Using the symbolic expression of the response it is possible to calculate criteria (17) and (18) for an alternative and (20), (21) and (22) for the solution found.

4. Performance investigation

The first order reduction problem we consider is related to the SISO LTI system model

$$G(s) = \frac{s^3 + 7 \cdot s^2 + 24 \cdot s + 24}{s^4 + 10 \cdot s^3 + 35 \cdot s^2 + 50 \cdot s + 24} \quad (23)$$

which is a 4th order linear differential equation given, for example, in [6]. The linear dynamic model of the 2nd order, (9), is to be found as a solution of the extremum problem (17) and $u(t)=H(t)$. The transformation in (1) was used to calculate the fitness of an individual; the settings of COBRA are given above. After 100 algorithm launches, the following solution was found:

$$G_m^*(s)=\frac{0.761968643\cdot s+1.73503772}{s^2+2.639490066\cdot s+1.73503772} \tag{24}$$

In Figure 1 the higher order model response and the lower order model response are given. The best solution found has a fitness value of 0.99914635, the worst solution fitness value is 0.99157284, the mean fitness value of the solution found is 0.99856314 and the variance of the fitness function value is 0.00102815. The values of criteria (20), (21) and (22) are given in Table 1, where the proposed approach is compared to [6], [5] and [3].

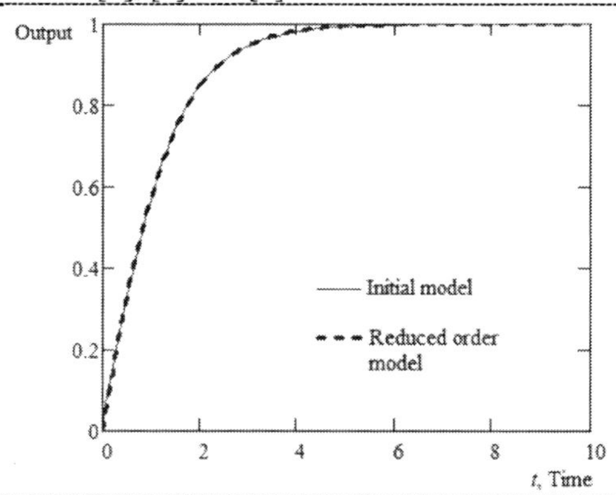

Figure 1. The unit-step function response of the reduced order model (24) and the initial model (23).

Table 1. Criteria values.

	Criterion		
Approach	I_1, (19)	I_2, (20)	I_3, (21)
Proposed	$7.408\cdot10^{-5}$	$7.458\cdot10^{-5}$	$6.09\cdot10^{-3}$
Narval et al.	$9.98\cdot10^{-5}$	$1.6\cdot10^{-4}$	$8.52\cdot10^{-4}$
Desai et al.	$2.178\cdot10^{-4}$	$2.01\cdot10^{-4}$	$4.3\cdot10^{-3}$
Parmar et al.	$4.192\cdot10^{-4}$	$4.19\cdot10^{-4}$	$1.85\cdot10^{-2}$

As can be seen, the proposed approach outperforms the approaches presented in Table 1, but not in terms of criterion I_3 values.

The bode diagram shows the frequency response of the initial and the reduced model and is given in Figure 2 and Figure 3 for the amplitude and the phase, respectively.

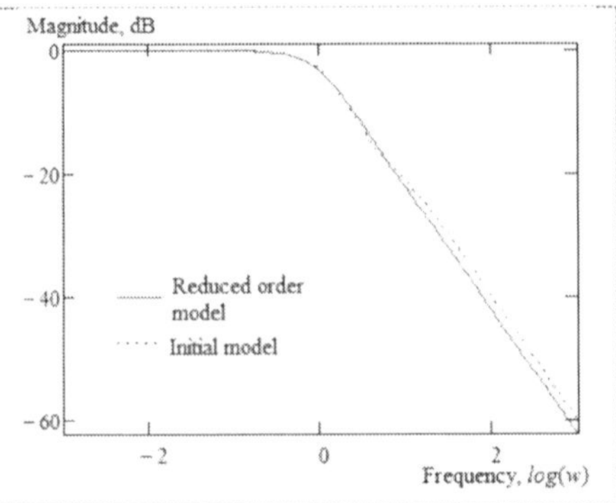

Figure 2. The unit-step function response magnitude of the reduced order model (24) and the initial model (23).

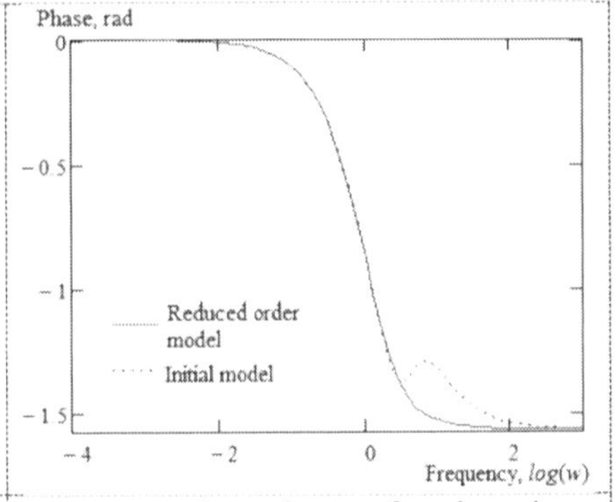

Figure 3. The unit-step function phase response of the reduced order model (24) and the initial model (23).

The same investigation was performed for the MIMO LTI system model, which is determined by the following equation:

$$H(s)=\begin{pmatrix} \dfrac{2\cdot(s+5)}{(s+1)\cdot(s+10)} & \dfrac{s+4}{(s+2)\cdot(s+5)} \\ \dfrac{s+10}{(s+1)\cdot(s+20)} & \dfrac{s+6}{(s+2)\cdot(s+3)} \end{pmatrix}. \tag{25}$$

It must be emphasized that there is a limited number of decimal places, but the accuracy can be increased by calculating the 2nd element in total for each nominator using formula (8) and knowing that $a_0^s=1$, $a_1^s=0.4$, $a_2^s=0.5$ and $a_3^s=1$.

As can be seen, the form of its representation in (7) would give us a denominator of the 6th order and nominators of the 5th order. The second order model, (10), is to be identified. Extremum problem (18) for $u(t)=H(t)$ was solved in a similar way 100 times. The responses are given in Figure 4, Figure 5, Figure 6 and Figure 7 for (1,1) matrix (6) component, (1,2), (2,1) and (2,2), respectively. The best solution found has a fitness value of 0.98445642, the worst solution fitness value is 0.98052264, the mean fitness value of the solution found is 0.98427018 and the variance of the fitness function value is 0.00045411. The values of criterion (20) for the proposed approach and the other approaches are given in Table 2 for different equations – matrix components. In Table 2 the proposed approach was compared with [6], [5], [12] and [13]. The same was performed for criteria (21) and (22), Table 3 and Table 4, respectively. The following solution in the form of (10) was found:

$$\begin{gathered} G_m^*(s)=\frac{1}{D_m^*}\cdot\begin{pmatrix} N_m^{*1.1} & N_m^{*1.2} \\ N_m^{*2.1} & N_m^{*2.2}\end{pmatrix}, \\ D_m^*(s)=s^2+2.3579688\cdot s+1.2967903, \\ N_m^{*1.1}(s)=1.1830408\cdot s+1.2967903, \\ N_m^{*1.2}(s)=0.7282982\cdot s+0.51871609, \\ N_m^{*2.1}(s)=0.5561906\cdot s+0.6483951, \\ N_m^{*2.2}(s)=1.4883673\cdot s+1.2967903. \end{gathered} \tag{26}$$

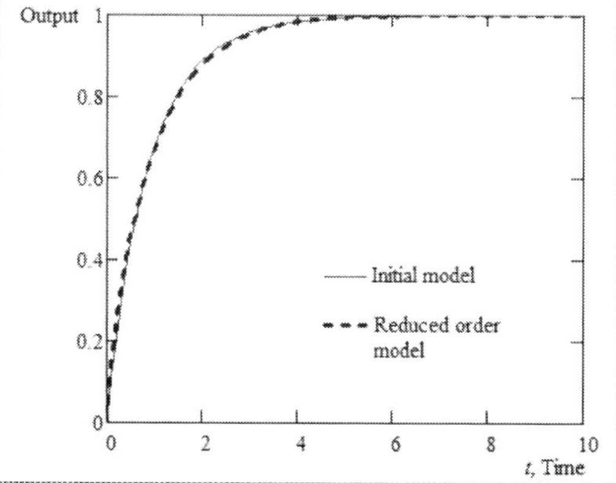

Figure 4. The unit-step function response of the reduced order model (26) and the initial model (25), (1,1)- component.

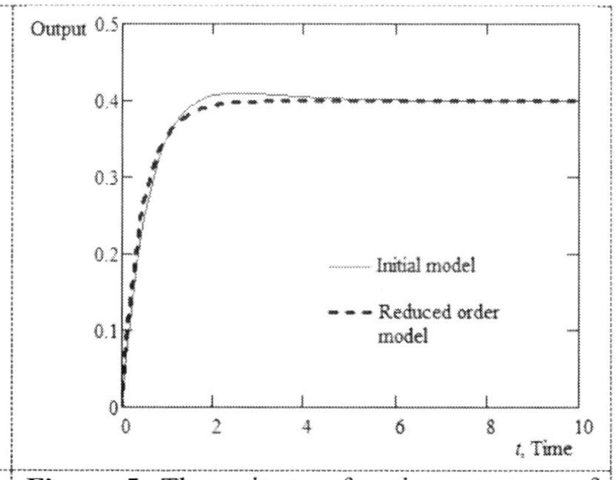

Figure 5. The unit-step function response of the reduced order model (26) and the initial model (25), (1,2)- component.

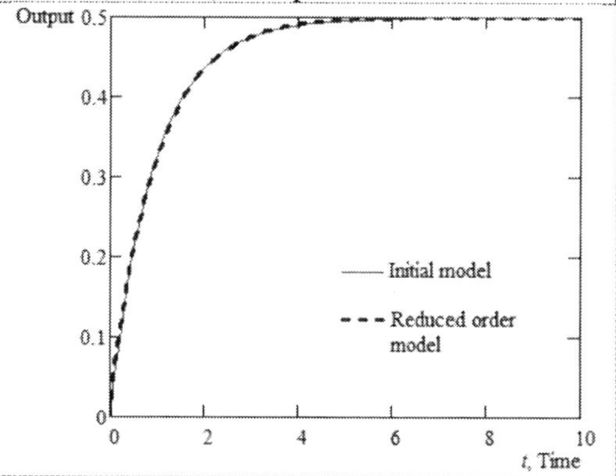

Figure 6. The unit-step function response of the reduced order model (26) and the initial model (25), (2,1)- component.

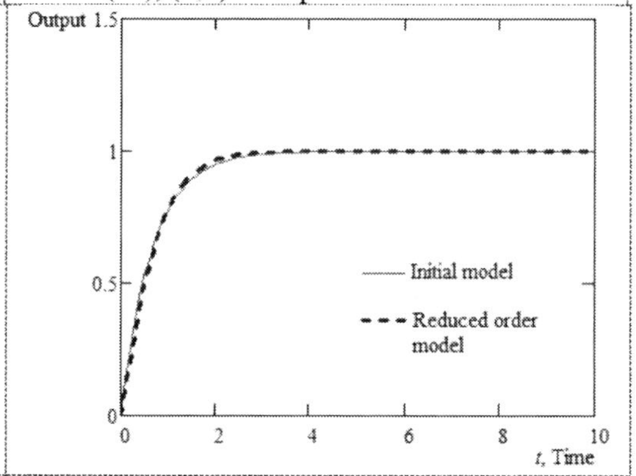

Figure 7. The unit-step function response of the reduced order model (26) and the initial model (25), (2,2)- component.

Table 2. Criterion (20) values for different approaches.

	System component			
Approach	1,1	1,2	2,1	2,2
Proposed	$8.612 \cdot 10^{-4}$	$7.226 \cdot 10^{-4}$	$6.278 \cdot 10^{-5}$	$1.517 \cdot 10^{-3}$
Narval et al.	$1.124 \cdot 10^{-3}$	$7.312 \cdot 10^{-4}$	$1.942 \cdot 10^{-4}$	$1.117 \cdot 10^{-3}$
Desai et al.	$6.72 \cdot 10^{-3}$	$5.918 \cdot 10^{-3}$	$9.58 \cdot 10^{-4}$	$5.896 \cdot 10^{-3}$
Prasad et al.	$3.069 \cdot 10^{-2}$	$2.56 \cdot 10^{-4}$	$2.62 \cdot 10^{-1}$	$2.168 \cdot 10^{-2}$
Prasad and Pal	$1.365 \cdot 10^{-1}$	$2.446 \cdot 10^{-3}$	$4.029 \cdot 10^{-2}$	$6.79 \cdot 10^{-2}$

The results in Table 2 show that the proposed approach is an efficient tool for solving the reduced order model parameter identification problem. The sum of the error values, which is the sum of the values in a row, for the proposed approach is less than the sum of the values for any other approach in Table 2.

Table 3. Criterion (21) values for different approaches.

	System component			
Approach	(1,1)	(1,2)	(2,1)	(2,2)
Proposed	$2.082 \cdot 10^{-3}$	$2.1 \cdot 10^{-2}$	$5.537 \cdot 10^{-4}$	$4.136 \cdot 10^{-3}$
Narval et al.	$1.08 \cdot 10^{-2}$	$3.68 \cdot 10^{-2}$	$8.7 \cdot 10^{-2}$	$1.65 \cdot 10^{-2}$
Desai et al.	$3.4 \cdot 10^{-2}$	$2.17 \cdot 10^{-2}$	$9.5 \cdot 10^{-3}$	$2.47 \cdot 10^{-2}$

The results in Table 3 confirm the efficiency of the proposed approach and that it outperforms the other approaches, since it gives more accurate models; the criterion (21) values provide both higher and lower order model response closure for all the MIMO LTI model components.

Table 4. Criterion (22) values for different approaches.

	System component			
Approach	(1,1)	(1,2)	(2,1)	(2,2)
Proposed	$4.3 \cdot 10^{-2}$	$3.4 \cdot 10^{-2}$	$3.1 \cdot 10^{-2}$	$2.9 \cdot 10^{-2}$
Narval et al.	$1.72 \cdot 10^{-2}$	$3.62 \cdot 10^{-2}$	$1.62 \cdot 10^{-2}$	$2.3 \cdot 10^{-3}$
Desai et al.	$2.49 \cdot 10^{-2}$	$4.87 \cdot 10^{-2}$	$2.43 \cdot 10^{-2}$	$3.7 \cdot 10^{-3}$

The proposed approach improves the solution properties: integral square error, (20), and relative integral square error, (21) , but it does not outperform the other approaches in terms of criterion (22) values. The reason for this is the problem statement, which provides the following aim: to receive the model that has a response on the unit-step function that is maximally close to the initial model response on that function. In reality, a different input function corresponds to a different extremum problem with its extremum points. Nevertheless, the proposed approach criterion (22) values are close to the values for the other approaches.

The bode diagram shows the difference between the initial and the reduced model in the frequency domain and is given in Figures 8-9 for the (1,1) MIMO system component, Figures 10-11 for the (1,2) component, Figures 12-13 for the (2,1) component and Figures 14-15 for the (2,2) component.

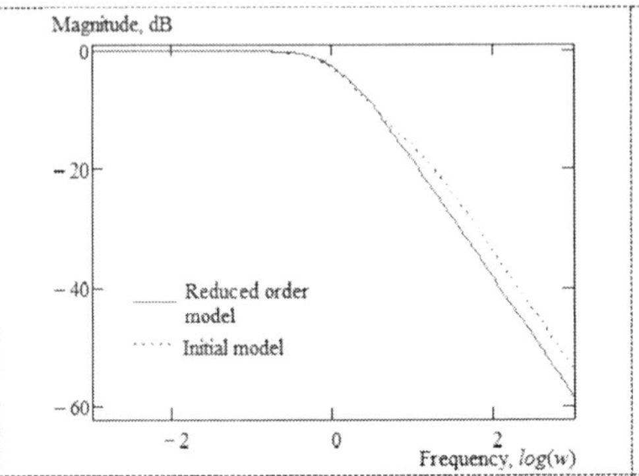

Figure 8. The unit-step function response magnitude of the reduced order model (26) and the initial model (25), (1,1)- component.

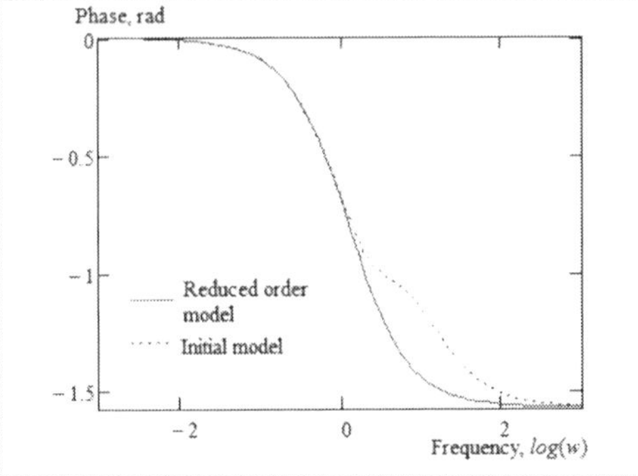

Figure 9. The unit-step function phase response of the reduced order model (26) and the initial model (25), (1,1)- component.

IOP Conf. Series: Materials Science and Engineering **173** (2017) 012014 doi:10.1088/1757-899X/173/1/012014

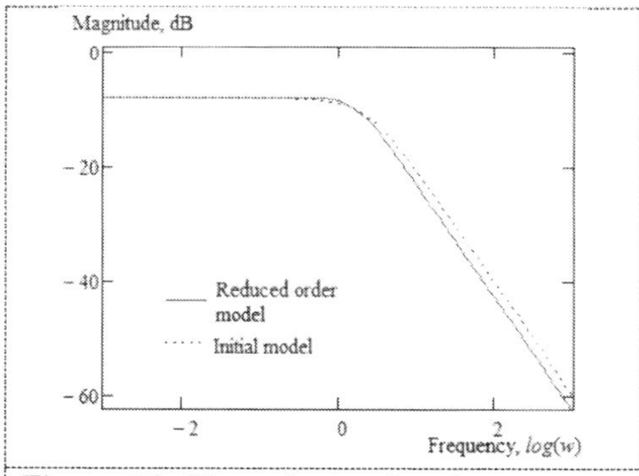

Figure 10. The unit-step function response magnitude of the reduced order model (26) and the initial model (25), (1,2)- component.

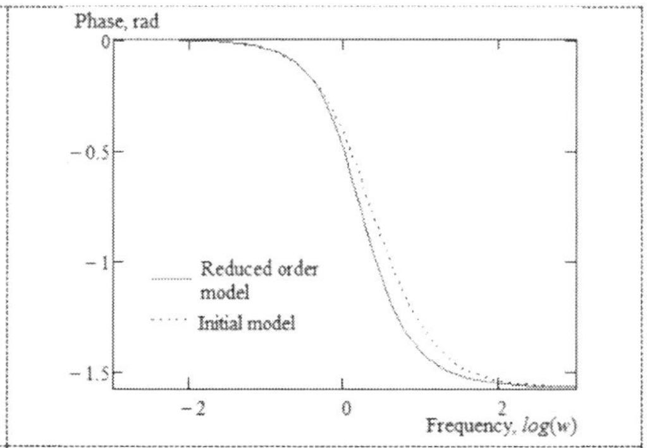

Figure 11. The unit-step function phase response of the reduced order model (26) and the initial model (25), (1,2)- component.

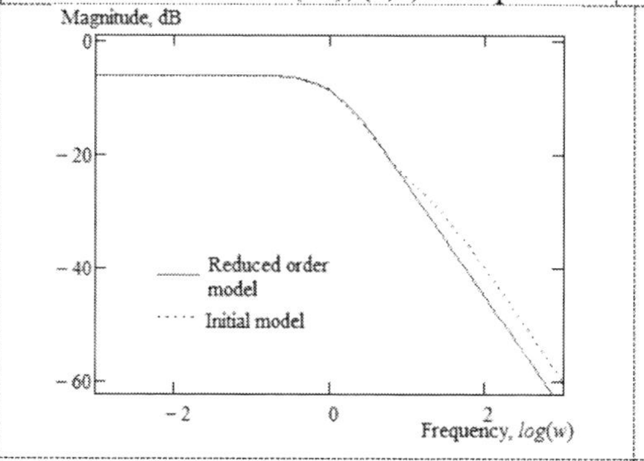

Figure 12. The unit-step function response magnitude of the reduced order model (26) and the initial model (25), (2,1)- component.

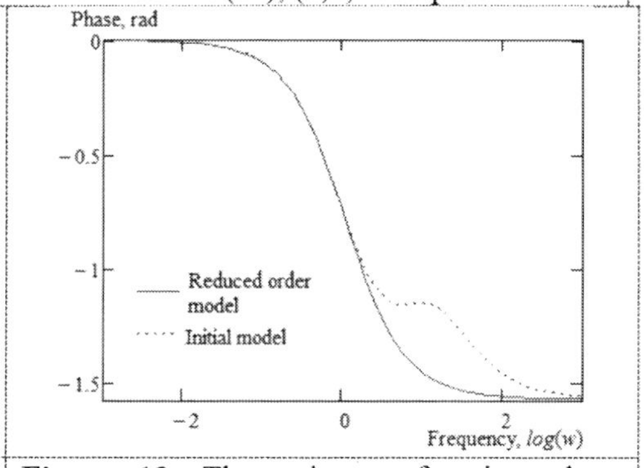

Figure 13. The unit-step function phase response of the reduced order model (26) and the initial model (25), (2,1)- component.

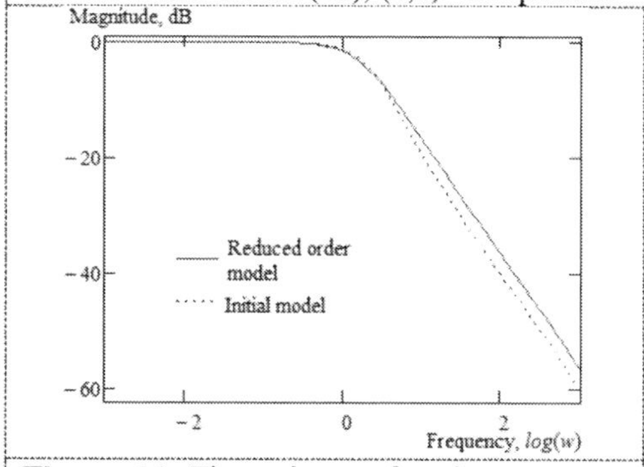

Figure 14. The unit-step function response magnitude of the reduced order model (26) and the initial model (25), (2,2)- component.

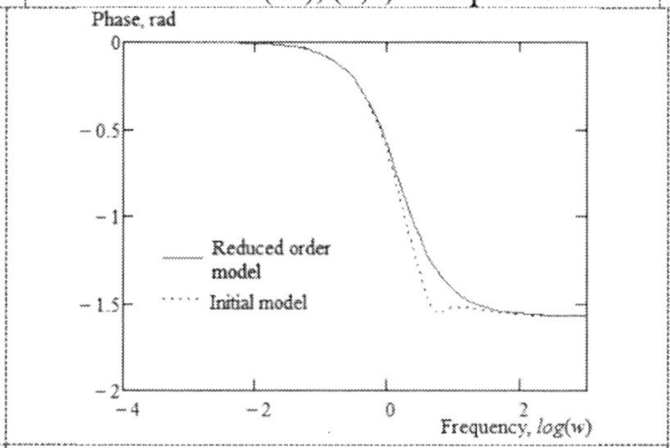

Figure 15. The unit-step function phase response of the reduced order model (26) and the initial model (25), (2,2)- component.

5. Conclusion

The order reduction problem for the LTI, SISO and MIMO systems was successfully solved in this study by applying COBRA to the extremum problem, based on asymptotic stability and its equivalence to the higher order LTI system model. The proposed approach significantly outperformed other approaches in terms of the relative integral square error value for the unit-step function response. Also the integral square error was the least for the first problem and the sum of the integral square errors for the second problem was the least, compared to other approaches. By this we can conclude that COBRA and the reduced order model determination improved the solutions found earlier. The determination rule allows a stable LTI system to be achieved, whose asymptote is equal to the desired value.

The COBRA statistics prove that it is a reliable technique, since the estimation of the integral square error is sufficiently low, which means that the expected value for the fitness function is close to its maximum value, so also to its worst value. Moreover, it can be seen that fitness function variance is low for the both problems.

Since the 2nd order LTI system model response can be transformed into a symbolic form, the future work is related to implementing the integral square error directly to the fitness function. It would solve a problem by estimating the main criteria with its numerical analogue: different estimations often give different results.

Another important task is to use as a main criterion the distance between responses of the higher and lower orders systems on the impulse function. This is necessary to compare the solutions and their properties in the case of a different main criterion.

Acknowledgements
This research is supported by the Russian Foundation for Basic Research, within project № 16-01-00767.

References

[1] Akhmedova Sh A, Semenkin E 2013 Co-operation of biology related algorithms *IEEE Congress on Evolutionary Computations*
[2] Chen T C, Chang C Y, Han K W 1979 Reduction of transfer functions by the stability equation method *Journal of Franklin Inst.* **308 (4)** pp 389-404
[3] Parmar G, Prasad R, Mukherjee S 2007 Order reduction of linear dynamic systems using stability equation method and GA *International Journal of computer and Infornation Engeneering* **1(1)** pp 26-32
[4] Ramesh K, Nirmalkumar A, Gurusamy G 2011 Order Reduction of LTIV continuous MIMO system using Stability Preserving approximation method *International Journal of Computer Applications* **36(8)** pp 1-8
[5] Desai S, Prasad R 2013 A novel order diminution of LTI systems using Big Bang Big Crunch optimization and Routh approximation *Appl. Math. Model.* **37** pp 8016-8028
[6] Narwal A, Prasad B R 2016 A Novel Order Reduction Approach for LTI Systems Using Cuckoo Search Optimization and Stability Equation *IETE Journal of Research* **62(2)** pp 154-163
[7] Kennedy J, Eberhart R 1995 Particle swarm optimization *IEEE International Conference on Neural Networks*
[8] Yang Ch, Tu X, Chen J 2007 Algorithm of marriage in honey bees optimization based on the wolf pack search *International Conference on Intelligent Pervasive Computing*
[9] Yang X S 2009 Firefly algorithms for multimodal optimization *The 5th Symposium on Stochastic Algorithms, Foundations and Applications*
[10] Yang X S, Deb S 2009 Cuckoo Search via Levy flight *World Congress on Nature & Biologically Inspired Computing*
[11] Yang X S 2010 A new metaheuristic bat-inspired algorithm. Nature Inspired Cooperative Strategies for Optimization *Studies in Computational Intelligence* **284**
[12] Prasad R, Pal J, Pant A K 1995 Multivariable system approximation using polynomial derivatives *J. Inst. Eng. India Part E Electric. Eng. Div.* **76** pp 186-194
[13] Prasad R, Pal J 1991 Use of continued fraction expansion for stable reduction of linear multivariable systems *J. Inst. Eng. India Part E Electric. Eng. Div.* **72** pp 43-47

IOP Conf. Series: Materials Science and Engineering **173** (2017) 012015 doi:10.1088/1757-899X/173/1/012015

Restart Operator Meta-heuristics for a Problem-Oriented Evolutionary Strategies Algorithm in Inverse Mathematical MISO Modelling Problem Solving

I S Ryzhikov, E S Semenkin

Russia, Krasnoyarsk, Siberian State Aerospace University

E-mail: ryzhikov-88@yandex.ru

Abstract. This study is focused on solving an inverse mathematical modelling problem for dynamical systems based on observation data and control inputs. The mathematical model is being searched in the form of a linear differential equation, which determines the system with multiple inputs and a single output, and a vector of the initial point coordinates. The described problem is complex and multimodal and for this reason the proposed evolutionary-based optimization technique, which is oriented on a dynamical system identification problem, was applied. To improve its performance an algorithm restart operator was implemented.

1. Introduction

It is widely known that adequate mathematical models are required for decision-making problem solving. This type of model is the only tool that is applicable to control problems, prediction problems and the investigation of properties. However, the difficulty is that many of the real processes cannot be mathematically determined, since these processes are complex and there is no information or existing theory on how they behave and why. In this case, inverse mathematical modelling problems are used.

Modern cybernetic techniques provide a variety of tools to solve regression problems, such as fuzzy approximation, artificial neural network modelling, genetic programming approximation and nonparametric statistics. All of the above-mentioned methods are essentially based on the reduction of a mathematical modelling problem to a parameter identification problem and of a parameter identification problem to an optimization problem. The space of the optimization problem is determined by the model representation and its bijective mapping to this space. The space of alternatives becomes more complex with growing uncertainty and a lack of information. Structural uncertainty and stochastic uncertainty are often found in inverse mathematical model problems and so the adequacy of solutions can be improved by reducing the effect of such uncertainty.

For this reason, many of today's optimization algorithms are problem-oriented. In a general case, the parameter identification problem is a black-box optimization problem. Such diverse types of extremum-seeking algorithms as evolution-based and natural-inspired algorithms have proved their high performance and reliability for black-box optimization. Therefore, various modifications are made to the stochastic extremum-seeking heuristics to improve their performance for solving some class of problems.

The described optimization problem classes can be completely different, which is why it is important to develop a meta-heuristic that could be applicable to distinct extremum seeking algorithms without implementing some operators deep into the algorithm schemes or designing their interaction with algorithm operators. One of these approaches is based on the restart idea. The first restart operator implementation is described in studies [1], [2], [3] and [4]. All of the proposed approaches use different criteria to check if the restart condition is met and these are also different from the one in this study. In the study [4] it is shown that changing the initial population at each restart allows better results to be achieved, which is why in the current work we also investigate the restart operator, which changes the initial population generation settings.

Back to the inverse mathematical modelling, each problem reduction allows a black-box optimization problem class to be assigned. As was mentioned above, there are many various approaches, but many of them are not suitable in a case where a dynamical model in symbolic form is required. Models in symbolic form are increasingly used and applied to the further study of such systems, their control and prediction, due to the high rate of development of these theories. This type of model is more flexible and useful. The approaches listed above along with new black-box optimization algorithms can be used to approximate the dynamical system output with some known input reaction, but this approximation itself cannot be used in a general way to solve control or prediction problems, for example. Moreover, stochastic uncertainty renders many approximation methods useless, unless the sample size is large and the sample is not flat, which cannot be guaranteed.

To solve this problem, a specific approach was developed. The mathematical model is in the form of a linear differential equation, whose order and coefficients are to be estimated [5]. The proposed approach was efficient in solving the identification problems in the case of the system being observed from its steady state. Another problem statement assumes the equation order to be known, but the system state coordinates by the time the observation begin are unknown [6]. The next improvement is associated with identifying the order, the coefficients and the initial point [7]. The idea of solving the identification problem for the dynamical systems by means of bio-inspired computation was first described in [8], where the system structure is assumed to be known. In this study and in the studies [9] and [10], where the coefficients of second order differential equation are estimated, a genetic algorithm was used. The same problem was also solved using the big bang big crunch and cuckoo optimization algorithms in the studies [11] and [12], respectively. The particle swarm optimization algorithm was applied to the linearized model identification is the study [13]. In this study, the model searches in the form of a linear differential equation, whose coefficients, order and initial value are unknown, so the problem definition is close to real system modelling.

In this study we consider a system with multiple inputs and a single output in the case of all the inputs being additive and known. The experimental results proved that the proposed approach is reliable and useful in solving problems of such definition. It is also shown that an algorithm with a meta-heuristic allows a better model with a higher probability to be achieved.

2. Single input multiple output dynamical system identification

Since the considered inverse mathematical modelling problem is related to dynamical systems, it is supposed that the model is a differential equation. In this study we assume that the dynamical system is linear or quasilinear: in some neighbourhood of the system output reaction with some inputs the dynamical system is linear and, therefore, the model appears in the form of a linear differential equation.

The information about the system consists of the measurements of its output, which are represented with a sample of size s, $Y=\{y_i, i=\overline{1,s}\}$, taken at time points $T=\{t_i, i=\overline{1,s}\}$ and the control functions $u_j(t): R^{+}\bigcup\{0\}\rightarrow R$, $j=\overline{1,n_{inp}}$, where n_{inp} is the number of input functions. The single input and multiple output systems are described, so $y_i \in R$. It is assumed that the object is determined by the linear differential equation

$$\sum_{i=0}^{n_{obj}} a_i \cdot x^{(i)}(t) = \sum_{j=0}^{n_{inp}} b_j \cdot u_j(t), \tag{1}$$

where n_{obj} is a differential equation order; $a_{n_{obj}} \neq 0, a_i, b \in R$ are the coefficients and $x^{(i)}(t)$ is the i-th derivative ($x^{(0)}(t)$ is a system state $x(t)$). Therefore, the observation data can be determined by the expression

$$y_i = x(t_i) + \xi_i, \tag{2}$$

ξ_i is a uniformly distributed random value, $\xi \sim U(-L_n, L_n)$ (L_n is a noise level, which is unknown). In the current study it is assumed that $E(\xi) = 0$ and $D(\xi) < \infty$.

Let the model of the system (1) be represented by the following differential equation:

$$\sum_{i=0}^{m} \tilde{a}_i \cdot \hat{x}^{(i)}(t) = \sum_{j=0}^{n_{inp}} \tilde{b}_j \cdot u_j(t), \tag{3}$$

where all the parameters are to be identified. On the basis of the data we have, the parameters could be found as a solution of the optimization problem

$$\sum_{i=1}^{s} \left\| y_i - \hat{x}(t_i) \right\| \to \min_{\tilde{a} \in R^{m+1}, \tilde{b} \in R^{n_{inp}}, m} \tag{4}$$

here $\|\cdot\|$ is an any Euclidean norm.

The extremum problem (4) is not formalized and cannot be solved, since the model state is not determined. The state function $\hat{x}(t)$ is a solution of the Cauchy problem for some initial value. Since the described systems are time invariant, as all systems are that can be determined with stationary differential equations, it makes no difference what the value of the starting point is. Without loss of generality, let the initial value be $\hat{x}(0)$, then

$$\sum_{i=0}^{m} \tilde{a}_i \cdot \hat{x}^{(i)}(t) = \sum_{j=0}^{n_{inp}} \tilde{b}_j \cdot u_j(t),\ \hat{x}(0) = v \tag{5}$$

would determine the output function of the model. To avoid linear dependence of the coefficients and keeping in mind that, by definition, the order is determined by the variable m, so $\hat{a}_m \neq 0$, the first coefficient on the left side of the equation (5) can be excluded. For this purpose, the expression (5) after division by $\tilde{a}_m$ can be denoted as

$$\hat{x}^{(m)}(t) + \sum_{i=0}^{m-1} \hat{a}_i \cdot \hat{x}^{(i)}(t) = \sum_{j=0}^{n_{inp}} \hat{b}_j \cdot u_j(t),\ \hat{x}(0) = v, \tag{6}$$

where $\hat{a}_i = \dfrac{\tilde{a}_i}{\tilde{a}_m}$ and $\hat{b}_j = \dfrac{\tilde{b}_j}{\tilde{a}_m}$, $j = \overline{1, n_{inp}}$.

IOP Conf. Series: Materials Science and Engineering **173** (2017) 012015 doi:10.1088/1757-899X/173/1/012015

Some constrains are added to limit the search space. Let M be the maximum order of the differential equation (6), so this parameter is taken to be no less than n_{obj}, (1). Now the coefficients and the initial value of the problem (4) can be determined by three vectors $\hat{a} \in R^M$, $v \in R^M$ and $b \in R^{n_{inp}}$ and an integer $m \le M$, $m \in S(M), S(M) = \{k : k \in N, k \le M\}$. Thus, the extremum problem for the model (6) can be reduced to the following one

$$\sum_{i=1}^{s} \left\| y_i - \hat{x}(t_i) \right\| \to \min_{\hat{a},\hat{b},m} \tag{7}$$

We propose performing the estimation of all the required parameters with two vectors: the equation coefficients $\hat{a} \in R^{M+1}$ and the initial value coordinates $\hat{x}_0 \in R^M$.

Let A be some vector space $R^{\dim_R(A)}$, so using vectors $\hat{a}$, $\hat{x}_0$ and transformation

$$\begin{gathered} T(a \in A, n) : A \times N \to R^{\dim_R(A)-n}, \\ T(a \in A, n) = v : v_i = a_i, i = \overline{1, \dim_R(A) - n} \end{gathered} \tag{8}$$

one can define the parameters in the notation (3).

Let $\hat{x}(t)\big|_{\langle \hat{a},\hat{b}\rangle = \langle a,b\rangle, \hat{x}(0)=x_0}$ be a solution of the Cauchy problem (5) with parameters: m, $\hat{a} = T(a,m)$, $\hat{b}$ and $\hat{x}(0) = T(x_0, m)$. The conceptual formulation of the inverse mathematical modelling problem involves receiving the model from the data. In other words we need to find a solution of the extremum problem: m^*, $\hat{a}^*_{m^*} = P(\hat{a}^{opt}, m^*)$, $\hat{b}^*$ and $\hat{x}^*(0) = P(\hat{x}_0^{opt}, m^*)$, where

$$\begin{gathered} \hat{a}^{opt}, \hat{b}^{opt} \hat{x}_0^{opt} = \arg\min_{a \in R^M, b \in R^{n_{inp}}, x_0 \in R^M} I(a,b,x_0), \\ I(a,b,x_0) = \sum_{i=1}^{s} \left| y_i - \hat{x}(t_i)\big|_{\langle \hat{a},\hat{b}\rangle = \langle a,b\rangle, \hat{x}(0)=x_0} \right| \end{gathered} \tag{9}$$

According to (9), each alternative of this problem is an element of the product $R^M \times R^{n_{inp}} \times R^M \times S(M)$ or else $R^{2 \cdot M + n_{inp}} \times S(M)$.

We need to point out that the real input coefficients can be estimated if the inputs are linearly independent. Let us prove it. If $\exists i \in N : 1 \le i \le n_{inp}, \alpha_i \ne 0$ and $\forall i : 1 \le i \le n_{inp} \Rightarrow u_i(t) \ne 0$ and we have a solution of the equation $\sum_i \alpha_i \cdot u_i(t) = 0$, then from the equality $\sum_i \tilde{\alpha}_i \cdot u_i(t) = \sum_i \beta_i \cdot u_i(t)$ does not follow the equalities $\tilde{\alpha}_i = \beta_i, 1 \le i \le n_{inp}$, since $\sum_i \tilde{\alpha}_i \cdot u_i(t) + \sum_i \alpha_i \cdot u_i(t) = \sum_i \beta_i \cdot u_i(t)$ and $\sum_i (\tilde{\alpha}_i + \alpha_i) \cdot u_i(t) = \sum_i \beta_i \cdot u_i(t) \Rightarrow \tilde{\alpha}_i + \alpha_i = \beta_i$. The contradiction proves the statement.

Thus, the identification problem (4), whose solution is a dynamical system model in the form of a linear differential equation is reduced to an extremum problem (9). This problem requires a specific optimization tool that works both with real values and integers. It is easy to see that such an exhaustive search is computationally unfavourable. The problem statement makes it impossible to use various standard optimization algorithms, unless a modification of one or a collaborative technique or hierarchy that handles all the model determination features is applied. Since there is no limitation for real values, it is not justified to use pseudo Boolean optimization techniques. Finally, the required algorithm should work on the reduced search space without any information on the solution area localization.

3. A problem-oriented optimization algorithm and the restart operator

In our previous work, we have solved the problem of modifying some real-value optimization problems. The evolutionary strategies (ES) approach was used as the origin for the problem-oriented optimization technique. In the studies [5] and [7] it was shown that the algorithm with the proposed modifications has a higher level of performance than a standard algorithm with its tuned settings. We describe a (λ,μ)-ES algorithm and $\lambda=\mu$. A generalized analogue of the problem-oriented algorithm that makes it possible to solve the optimization problems on real value vector fields and integer value fields is given in the study [7]. By the word 'generalization' we mean that the integer values also have strategic parameters that take effect when objective genes mutate. Despite the fact that all the modifications increase the number of settings, their implementation makes the algorithm much more powerful.

The main feature of the evolutionary strategies algorithm is that the individual i matches a set, which is represented by an alternative and a strategy, so it has a structure: $i=\langle p_{obj},p_{str}\rangle$, $p_{obj}\in X$, $p_{str}\in R^{\dim(X)}$, where p_{obj} and p_{str} are the objective and strategic parameters, respectively and X is an alternative space. The objective parameters can be subdivided by the nature of variables, $p_{obj}=\langle i^{r}_{obj},i^{c}_{obj}\rangle$, a set of real variables i^{r}_{obj} and integers i^{c}_{obj}. For the fitness function evaluation, all the objective parameters are transformed into optimization problem variables: $i^{r}_{obj}\to\hat{a},\hat{b},v$ and $i^{c}_{obj}\to m$. The same for the strategic parameters: $p_{str}=\langle i^{r}_{str},i^{c}_{str}\rangle$. Now, let $\hat{x}(t)\big|_{\langle\hat{a},\hat{b}\rangle=i^{r}_{obj},i^{c}_{obj}}$ be a solution of the initial value problem (6) with the parameters transformed from objective parameters, then the fitness function can be evaluated as follows

$$fit(i)=\left(1+\sum_{i=1}^{s}\left\|y_i-\hat{x}(t_i)\big|_{i^{r}_{obj},i^{c}_{obj}}\right\|\right)^{-1}. \quad (10)$$

The first aim of this improvement is to increase the efficiency of the local optimization operator by applying a limitation to the search space. Here we briefly describe the improvements that were made. The evolutionary strategies algorithm is hybridized with another stochastic optimization algorithm, which is similar to the coordinate-wise optimization approach and has its own parameters. The main principle of this implementation is: randomly choose an individual, randomly choose its coordinate n^{l}_{2} times, and make for each coordinate n^{l}_{3} steps in a previously generated direction until the fitness value grows. Repeat the search n^{l}_{1} times, so that the same individual can be chosen several times, and the same for the coordinate. The step value is also the search setting and it can be fixed or be randomly generated with the selected limitation. However, every individual has a gene that determines the order of the differential equation and so the number of genes to be improved in each loop should be reduced. It means that if a chosen variable does not affect the equation, it must be avoided.

In this case, for every particular i^{c}_{obj} value, the local optimization operator works only for the following genes $\left(i^{r}_{obj}\right)_k, k:\left(i^{r}_{obj}\right)_k\to\hat{b}\vee\left(i^{r}_{obj}\right)_k\to\hat{a}_j\vee\left(i^{r}_{obj}\right)_k\to v_j$, where $j\le i^{c}_{obj}$. This modification sufficiently increases the speed to reach the local optimum.

The second improvement is a modification of the generalized mutation operator for the integers. In the current study, integers are the rank variable, not the nominal one. This is why the mutation could not just be to change the current variable value to a randomly generated integer, but rather to randomly greaten or lessen it by adding a random integer value. There is only one constraint: the integer variable after the mutation must fit into the boundaries.

Let a mutant individual be denoted as $\tilde{i}$. The random variables r_1, r_2 are distributed by the following laws: $P(r_1 = 1) = p^c_m$, $P(r^c_m = 0) = 1 - p^c_m$ and $P(r_2 = 1) = i^c_{str}$, $P(r_2 = 0) = 1 - i^c_{str}$. The parameter r_m is a controlling parameter for the mutation operator, if it is necessary to reduce the mutation. The strategic parameter mutates as for the standard evolutionary strategies algorithm and the current improvement is related to the objective parameter mutation. The random value $r_3 : P(r_3 = 0) = P(r_3 = 1) = \frac{1}{2}$, determines the mutation direction. If $i^c_{obj} = 1$, then $\tilde{i}^c_{obj} = i^c_{obj} + r_1 \cdot r_2$, if $i^c_{obj} = M$, then $\tilde{i}^c_{obj} = i^c_{obj} - r_1 \cdot r_2$ and if $1 < i^c_{obj} < M$:

$$\tilde{i}^c_{obj} = i^c_{obj} + r_1 \cdot r_2 \cdot (-1)^{r_3}. \tag{11}$$

The same mutation probability controlling parameter for real value variables is denoted as p^r_m, and the mutation for each of these variables takes place if a value uniformly distributed on $[0,1]$ is less than p^r_m.

All the algorithm settings were taken as they were tuned in the study [7].

The initial population of this algorithm is generated randomly. All the strategic parameters are generated uniformly on the interval $[0, 1]$, and the real value objective parameters are equal to 0. In previous work it was estimated to be the most efficient initial distribution for the algorithm without integers. The integer variables can take any value with the same probability, so

$$P(i^c_{obj} = j) = \frac{1}{M}, \forall 1 \le j \le M. \tag{12}$$

Thus, when the algorithm restarts, its initial population is generated according to the above and (16). In this study three more schemes of initial population generation were proposed. As was already mentioned in the introduction, we need a useful, effective and reliable tool for improving the algorithm performance, no matter what structure, properties, modifications and nature the algorithm has. The proposed meta-heuristic can be used irrespective of the type of algorithm it is applied to, be it a cooperation of distinct algorithms, coevolutionary algorithm or an algorithm with another meta-heuristic.

First of all, it is necessary to estimate if the algorithm is attracted to some local optimum and stacked in its neighbourhood. In terms of the evolution-based search, the algorithm stagnates. Nevertheless, since the optimization techniques are stochastic, there are always perturbations in the coordinates of alternatives. Therefore, in each generation there are different values for the maximum, average and minimum fitness function of the current population. The same is true for the population average or best alternative positions. Since there is still accuracy – and no longer such a contradiction when gathering information, the following characteristic is more preferable.

Let X be the alternative set. Let x^*_i be the alternative with the highest fitness function ever found during the work of the algorithm up to the i-th generation, so

$$x^*_i = \max_{j \le i} \left(\arg\max_{k < N^j_p} \left(fit\left(x^j_k\right)\right)\right), \tag{13}$$

where $j,k \in N$, N_p^j is the population size and $fit(\cdot): X \to R$ is a fitness function. As can be seen, $x_i^*, i = \overline{1,\infty}$ is a nondecreasing sequence. When alternatives localize near a specific optimum, the elements of this progression become closer to each other or even equal. This situation can last for a long time, during which the number of generations or objective function evaluations can be found.

In this case it can be seen that the algorithm stagnates and the computational resources are wasted. To avoid this problem the algorithm restart can be used. We denote by $H_1^{n_1}$ a set that contains the last n_1 alternatives with the highest fitness function values (1): $H_1^{n_1}(i) = \left\{x_j^*, j = \overline{i - n_1 + 1, i}\right\}$. The following criterion is used to determine the best solution fitness growth

$$R_1 = fit\left(H_1^{n_1}(i^*)_i\right) - fit\left(H_1^{n_1}(i^*)_1\right), \tag{14}$$

where i^* is the current generation. The parameter t_R^1 is a threshold level, thus the restart launches once the criterion (14) value becomes less than the threshold level value,

$$R_1 < t_R^1. \tag{15}$$

The first restart condition is based on two parameters: the number of generations n_1 and the threshold level t_R^1, which are to be tuned for every particular class of the optimization problem. In addition, there are still many other ways to implement the stagnation criterion but this question goes beyond this study and is to be investigated on a more general set of optimization problems.

The second restart operator requires the history of previous restarts. For this purpose let $H_2^{n_2}$ be a set of alternatives, which cause the algorithm restart. After each restart caused by condition (15) we would have

$$n_2 = n_2 + 1,\ H_2^{n_2} = H_2^{n_2} \bigcup x_i^*, \tag{16}$$

where x_i^* is a solution found, (13). The initial condition, when the algorithm has not ever restarted, is $n_2 = 0$, $H_2^{n_2} = \varnothing$.

After each generation, the meta-heuristic checks if the current best solution (13) is close to a solution that has already caused a restart. The following criterion is evaluated

$$R_2 = \max_{0<j\le n_2}\left(\left\|x_i^* - H_2^{n_2}{}_j\right\|\right), \tag{17}$$

where $\|\cdot\|: X \to R$ is a norm on X.

Now we check the condition based on the criterion (17) value for some taken threshold level $t_R^2 > 0$:

$$R_2 < t_R^2, \tag{18}$$

if the condition is fulfilled. the algorithm restarts.

The meta-heuristic works after each generation and if the history set is not empty, $H_2^{n_2} \neq \varnothing$, the condition (18) is checked. If the criterion (17) value is less than the threshold value t_R^2, the algorithm restarts and the set $H_1^{n_1}$ is deleted. In the other case, the inequality (15) is verified, and if the criterion (14) is less than the threshold value t_R^1, then the algorithm restarts and (16) evaluates.

In this paper the initial population generation after the restart increases the variable that controls the minimum differential equation order. After every restart, the minimum order value for the integer values of the initial population is increased by one with a probability $\frac{1}{i_{obj}^c}$. The purpose of this idea is to prevent the convergence to a lower order system with fewer parameters.

4. Experimental results

Let us denote criterion (9) divided by the sample size as C_1, which is a measure of fitting the observation data. Since the samples are made on the basis of known dynamical system outputs, it is preferable to compare the model output with the system's one. This criterion will strongly correlate with C_1, if and only if the sample is representative and the output is observed as it is: there is no data distortion and the sample is uniformly distributed. However, normally, real problems are identification problems with stochastic and parametric uncertainties. The criterion C_2 is the distance between the model output function and the real output: $C_2 = \int_0^T \left|x(t) - \hat{x}(t)\right| dt$. There are two more criteria, which are based on mathematical model parameters: $C_3 = \frac{n_{order}}{n_{runs}}$, where n_{runs} is the number of approach runs; n_{order} is the quantity of runs, as a result of which the real differential equation order is found; v_r is the real system initial value. The criterion C_3 is a probability estimation of receiving the model with the same order as for the real system equation.

And the last two criteria are useful in estimating how the right part of the equation (6) is identified. The criterion $C_4 = \left\|b - \hat{b}\right\|$ shows how accurate the identification result is and the criterion $C_5 = \int_0^T \left|\sum_{i=1}^{n_{inp}} \left(b_i \cdot u_i(t) - \hat{b}_i \cdot u_i(t)\right)\right| dt$ is an estimation of the closure of the summarized control input.

In this chapter the experimental results are given. Firstly, let us describe the sample forming and the test tasks. The dynamic system output is the Cauchy problem solution on $[0, T]$, as it corresponded to (6). This solution is discretised and represented as a set with s elements. Let $I^s = \{r_i^{\text{int}} \le N_s, \forall i < s\}$ be a set of randomly chosen different integers, so accordingly, the expression (2) is, $y_i = x\left(\frac{I_i^s \cdot T}{N_s}\right) + N(0, \sigma^2)$ and $t_i = \frac{I_i^s \cdot T}{N_s}$, where $i = \overline{1, s}$. In the current study it was suggested to compare the algorithms in the case of no noise.

Due to the results of previous work it was decided to use the following settings in the current investigation: 100 individuals for 100 populations, tournament selection, 3 parents for random crossover, modified mutation scheme, the mutation probability for the real variables $p_m^r = \frac{2}{2 \cdot N + 1}$, the mutation probability for the integer variables $p_m^c = 0.5$, $N_1^L = 40$ individuals for the local stochastic improvement, $N_2^L = 50$ genes and $N_3^L = 1, h_L^{\max} = 0.5$. The strategic parameters of the initial population were uniformly generated $U(0,1)$, and every objective parameter is equal to 0. The order limitation as in all the previous studies is taken as 10.

To examine the algorithm with meta-heuristics and its different settings: $n_1 \in \{5, 7, 9, 11, 15\}$, $t_R^1 \in \{0.005, 0.01, 0.05, 0.1\}, t_R^2 \in \{0.05, 0.1, 0.2, 0.5\}$, three different identification problems have been considered, which are used to estimate a similar algorithm performance for single input systems [7]. The list of differential equations that simulated the dynamic process is given in Table 1. The sample size is 100. For this experiment the variance of the noise is equal to 0, so the experimental data allows us to see the algorithm performance in a basic form. In Table 2 the different multiple inputs are given, so that performance investigation in the case of there being no data distortion would be based on tuning the meta-heuristic settings on the set of the following problems: each of the dynamical systems and each of the first three control functions, given in Table 2. All the functions are laniary independent, so there is the only one global of the criteria for the real system parameters.

Table 1. Dynamical systems.

Number	Differential equation of the object
1	$x'' + 2\cdot x' + x = u(t)$, $x_0 = (2\,0)^T, T = 12.5$
2	$x''' + x'' + 2\cdot x' + x = u(t)$, $x_0 = (2\,0\,0)^T, T = 12.5$
3	$x^{(4)} + x''' + 4\cdot x'' + 3\cdot x' + x = u(t)$, $x_0 = (2\,0\,0\,0)^T, T = 12.5$

Table 2. Considered inputs.

Number	Number of inputs	Additive control function
1	2	$\sqrt{t+1} + \sin(t)$
2	3	$\sqrt{t+1} + \sin(t) + 2\cdot\cos(2\cdot t)$
3	4	$\sqrt{t+1} + \sin(t) + 2\cdot\cos(2\cdot t) + \dfrac{\sin(0.5\cdot t)}{1+0.2\cdot t}$

The algorithm with minimum error C_1 and C_2 is the one with the settings: $c_1^r = 0.005$, $c_2^r = 0.05$, $n_t = 11$, and the criterion average value equal to 0.1587. Let us show how the examination results correlate with the performance estimation of the hybrid modified evolutionary strategies algorithm without the meta-heuristic. For this purpose we put on a chart the average criteria values for the algorithm without the meta-heuristic, the average criterion value for the algorithm with the best meta-heuristic settings. The following statistics are given for the criteria C_1, C_2, C_3, C_4 and C_5 on the Figures 1-5, respectively. On every Figure there are average values of criteria for problems with 2, 3 and 4 inputs and overall average value. As one can see, the algorithm with the proposed restart operator and its tuned settings outperforms the initial algorithm by all the criteria.

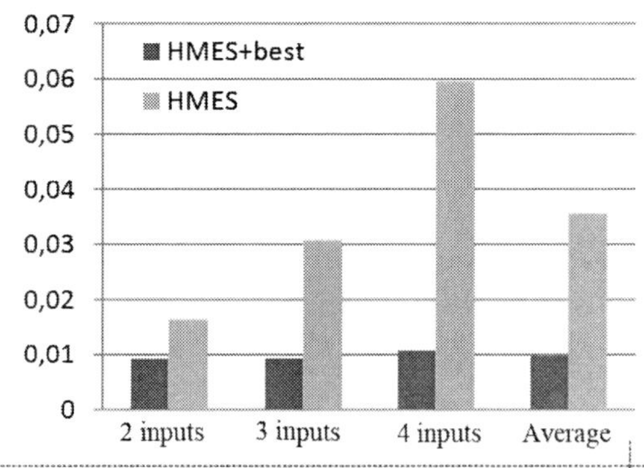

Figure 1. Average criterion C_1 values.

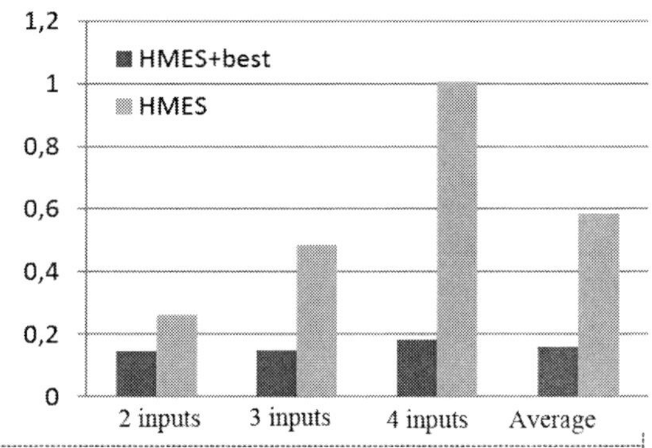

Figure 2. Average criterion C_2 values.

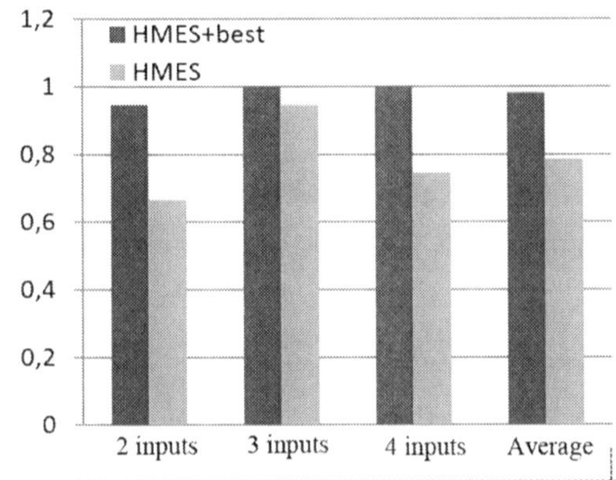

Figure 3. Average criterion C_3 values.

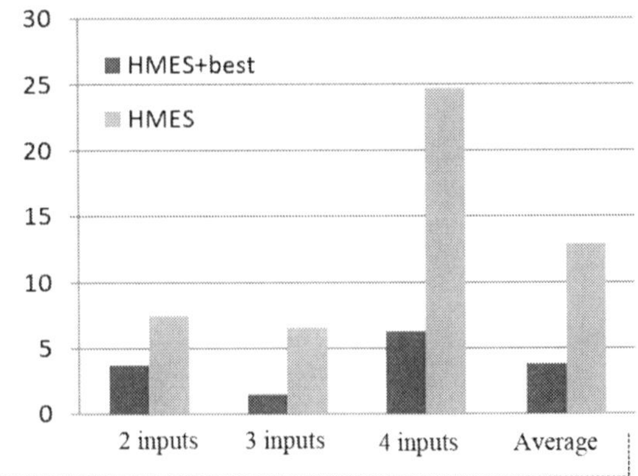

Figure 4. Average criterion C_4 values.

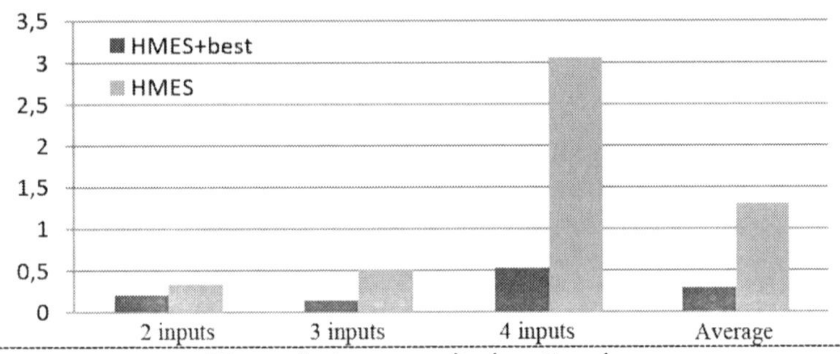

Figure 5. Average criterion C_5 values.

The examination results prove that the proposed meta-heuristic implemented into the problem-oriented algorithm results in an improved level of performance, so with the same computational effort the result that can be achieved using the proposed approach is more accurate: model fits the observations data more and so its output is closer to the system state. The same applies to the multiple inputs coefficients estimation and the sum of these input functions estimation.

5. Conclusion

In the current study the MISO system inverse mathematical modelling problem was considered. The problem was reduced to an optimization problem on real value vector space and integer value space. Each alternative determines the mathematical model in the form of a linear differential equation with and an additional part on the right for multiple inputs. The variables of the extremum problem determine the order of the differential equation, its coefficients, coefficients of multiple inputs and the initial value coordinates.

We proposed an optimization algorithm, which is a problem-oriented modification of the hybrid generalized evolutionary strategies algorithm supplemented by a restart operator meta-heuristic. The examination results proved its high performance and higher efficiency in solving the considered class of optimization problems, compared with the problem-oriented algorithm with no meta-heuristic implementation.

Further studies are correlated with developing an inverse mathematical modelling of the dynamic system along with developing extremum seeking approaches for the variables of a different nature and meta-heuristics for improving algorithm performance.

The MISO inverse modelling problem is the previous step to modelling MIMO dynamical systems, since each MIMO system state can be described as a MISO system with multiple inputs, which are the control inputs and states and its derivatives of other system states.

Acknowledgements
The reported study was funded by Russian Foundation for Basic Research, Government of Krasnoyarsk Territory, Krasnoyarsk Region Science and Technology Support Fund to the research project №6-41-240822.

References
[1] Beligiannis G N, Tsirogiannis G A, Pintelas P E 2014 Restartings: a technique to improve classic genetic algorithms' performance *Optim.* **1** pp 112-115
[2] Dao S D, Abhary K, Marian R M 2014 Optimization of partner selection and collaborative transportation scheduling in Virtual Enterprises using GA *Expert Syst. Appl.* **41(15)** pp 6701-6717
[3] Fukunaga A S 1998 Restart Scheduling for Genetic Algorithm *Lecture Notes of Computer Science* **1498** pp 357-369
[4] Loshchilov I, Schoenauer M, Sebag M 2012 Alternative Restart Strategies for CMA-ES *Lecture Notes in Computer Science* **7491** pp 296-305
[5] Ryzhikov I, Semenkin E 2013 Evolutionary Strategies Algorithm Based Approaches for the Linear Dynamic System Identification *Lecture Notes in Computer Science: Adaptive and Natural Computing Algorithms* **7824** pp 477-483
[6] Ryzhikov I, Semenkin E, Akhmedova S 2016 Linear ODE coefficients and initial conditions estimation with co-operation of biology related algorithms *Advances in Swarm Intellegence. Lecture Notes in Computer Science* **7824** pp 228-235
[7] Ryzhikov I, Semenkin E, Panfilov I 2016 Evolutionary optimization algorithms for differential equation parameters, initial value and order identification *Proceedings of the 13th International Conference on Informatics in Control, Automation and Robotics* **1** pp 168-176
[8] Sersic K, Urbiha I 1999 Parameter Identification Problem Solving Using Genetic Algorithm *Proceedings of the 1. Conference on Applied Mathematics and Computation* pp 253–261
[9] Parmar G, Prasad R, Mukherjee S 2007 Order reduction of linear dynamic systems using stability equation method and GA *International Journal of computer and Infornation Engeneering* **1(1)** pp 26-32
[10] Ramesh K, Nirmalkumar A, Gurusamy G 2011 Order Reduction of LTIV continuous MIMO system using Stability Preserving approximation method *International Journal of Computer Applications* **36(8)** pp 1-8
[11] Desai S, Prasad R 2013 A novel order diminution of LTI systems using Big Bang Big Crunch optimization and Routh approximation *Appl. Math. Model.* **37** pp 8016-8028
[12] Narwal A, Prasad B R 2016 A Novel Order Reduction Approach for LTI Systems Using Cuckoo Search Optimization and Stability Equation *IETE Journal of Research* **62 (2)** pp 154-163
[13] Naiborhu J, Firman, Mu'tamar K 2013 Particle Swarm Optimization in the Exact Linearization Technic for Output Tracking of Non-Minimum Phase Nonlinear Systems *Applied Mathematical Sciences* **7(109)** pp 5427-5442

Network intrusion detection by the coevolutionary immune algorithm of artificial immune systems with clonal selection

T Salamatova, V Zhukov

Reshetnev Siberian State Aerospace University, 31, Krasnoyarsky Rabochy Av., Krasnoyarsk, 660037, Russian Federation

E-mail: salamatova_ta@mail.ru

Abstract. The paper presents the application of the artificial immune systems apparatus as a heuristic method of network intrusion detection for algorithmic provision of intrusion detection systems. The coevolutionary immune algorithm of artificial immune systems with clonal selection was elaborated. In testing different datasets the empirical results of evaluation of the algorithm effectiveness were achieved. To identify the degree of efficiency the algorithm was compared with analogs. The fundamental rules based of solutions generated by this algorithm are described in the article.

1. Introduction

Nowadays the qualitative and quantitative characteristics of network attacks on the informational objects are constantly changing. Under such conditions the role of intrusion detection systems (IDS) for information security is steadily becoming more and more important. IDS are becoming an integral part of the security infrastructure in public and commercial organizations (figure 1). In some cases applying IDS is a must, and it is governed by the requirements of legal acts and methodological documents.

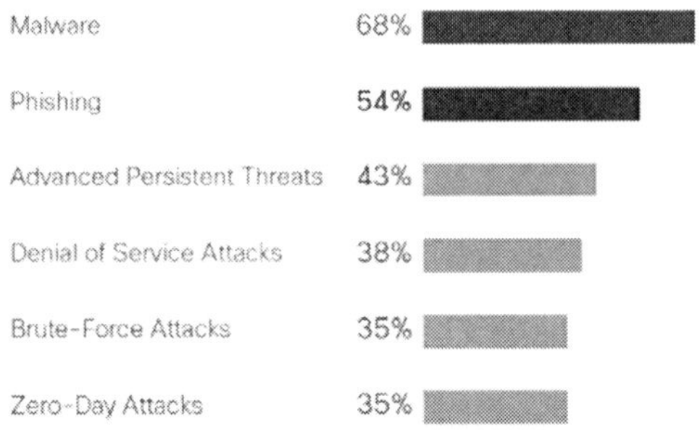

Figure1. ExternalChallengesFaced 2015.

Comparative analysis of IDS showed that most of the Russian IDS use only signature methods for data analysis. The ability to update the fundamental rules database is available only for the developers.Users must buy such updates, they do not have the possibility to create and automatically

IOP Conf. Series: Materials Science and Engineering 173 (2017) 012016 doi:10.1088/1757-899X/173/1/012016

generate their own fundamental rules. IDS based only on signature methods do not allow detecting unknown threats of information security («zero day» attacks). As practice shows such IDS do not provide the required level of security for IS. This fact increases information risks and, as a consequence, financial losses. A possible way out of this situation is using adaptive heuristic intrusion detection algorithms in addition to IDS. It will allow not only detecting «zero day» attacks, but also automating a local content (update) of a fundamental rules base for information systems (IS). A heuristic algorithm built on the basis of computer models of artificial immune systems (which is based on the clonal selection methods and the theory of acquired immunity) is considered in this paper.

The paper consists of several sections. The 2nd section of this article is devoted to artificial immune systems. The modified algorithm of artificial immune systems (AIS) with clonal selection is discussed in Section 3. Section 4 evaluates its effectiveness (test and real data were evaluated). Section 5 is devoted to comparative analysis with similar algorithms according to the degree of their efficiency. Section 6 shows the use of generated information structures in the fundamental rules of intrusion detection systems. Finally, section 7 reflects the main results and suggestions for further research.

2. Artificial immune system

AIS is a complex adaptive structure which simulates some aspects of the immune system functioning in protecting organisms against external microorganisms (antigens (Greek *anti* – against, genes – generating, born)) by protective cells of the immune system (antibodies (Greek *anti* – against) or detectors). A characteristic which describes the strength of antibodies and antigens interaction is defined as the affinity (Latin *affinis* – related). Presently, such computational models of immune systems as clonal selection algorithms [2, 3], negative selection [4] and immune network algorithms are of interest [5, 6].

The application of clonal selection algorithm based on the methods of clonal selection and the theory of acquired immunity is inspected in the paper. Clonal-selection theory was formulated in 1957 by Frank Macfarlane Burnet in Australia and D. Talmagein in the USA independently of each other. This theory explains how the immune system resists alien macromolecules (antigens). Each individual's antibody-producing cell system contains all the information necessary for the synthesis of any antibody (detector) before meeting the antigen. The antigen does not give information to these cells; it simply selects cells which synthesize antibodies corresponding to this antigen, and encourages them to multiply and generate such antibodies. Cells synthesizing such type of antibodies belong to one clone which consists of all the descendants of the parent cell (this parent cells acquired the genetic ability to respond to the antigen in a random process). This clone is numerically small before the antigen appearance. The presence of the antigen stimulates clone reproduction; it is able to synthesize the corresponding antibodies. The better this antigen will be recognized (affinity), the greater amount of offspring (clones) will be generated.

The heuristic algorithm built on the basis of synthesis of AIS clonal selection algorithm as well as evolutionary and coevolutionary strategies are presented in this work.

3. Modification algorithm of artificial immune systems with clonal selection

During the development and research of clonal selection algorithm AIS [7] it was determined that the algorithm can detect deliberate changes in the monitored data, stabilize the average number of errors of type I and II, and make them dependent on the resources allocated to the algorithm. However, it became clear that in the process of results averaging the error rate of the first and second type was quite high. The problem of forming a representative set of high affinity detectors has been determined. It became necessary to introduce additional mechanisms for solving the problem. It was proposed to replace the mechanism of detector reproduction and mutation (which was the standard for clonal selection algorithm) with the external optimization procedure. This procedure is based on the application of evolutionary algorithms strategy [8]. The results of experiments with evolutionary immune clonal selection algorithm showed that the proposed modifications reduced the number of type I errors and nullified type II errors. However, the application of evolutionary strategies resulted in the problem of identification of its optimal settings, such as selection, recombination and mutation.

To solve the problem of automated index selection and setting the immune evolutionary algorithm parameters, the use of coevolutionary strategy was proposed (*coevolution*: Latin. *co* – compatibility, consistency, *volutio* – deployment). Coevolutionary immune algorithm with clonal selection AIS (coevolutionary algorithm) presents some independent, self-acting evolutionary immune algorithms with clonal selection AIS. These algorithms have different settings in competition for granted (limited) computing resource.The generalized scheme of coevolutionary algorithm is presented in figure 2.

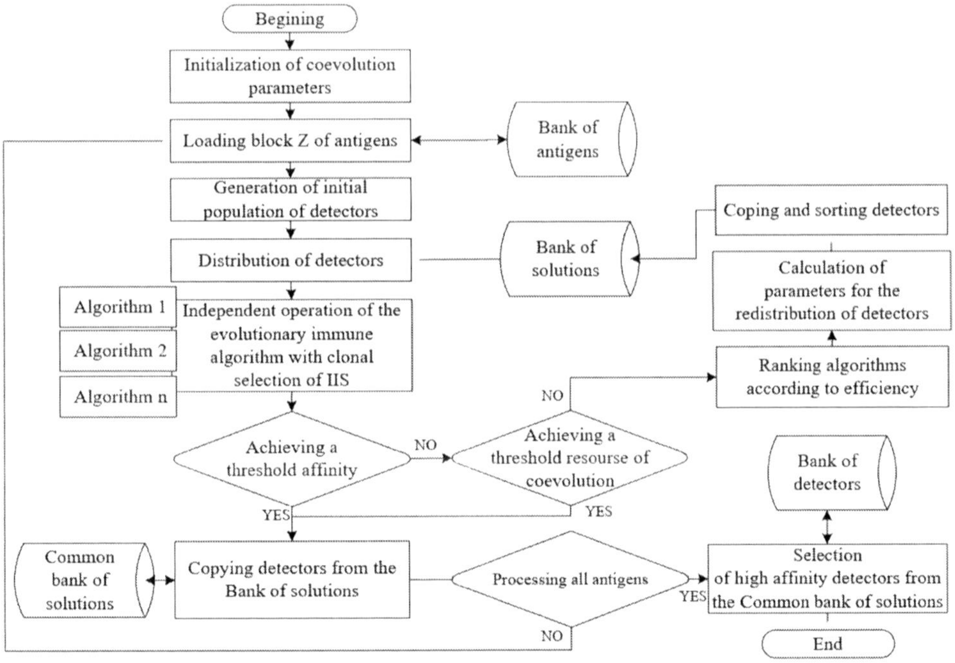

Figure 2. Flowchart of generation of detectors by coevolutionary immune algorithm with clonal selection AIS.

Thanks to the coevolutionary strategy the necessity of a parameters random selection and / or the need of exhaustive combinations search during evolutionary immune algorithm separate testing have been obviated.

4. Experiments and Results

Empirical results of evaluation of coevolutionary algorithm effectiveness (on test data) were introduced [7, 8]. These results indicate that the developed algorithm can detect real data intentional changes (malicious traffic patterns). Also, the algorithm makes it possible to create information structures for automatic construction of fundamental rules (signatures) in the specified format. Effectiveness evaluation was conducted on a set of test data from the public database of network traffic *KDD Cup 1999* (database «*KDD*'99») [10], collected by the Information and Computer Science University of California. The number of detected antigens in the test multiplicity (%), the number of the first and second type errors (%), the ratio of average number of the first and second type errors to the number of detectors and the ratio of abnormal events (antigens) database size to the size of the created detector database were used as evaluation criteria of evolutionary algorithm efficiency.

The next logical step is to evaluate the developed algorithm efficiency in the network traffic «real data» containing or not containing malicious content. Network analyzers (e.g. *CommView, Wireshark*) using a common format *Packet Capture* (*pcap*) as a storage format of the intercepted traffic are typically applied to create the dumps of network traffic. Multiplicity of data for research (patterns of network traffic in *pcap* format from international cyber-training Locked Shields [11] and DEF CON competition [12]) was formed as a result of open source studying. Figure 3 shows some part of effectiveness evaluation results of the coevolutionary algorithm for antigens unit |Z|=10 when tested on real samples of malicious and the «normal» traffic *pcap* format (*REAL* column). Previously reached on the test base «*KDD*'99» results are compared (*TEST* column). D_{best} is multiplicity of detectors recorded into the detectors database. A is a set of antigens (abnormal events). E is test multiplicity. The results of the experiments averaged over multiple startups show that the percent of detection of malicious samples in the test data and the level of Type II error (100% and 0% respectively) remained the same. The level of the type I error has increased by 1% on the average. The average number of type I errors per one detector remained at the same level as that obtained when tested on the test database «*KDD*'99».

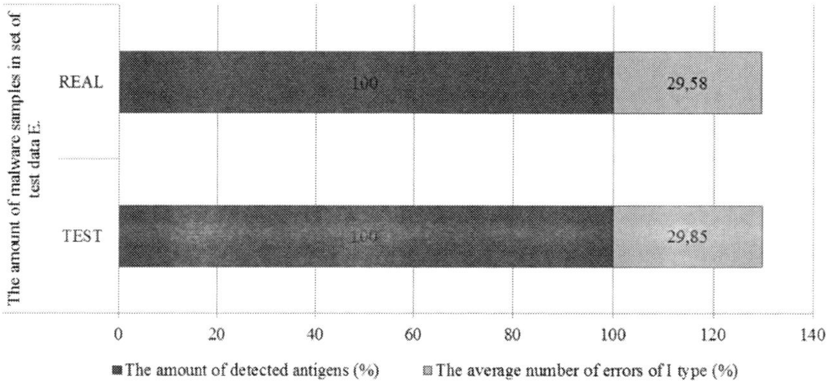

Figure 3. The percentage ratio of the amount of detected antigens and the average number of type I errors by varying the set of test data E and $|D_{best}|=|A|\cdot 15\%$.

Figure 4 shows the size of generated detectors database (Bank of detectors) for the given size of abnormal events database (Bank of antigens) and normal events database (Bank of normal events).

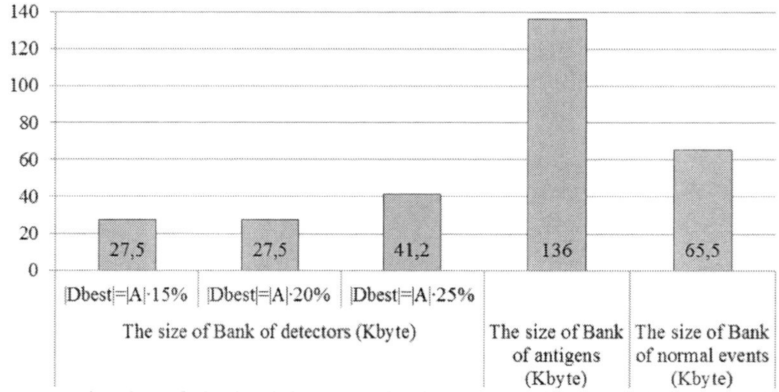

Figure 4. The size of obtained detectors database at specified test databases of normal and abnormal events.

Thus, the ratio of the abnormal events database to detectors database is about 3:1.

5. Comparison of the results of coevolutionary algorithm research with the other results.
To evaluate the effectiveness of the coevolutionary algorithm developed by the authors it was decided to conduct a comparative analysis of the results of the algorithm and the results of other algorithms (running on the test set «*KDD*'99»base).

Shirazi H. M. with colleagues proposed an IDS model where collected data analysis is carried out by the algorithm based on Memetic algorithm and Bayesian networks [13]. Also, the authors additionally solve the classification of attacks problem in the research. The model was tested by the «*KDD*'99» database. *Detection Rate* (*DR*) and *False Acceptance Rate* (*FAR*) were used as the criteria of effectiveness calculated as follows:

$$DR = \frac{TP}{TP + FN} \cdot 100\ \%,\ FAR = \frac{FP}{FP + TN} \cdot 100\ \%, \tag{1}$$

where *TP* (*True Positive*) is the situation when a signature is fired properly as soon as an attack is detected and an alarm is generated; *FP* (*False Positive*) is the situation when normal traffic causes the signature to raise an alarm; *TN* (*True Negative*) is the situation when normal traffic does not cause the signature to raise an alarm; *FN* (*False Negative*) is the situation when a signature is not fired when an attack is detected.

Figure 5 presents a comparative analysis of the coevolutionary algorithm with the results of research of Shirazi H. M. and other results of scientific researches and works solving a similar problem of intrusion detection mentioned by Shirazi H. M.

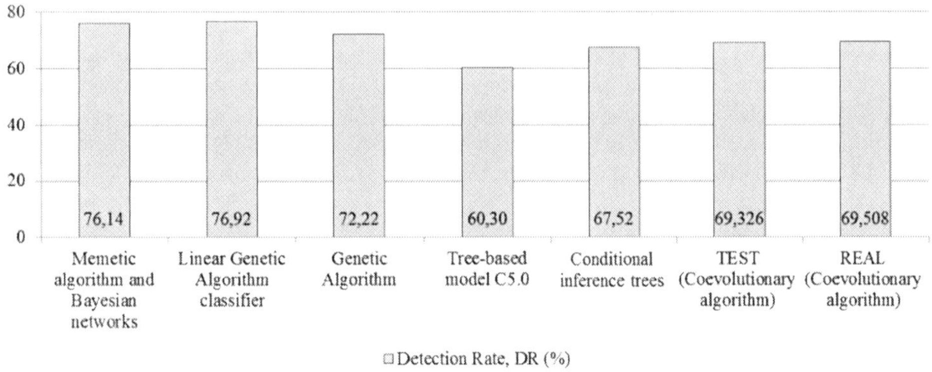

Figure 5. Comparison of the research results of Shirazi H. M. with the other results and the coevolutionary algorithm.

Summarizing the results of research one can draw the conclusion that the percentage ratio of the average number of I and II types errors is comparable with analogs built on different methods of artificial intelligence.

6. The format of the decision rule
The detectors generated by the research coevolutionary algorithm can be used for creation of the IDS decision rules and, in the future, in some other means and systems of information protection (e.g., antivirus software). A lot of modern IDS support the «Snort» signature format of Cisco Company [14], so it is possible to specify the location of the generated coevolutionary algorithm detector as the example of its constructing.

«Snort» signatures (the decision rules) consist of two parts: a rule header and a body (parameters) of a rule. A rule header contains a description of the action, a data transfer protocol, IP-address, a network mask, a source and destination ports. Rule parameters store a warning message and inform what part of the detected packet should be processed in case of rules activation. There are four categories of rule parameters in the modern «Snort»: *general, payload, non-payload* and *post-detection*. During the analysis of the interaction of the coevolutionary algorithm detectors and the format of «Snort» signatures it was found that the generated detectors can replace the last three parameters in «Snort» signature structure (figure 6).

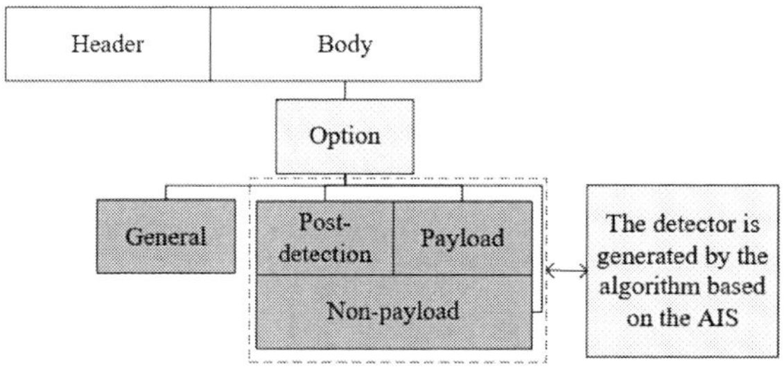

Figure 4. The detector location in the IDS «Snort» signature format.

For example, there is a signature (figure 7):
alert tcp any any <> *any* 80 (*msg*: «*MD5 Alert*»; *protected_content*: «293C9EA246FF9985DC6F62A650F78986»; *hash*:*md5*; *offset*:0; *length*:4;).

In this case, there are two types of parameters involved in the body of the rule: general (*msg*) and payload (*protected_content, hash, offset, length*). The *msg* option sets the rules *alert* and *log* a message which is linked to the packages in the dump or is displayed in triggered *alert*. Keyword *protected_content*as as well as *content* search the matches in the lines of certain information bits but the search is performed in hashed packets whose content is compared with a predetermined value in the rule. Keyword *hash* is used to specify the type of hash (*MD*5, *SHA*256 и*SHA*512). Keyword *offset* allows specifying a search range set value (value from -65535 to 65535). Keyword *length* is used to indicate the initial length of hashed content specified in the rule (value from 0 to 65536).

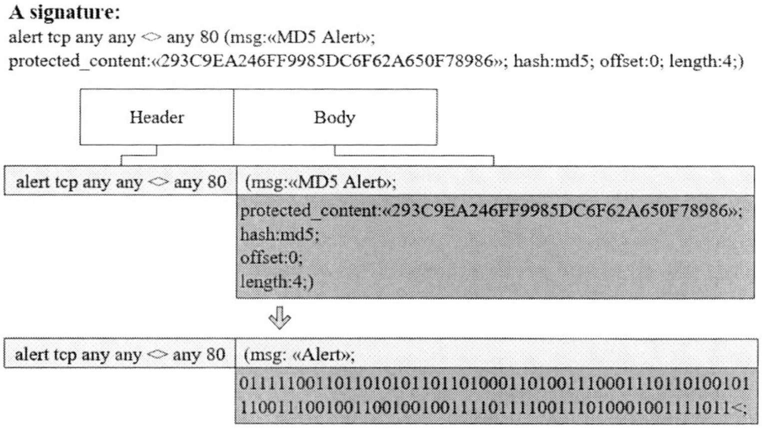

Figure 5. The synthesis submission the IDS «Snort» signature and the generated detectors.

The analysis of the individual components of the signature parameters shows that the rule demonstrates text search «*HTTP*» in packages of length of 4 bytes at zero bias. The detection of the text will activate «*MD5 Alert*» warning.

In using the information structure of the detector generated by the training coevolutionary algorithm previously the signature will acquire the following form:
alerttcp any any<>any 80 (*msg*: «*MD5 Alert*»; 0111110011011010101101101000110100111000111011010010111001110010011001001001111011 1100111010001001111011<;),
where the last value reflects the affinity value whose exceeding is the rule activation condition (in this case "<" is 60 encoded value in the Unicode format). It isn't necessary to set parameters of *payload* type using this signature.

Today it is possible to point out the foreign solutions IDS such as «Suricata», «HPE TippingPoint Next Generation Intrusion Prevention Systems», «The Huawei Network Intelligent Protection system», «IBM Security Network IPS», «Cisco IPS», «McAfee Network Security Platform v.8.x» and Russian solutions such as «ViPNet IDS», «User Gate Proxy&Firewall 6.0 VPNGOST», «Dionis-NX» of Factor-TC company, «Outpost» and «Attack Detection «Continent» which support the «Snort» signature format (import and /or export of decision rules).

7. Conclusion

AIS were selected among the existing and rapidly developing methods for the development of a model-algorithmic support system of the heuristic detection methods for IDS. AIS have such important properties as adaptability, self-regulation and self-learning, due to which IDS allow making up the base of decision rules («learning mode») automatically. These data bases will be periodically updated as long as experience is accumulated («operation mode») that in the end could help to reduce the cost of creating the bases of IDS decision rules and updating them.

In general coevolutionary algorithm proved its efficiency both in the test and "real" data sets as well as the problem of the automatic creation of information structures for the construction of rules to build the IDS decision rules base. However it is necessary to continue research of the coevolutionary algorithm aimed at reducing the errors level of type I (improve the quality of the decisions) and optimizing the algorithm for the rate of its convergence and resource. Further, the authors plan to submit comparative analysis of the research results with the results of other algorithms on the «real» data set and to develop recommendations for improving the algorithm.

References

[1] Annual Security Report Available at: http://www.cisco.com/c/m/en_us/offers/sc04/2016-annual-security-report/index.html
[2] Burnet F 1957 *Australian Journal of Science* **20** pp 67-69
[3] Talmage D 1957 *Annual Review of Medicine* **8** pp 239-256
[4] Forrest S, Perelson A 1994 Self-nonself discrimination in a computer *Proceedings of the IEEE Symposium on Research in Security and Privacy* pp 202-212
[5] Jerne N 1974 Towards a network theory of the immune system *Annales d'Immunologie* **125** pp 373-389
[6] Kim J and Bentley P 2001 The Artificial Immune System for Network Intrusion Detection: An Investigation of Clonal Selection with Negative Selection Operator *Proceedings of the Congress on Evolutionary Computation* pp 1244-1252
[7] Zhukov V and Salamatova T 2014 The detection of network intrusion by evolutionary immune algorithm with clonal selection *Vestnik SibGAU* **4 (56)** pp 41-47
[8] Zhukov V and Salamatova T 2014 Detection of information security incidents modified algorithm of artificial immune system with clonal selection *In the World of Scientific Discoveries* **6.1 (54)** pp 497-517
[9] Salamatova T 2015 The determination of competitive search strategy in the coevolutionary immune algorithm with clonal selection *Materials XIX International scientific conference Reshetnev reading* **2** pp 296-298
[10] KDD Cp 99 Intrusion detection data set Available at: http://kdd.ics.uci.edu/

[11] NATO Cooperative Cyber Defence Centre of Excellence Available at: https://ccdcoe.org/index.html
[12] DEF CON Communications Available at: https://defcon.org/
[13] Shirazi H and Namadchian A 2012 Combined anomaly base intrusion detection using memetic algorithm and Bayesian networks. *International journal of machine learning and computing* **2 (5)** pp 706-710
[14] Snort. Available at: https://snort.org/

Developing of discrimination experiment to find most adequate model of plant's multi-nutrient functional response

M Yu Saltykov[1], **S I Bartsev**[1]

[1] Institute of Biophysics SB RAS, Russia, Krasnoyarsk, Akademgorodok 50-50

E-mail: saltykoff.mixail@yandex.ru

Abstract. To create reliable Closed Ecological Life Support Systems (CELSS) it is necessary to have models which can predict CELSS dynamic with good accuracy. However it was shown that conventional ecological models cannot describe CELSS correctly if it is closed by more than one element. This problem can be solved by means more complex models than conventional ones - so called flexible metabolism models. However it is possible that CELSS also can be described correctly in "semi-conventional" framework – when only one trophic level is described by flexible metabolism model. Another problem in CELSS modeling is existence of different and incompatible hypotheses about relationships between plants growth rate and amounts of nutrients (functional responses). Difficulty of testing these hypotheses is associated with multi-nutrient dependency of growth rate and comprehensive experimental studies are expensive and time-consuming. This work is devoted to testing the hypothesis that "semi-conventional" approach is enough to describe CELSS, and to planning the discrimination experiment on selecting correct type of the plant's functional response. To do that three different models of plants (one flexible and two conventional) were investigated both in the scope of CELSS model, and in hemostat model. Numerical simulations show that each of the models has typical patterns which can be determined in experiment with real plants.

1. Introduction

The usage of Closed Ecological Life Support Systems (CELSS) in space is possible only if the range of working regimes is estimated correctly. It is obvious that estimation of CELSS reliability has to be based on CELSS testing at different working conditions and external influences to simulate dangerous situations (similar to car crash-tests). However "crash-tests" of full-scale CELSS with humans are impossible due to obvious reasons.

This problem can be solved by mathematical models. In that case parameters of system can be measured in non-catastrophic experiments, after that estimation of working range of CELSS is conducted by an adequate model. Since there is no uniform method of CELSS models creation, some semiformal procedure of their creation has to be developed if we want to have a technology (not art) of CELSS creation.

To simulate CELSS it's necessary to find mathematical description which can describe processes in the system correctly. It's obvious that biological components of CELSS are hardest for description.

A number of experimental papers which are devoted to biological part of CELSS exist now [1,2,3,4]. This articles show some dynamical properties of ecosystems, at least near steady state. Articles on CELSS modeling [5,6,7,8,9,10] are devoted to the description of steady state of CELSS. In the same time experimental and model analysis of transit and critical regimes of CELSS had not conducted.

Creation of adequate dynamic models of CELSS is connected with steady-state stability analysis and analysis of system sensibility to big external disturbances. Since CELSS as usual includes several species the objective of ensuring their consistent and stable operation raises. This objective is not trivial since biochemical compound and required proportions of chemical elements consumption (similar to stoichiometry in chemistry) are different for each species. It has previously been shown [9] that if biological part of CELSS built in tradition Lotka-Volterra framework with fixed proportions of substrate consumption then in general case the ecosystem cannot be closed with respect to more than one element. Traditional ecological models can be used if there is no switching between limiting elements during the life-time of ecosystem. But our task is study of transition and, especially, critical processes and there is not any reason to exclude possibility of limiting element changing. It means that it is necessary to use models with different structure then traditional one to describe CELSS properly.

As was shown earlier [9,10] the problem of closing can be resolved by using models with "flexible" metabolism. In these models organisms can change their diet in accordance with their needs and nutrients distribution. Models with similar properties of biological components (humans and plants) were used for calculations of optimal stationary configurations of CELSS [6,8]. However these models were not studied in dynamic case in these works. Dynamic flexible models were studied in [9,10] but these models are rather complex and have number of hardly-estimated parameters. It makes the question on the necessity to use "flexible" approach to describe all species in CELSS or only part of them very actual.

Adequate description of real CELSS is associated with another important objective – selecting proper trophic function of plants included. This task becomes fundamentally important to build realistic scenarios at extreme situations.

In accordance with well-known Liebig's law of the minimum [11] plant's growth rate is depended by element which is insufficient. Since "to be insufficient" is Boolean (can be only "yes/true" or "no/false") property distinctive feature of models based on Leibeg's law is discontinues functional response to changes of substrate concentration. At the same time in practical applications including models of CELSS [5] smooth multiplicative function was used as functional response. Therefore there are at least two widely used versions of mathematical descriptions of nutrient consumption by plants. In this paper for the sake of brevity the model with discontinuous trophic function is called Liebig's models and smooth - Mitscherlich's models [12].

This work has the following goals:

1) to test the hypothesis that for the closure of the ecosystem with respect to several biogenic elements only one "flexible" trophic level is sufficient;
2) to reveal characteristic features of the dynamics of Closed Ecological Systems for different functional responses (functions, which describe dependence between growth rate and nutrients concentrations);
3) to formulate conditions of discrimination experiments to choose the most adequate type of functional response.

Our goal at this stage is not to get new fundamental results, but developing tools for creation and, which is more important, practical usage of CELSS models.

2. Models description

As a real prototype of models of CELSS we consider a system that ensures the crew only by plant food since introducing animals into CELSS is very complex objective, and the data on the functioning of such systems is not sufficient [4]. Then CELSS can be decomposed into fundamentally different mass components:

1) Biogenic elements in the inorganic form
2) Photoautotrophs (plants) that consume inorganic nutrients.

3) Inedible parts of plants and organic waste produced by man.

A feature of CELSS as an ecosystem is that the biomass of the consumer (human) has to be almost constant. This greatly simplifies the model by excluding dynamics of human biomass. Furthermore it is possible to consider human metabolism as "black box" - it consumes plants at a constant rate, and disposes of wastes containing biogenic elements in the same amount they were consumed (a necessary condition of organism homeostasis and the steady state of the ecosystem). The details of human diet are not considered here.

Although in the CELSS model the rate of consumption and waste production of crew can be considered as constants, for a more accurate description of extreme situations it makes sense to use a variant of the well-known function Mono with small semi-saturation constants:

$$H_i(X_1,\ldots,X_n)=\frac{H_i\prod_j X_j}{\prod_j (h+X_j)},\ j=1\ldots n \tag{1}$$

where H_i is rate of consumption of plant i at normal conditions, h – semi saturation constant and X_j is biomass of plant species i.

This function provides practically constant specific consumption rate of photoautotrophs X_j for all $X_j >> h$. If the biomass of one of the plant species decreases to h then human consumption of all plant species simultaneously decreases. It is the simplest way to describe the influence of non-balanced diet.

In reality human organism can adapt to the deficiency of one of the foodstuff. However in the scope of objective - the construction of a minimal model of CELSS providing closure - this simplification is acceptable since it gives the evaluation of the lower boundary of CELSS stability.

In this work the processing detritus into inorganic substances is supposed to be performed by chemical reactor. Obviously reactor is not sensitive to the absence of any detritus nutrient element unlike bacteria decomposers. However the reactor has a limited capacity and therefore can be saturated by substrates. On this basis the reactor can be described quite adequately by following system of equations:

$$\begin{cases}\dfrac{dC}{dt}=\dfrac{RD_C}{(r+D_C+D_N+D_P)}\\ \dfrac{dN}{dt}=\dfrac{RD_N}{(r+D_C+D_N+D_P)}\\ \dfrac{dP}{dt}=\dfrac{RD_P}{(r+D_C+D_N+D_P)}\end{cases}, \tag{2}$$

where R is maximum detritus processing rate; D_C, D_N and D_P are element content of the detritus; C, N and P are biogenic elements (carbon, nitrogen and phosphorus) in inorganic form.

General view of the model of CELSS closed by three biogenic elements including three types of producers taking into account the above adjustments is as follows:

$$\begin{cases} \frac{dC}{dt} = \frac{RD_C}{(r + D_C + D_N + D_P)} - \sum_{i=1}^{i=3} f_{Ci}(C,N,P,X_i) \\ \frac{dN}{dt} = \frac{RD_N}{(r + D_C + D_N + D_P)} - \sum_{i=1}^{i=3} f_{Ni}(C,N,P,X_i) \\ \frac{dP}{dt} = \frac{RD_P}{(r + D_C + D_N + D_P)} - \sum_{i=1}^{i=3} f_{Ni}(C,N,P,X_i) \\ \frac{dX_1}{dt} = g_1(C,N,P,X_1) - \frac{H_1 X_1 X_2 X_3}{(h + X_1)(h + X_2)(h + X_3)} \\ \frac{dX_2}{dt} = g_2(C,N,P,X_2) - \frac{H_2 X_1 X_2 X_3}{(h + X_1)(h + X_2)(h + X_3)} \\ \frac{dX_3}{dt} = g_3(C,N,P,X_3) - \frac{H_3 X_1 X_2 X_3}{(h + X_1)(h + X_2)(h + X_3)} \\ \frac{dD_C}{dt} = \frac{W_1 X_1 X_2 X_3}{(h + X_1)(h + X_2)(h + X_3)} - \frac{RD_C}{(r + D_C + D_N + D_P)} \\ \frac{dD_N}{dt} = \frac{W_2 X_1 X_2 X_3}{(h + X_1)(h + X_2)(h + X_3)} - \frac{RD_N}{(r + D_C + D_N + D_P)} \\ \frac{dD_P}{dt} = \frac{W_3 X_1 X_2 X_3}{(h + X_1)(h + X_2)(h + X_3)} - \frac{RD_P}{(r + D_C + D_N + D_P)} \end{cases} \quad (3)$$

where $f_{Ci}(C,N,P,X_i)$, $f_{Ni}(C,N,P,X_i)$, $f_{Pi}(C,N,P,X_i)$ and $g_i(C,N,P,X_i)$ are functional responses; W_i is human dissimilation rate of element i.

In this work three different types of functional response of plant species were investigated:

"Inner regulatory pool" (IRP) which was considered in [9,10] early:

$$\begin{aligned} f_{Ci}(C,X_i) &= k_{Ci} C (c_i^{max} X_i - c_i)(X_i^{max} - X_i) \\ f_{Ni}(N,X_i) &= k_{Ni} N (n_i^{max} X_i - n_i)(X_i^{max} - X_i) \\ f_{Pi}(N,X_i) &= k_{Pi} P (p_i^{max} X_i - p_i)(X_i^{max} - X_i) \\ g_i(C,N,P,X_i) &= \mu_i c_i n_i p_i \end{aligned} \quad (4)$$

where k_{Ci}, k_{Ni} and k_{Pi} are specific rates of element absorption by plant, X_i^{max} are maximum plant species biomass, c_i^{max} n_i^{max} and p_i^{max} are maximum concentrations of biogenic elements in the inner regulatory pools. Variables c_i n_i and p_i are amounts of elements in inner regulatory pools of plant and can be calculated by following ordinary differential equations system:

$$
\begin{cases}
\dfrac{dc_i}{dt} = k_{Ci}C\left(c_i^{max}X_i - c_i\right)\left(X_i^{max} - X_i\right) - A_{ci}\mu_i c_i n_i p_i - \dfrac{c_i}{X_i}\dfrac{H_i\prod_j X_j}{\prod_j (h + X_j)} \\
\dfrac{dn_i}{dt} = k_{Ni}N\left(n_i^{max}X_i - n_i\right)\left(X_i^{max} - X_i\right) - A_{ni}\mu_i c_i n_i p_i - \dfrac{n_i}{X_i}\dfrac{H_i\prod_j X_j}{\prod_j (h + X_j)} \\
\dfrac{dp_i}{dt} = k_{Pi}P\left(p_i^{max}X_i - p_i\right)\left(X_i^{max} - X_i\right) - A_{pi}\mu_i c_i n_i p_i - \dfrac{p_i}{X_i}\dfrac{H_i\prod_j X_j}{\prod_j (h + X_j)}
\end{cases}, \quad (5)
$$

where μ_i is specific rate of element consumption by the metabolism of i-th plant and $A_{ci}+A_{ni}+A_{pi}=1$ are normalized stoichiometric coefficients of plants or proportions of carbon, nitrogen and phosphorous therein. Summands $\dfrac{c_i}{X_i}\dfrac{H_i\prod_j X_j}{\prod_j (h + X_j)}$, $\dfrac{n_i}{X_i}\dfrac{H_i\prod_j X_j}{\prod_j (h + X_j)}$ and $\dfrac{p_i}{X_i}\dfrac{H_i\prod_j X_j}{\prod_j (h + X_j)}$ describe decreasing of biogenic elements in their pools due to the consumption of plants by the crew.

Another variant of plant metabolism description was Mitscherlich's model:

$$
\begin{cases}
f_{Ci}(C, X_i) = A_{ci}g_i(C, N, P, X_i) \\
f_{Ni}(C, X_i) = A_{Ni}g_i(C, N, P, X_i) \\
f_{Pi}(C, X_i) = A_{Pi}g_i(C, N, P, X_i) \\
g_i(C, N, P, X_i) = \dfrac{v_i CNP}{(V_{Ci} + C)(V_{Ni} + N)(V_{Pi} + P)}\left(X_i^{max} - X_i\right)X_i
\end{cases}, \quad (6)
$$

where v_i is maximum growth rate and V_{Ci}, V_{Ni}, V_{Pi} are semi saturation constants.

The third variant was classic Liebeg's model:

$$
\begin{cases}
f_{Ci}(C, X_i) = A_{ci}g_i(C, N, P, X_i) \\
f_{Ni}(C, X_i) = A_{Ni}g_i(C, N, P, X_i) \\
f_{Pi}(C, X_i) = A_{Pi}g_i(C, N, P, X_i) \\
g_i(C, N, P, X_i) = min\left(\dfrac{v_{Ci}C}{(V_{Ci} + C)}, \dfrac{v_{Ni}N}{(V_{Ni} + N)}, \dfrac{v_{Pi}P}{(V_{Pi} + P)}\right)\left(X_i^{max} - X_i\right)X_i
\end{cases} \quad (7)
$$

As can be seen from the equations of Mitscherlich's and Liebig's models the consumption of biogenic elements by plant is proportional to the rate of growth of the organism, and the coefficients of proportionality are constant. As shown in [9] ecosystem models built **only on** the base of these principles cannot be closed with respect to more than one element. However in all three models plants are consumed by human, and it is considered that human consume plants in proportion which allow him to live. In this way, human is "flexible" part of models but mechanism of its flexibility is not described now. Consequently there is reason to expect closure achieving when using Liebig's and Mitscherlich's models.

Constants and variables from all three models are represented in Table 1.

Table 1. Model's constants and variables.

constant or variable	units	short description
H_i	g*day^{-1}	rate of consumption of plant by human at normal conditions
h	g	semisaturation constant
R	g*day^{-1}	maximum detritus processing rate
r	g	semisaturation constant
k_{Ci}, k_{Ni}, k_{Pi}	g^{-2}*day^{-1}	specific rates of element absorption by plant
c_i^{max} n_i^{max}, p_i^{max}		maximum concentrations of biogenic elements in the inner regulatory pools
A_{ci}, A_{ni}, A_{pi}		normalized stoichiometric coefficients of plants or proportions of carbon, nitrogen and phosphorous therein
μ_i	g^{-2}*day^{-1}	specific rate of element consumption by plant's the metabolism
v_i	g^{-1}*day^{-1}	maximum growth rates (Mitscherlich)
V_{Ci}, V_{Ni}, V_{Pi}	g	semi-saturation constants
W_i	g*day^{-1}	rate of waste of plant by human at normal conditions
X_i^{max}	g	maximum plant's biomass
v_{Ci}, v_{Ni}, v_{Pi}	g^{-1}*day^{-1}	maximum growth rates (Liebig)
C, N, P	g	biogenic elements (carbon, nitrogen and phosphorus) in inorganic form
X_i	g	plant's biomass
c_i, n_i, p_i	g	amounts of elements in inner regulatory pools
D_C, D_N, D_P	g	element content of the detritus

For a correct comparison of these three models production characteristics of the simulated plants were made practically equal. Equations were solved with Scilab 5.4.1 with default method of ode() function (automatically selects between nonstiff predictor-corrector Adams method and stiff Backward Differentiation Formula (BDF) method) with default relative and absolute estimated errors (10^{-5} for relative one and 10^{-7} for absolute).

3. Results

The study was conducted by numerical simulations of CELSS models with random set of plant parameters. Some parameters of plants and the ranges of variation for others are shown in Table 1. Uniform distribution was used in the random number generator.

Table 2. Basic parameters of the plants in the test model of CELSS.

Parameters\Abstract plants	Plant1	Plant2	Plant3
A_{ci}	0.9883	0.9912	0.9926
A_{ni}	0.0105	0.0077	0.0062
A_{pi}	0.0012	0.0011	0.0012
c_i^{max}	0.1	0.1	0.1
n_i^{max}	0.01	0.01	0.01
p_i^{max}	0.001	0.001	0.001

k_{Ci}, k_{Ni}, k_{Pi} (g^{-2}*day^{-1})	1 – 2
μ_i (g^{-2}*day^{-1})	$2*10^6 - 3*10^6$
x_i^{max} (g)	10 – 20

The stoichiometric ratios are chosen close to the real [13].

Typical results for all 100 numerical simulations of the plants dynamics in CELSS model with parameters ranges from table 2 are presented in Figure 1.

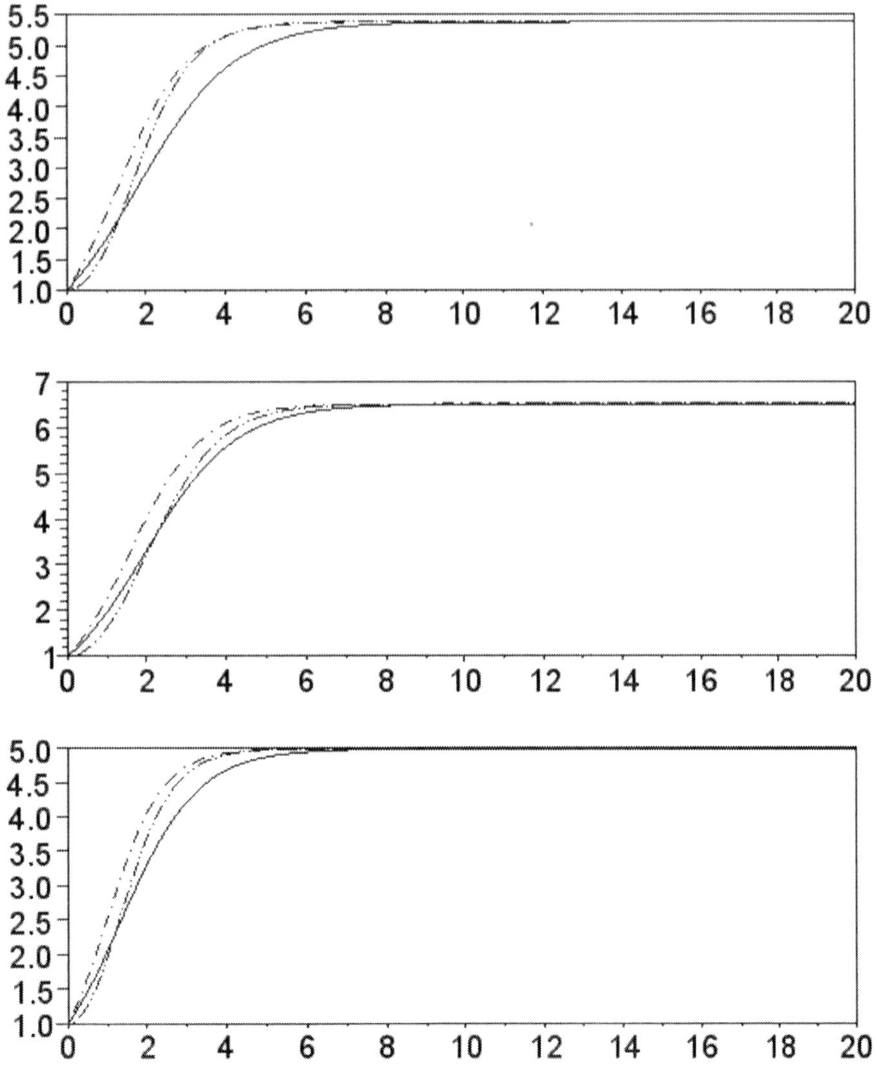

Figure 1. Dynamics of plant components of CELSS models. Line with colons - IRP, dot-dash - Mitscherlich model, solid - model Liebig.

From the results of numerical experiments it can be concluded that the closure of system (steady state of plant biomass is assumed) is possible for all three models in a wide range of parameters.

Hence the hypothesis on the possibility of closure of the ecosystem within the models accounting the flexibility of the metabolism of only one trophic level is confirmed.

3.1. Description of the experiment to determine the type of functional response of several variables

The experiment in the CELSS is rather complicated and expensive. However as shown in Figure 1 the differences in plant biomass dynamics for different models mean differences in the type of functional response of plant to changes of the amount of nutrients. So we have an ability to differentiate the type of plant functional response in experiments. The measurements of the growth rate of plants can be conducted in a phytotron with intensive carbon dioxide flow. At sufficiently high rate of CO_2 input its concentration in the phytotron can be considered as constant. If influx of carbon dioxide is turned out its concentration (measured in real-time by gas analyzer) begins to decline as a result of plants life. It is possible to determine the specific rate of consumption of carbon dioxide by plant from the curve of reduction of carbon dioxide concentration.

In this work numerical simulation of experiment on determination of the trophic functions of plants was conducted. To do that values of functions which describe consumption by the mechanisms of IRP, Liebig's, and Mitscherlich's models with varying amounts of nutrients were calculated. Four "computer experiments" were conducted as illustration. In each experiment the values of N and P were fixed while C concentration was varied. The dependence of growth rate on C concentration is presented by graph. The values of N and P in each of the experiments were taken as follows:

Table 3. Changing of amounts of nitrogen and phosphorous in experiments.

Experiment	N	P
1	N_{min}	P_{min}
2	N_{min}	P_{max}
3	N_{max}	P_{min}
4	N_{max}	P_{max}

where N_{min}, P_{min}, N_{max} *and* P_{max} are the smallest and largest amount of nitrogen and phosphorus used in these experiments. Values $N_{min}=1$, $P_{min}=0.1$, $N_{max}=7$, $P_{max}=0.7$ were used in computer experiments Selection of these values of nitrogen and phosphorus content is similar with nitrogen-phosphorous ratio in real plants.

Results of computer experiments are shown on Figure 2. Computer experiments consisted of identifying "concentration of CO_2 - specific growth rate" relation for 10 randomly generated sets of model parameters of IRP and the corresponding parameters of Liebig's and Mitscherlich's models. For better clarity the figure shows the typical curves for each of the three models. The next parameters of plants were varied: μ_i in the range ($2*10^6$-$3*10^6$ day^{-1}*g^{-2}) and k_{Ci}, k_{Ni} k_{Pi} in the range ($1-2$ day^{-1}*g^{-2}).

As can be seen from Figure 2, IRP, Liebig's, and Mitscherlich's models have following characteristic patterns which can be detected experimentally:

1) in the framework of the IRP and Liebig's models maximum growth rate are almost independent of the concentrations of nitrogen, but it depends on the concentration of phosphorus if nitrogen-phosphorous ratio;

2) for Liebig's model the break of the curve when entering the almost horizontal plateau are typical;

3) in Mitscherlich's model strong dependence of the upper level of "growth rate – CO2" curve on nitrogen and phosphorus is observed.

3.2. Description of the experiment to determine the type of functional response

The concentration of carbon dioxide in the phytotron with the flow of carbon dioxide is described by the following equation:

$$\frac{dS}{dt} = D(S_0 - S) - v(S), \tag{11}$$

where S is CO_2 concentration in the phytotron, S_0 is CO_2 input concentration, D is specific flow rate, $v(S)$ is rate of CO_2 consumption by plants. When the flow rate is much greater than the plant's rate of consumption then concentration of CO_2 in the phytotron is almost equal to input concentration. When flow is off then carbon dioxide concentration begins to fall in accordance with the equation

$$\frac{dS}{dt} = -v(S). \tag{12}$$

Obviously plant's carbon consumption rate can be reconstructed by the shape of the curve of CO2 concentration falling. In the case of short flow turn off decreasing of CO2 concentration can be considered as a linear.

There exists the possibility that with increasing CO2 concentration plants need some time to adaptation to a new concentration. To check existence of this period an additional experiment should be made:

0) Increasing of carbon dioxide input concentration.
1) Switching off CO_2 flow and measuring of CO_2 curve of fall.
2) Switching on CO_2 flow. CO_2 concentration return to initial level.
3) Repeating switching off after several tens of minutes.

If plants need no time for adaptation to the new CO2 concentration (both curves are identical) then experiments on growth curve determination can be produced through a series of pulsed CO2 injections with a gradual increase of its concentration. If plant adaptation exist then it is necessary to increase interval between switching off. Series of four curves corresponding to one of the three patterns shown in Figure 2 have to be the result of the experiment.

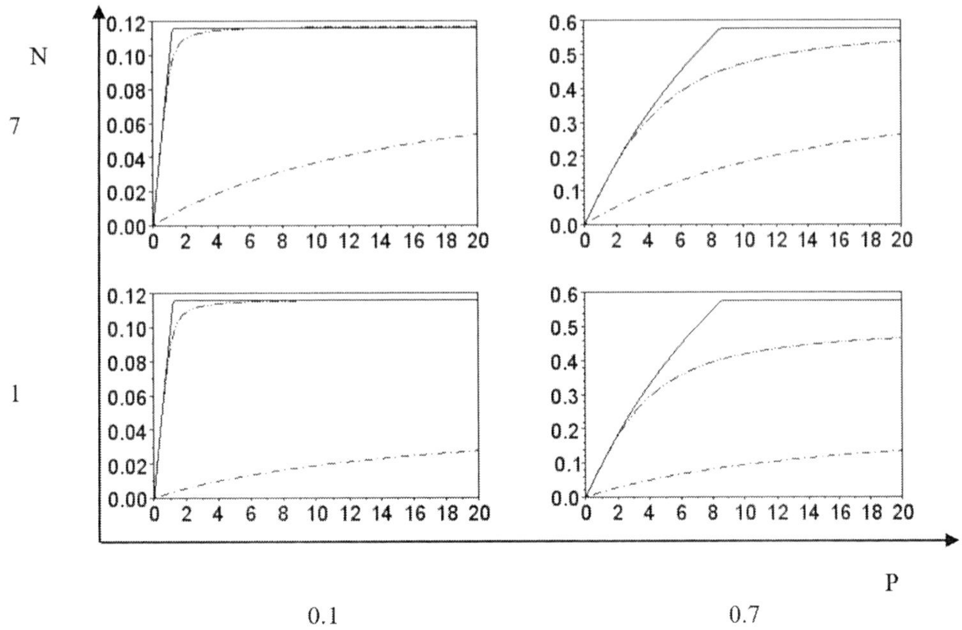

Figure 2. The results of computational experiments on the measurement of the food curve depending on the amounts of available nutrients. Line with colons is IRP, a line with a point is Mitscherlich's model, solid is Liebig's model.

4. Conclusion

In this work the possibility of CELSS closing by several biogenic elements was shown in the model with only one "flexible" species – human being. This result allows in some cases significantly simplify CELSS models.

To select the most adequate model for describing real plants searching for typical patterns for each of the models was conducted. It was shown that the investigated models demonstrate specific patterns allowing experimental discrimination of hypotheses:

1) In the framework of the IRP and Liebig's models maximum growth rate are almost independent on the concentrations of nitrogen.
2) In the Liebig's model growth curve has a break point and strictly horizontal "plateau".
3) In the Mitscherlich's model maximum growth rate in case of saturation by CO2 depends on the concentration of other nutrients.

Preliminary description of the experiment on the determination of the adequate way of describing the functional responses of plants according to one of the above patterns was made.

References

[1] Lehto K M, Lehto H J, Kanervo E A 2006 Research in Microbiology **157** 69–76
[2] Sychev V N, Levinskikh M A, Podolsky I G 2008 Acta Astronautica **63** 1119 – 1125
[3] Tikhomirov A A, Ushakova S A, Velichko V V, Tikhomirova N A, Kudenko Yu A, Gribovskaya I V, Gros J-B, Lasseur Ch 2011 Acta Astronautica **68** 1548–1554.
[4] Li M, Hu D, Liu H, Hu E, Xie B, Tong L 2013 Advances in Space Research 52 773–779.
[5] Rummel J D, Volk T 1987 Controlled Ecological Life Support System *Topical session of the XXVI COSPAR meeting* 57-65.
[6] Bartsev S I, Gitelzon J I, Lisovsky G M, Mezhevikin V V, Okhonin V A 1996 *Advances in Space Research* **39(8)** 617–622.
[7] Bartsev S I, Okhonin V A 1999 *Advances in Space Research* **24(3)** 407–412.
[8] Bartsev S I, Mezhevikin V V, Okhonin V A 2000 *Advances in Space Research* **26(23)** 323–326.
[9] Bartsev S I, 2004 *Advances in Space Research* **34(7)** 1509–1516.
[10] Saltykov M Yu, Bartsev S I, Lankin Yu P 2012 *Advances in Space Research* **49(2)** 223-229.
[11] de Baar H J W 1994 *Progress in Oceanography* **33(4)** 347-386.
[12] Mitscherlich E A 1909 *Landwirtschafliche Jahrbuecher* **38** 537-552.
[13] Sistla S A, Schimel J P 2012 *New Phytologist* **196** 68–78.

IOP Conf. Series: Materials Science and Engineering **173** (2017) 012018 doi:10.1088/1757-899X/173/1/012018

Determination of the periodicity and synchronization of anticipative agent based supply-demand model

A Škraba[1], M Bren[2], D Kofjač[1]

[1]University of Maribor, Faculty of Organizational Sciences, Cybernetics & Decision Support Systems Laboratory, Kidriceva cesta 55a, 4000 Kranj, Slovenia
[2]University of Maribor, Faculty of Criminal Justice and Security, Kotnikova ulica 8, 1000 Ljubljana, Slovenia

E-mail: andrej.skraba@fov.uni-mb.si

Abstract. The paper presents the transformation of cobweb model by including the anticipation about the supply and demand. Developed transformation leads to oscillatory behaviour. The periodic conditions of the model have been analytically determined by the application of z-transform. Periodic solutions of the system are presented in the form of an inverse Farey tree, where the Golden Ratio path could be observed. The table of periodic conditions is given up to period 8. The agent-based system was developed in order to show the possibility of controlling the system by varying the key parameter, which determines the frequency response of agents and their interaction. A note on application in the stock market has been provided.

1. Introduction

Controlling supply and demand in proper balance becomes more challenging in the new, unpredictable economy [4], [18], [15]. In the classical cobweb model the functions of demand Q_k^d and supply Q_k^s are specified in the following form:

$$Q_k^d = a + bP_k \quad (1)$$

$$Q_k^s = c + dP_{k-1} \quad (2)$$

where a, b, c and d are parameters of supply and demand specific to individual markets [21]. The price P_k and supply Q_k^s should be restricted to the positive values. In the cobweb model it is assumed that in any one time period producers supply a given amount (determined by the previous time period's price) and then the price adjusts so that all the products supplied are bought by customers. Initial cobweb equation could be restated by introducing additional time step and factors A, B, C and D:

$$P_{k+2} = \frac{d}{b}(A_k - (\frac{bB_k - c + a}{d})) \quad (3)$$

$$Q_{k+2}^s = \frac{d}{b}(C_k - a - \frac{b}{d}(D_k - c)) \quad (4)$$

with initial conditions:

$$P_{k+1} = \frac{p-a}{b} \tag{5}$$

$$P_k = \frac{bP_{k+1} + a - c}{d} \tag{6}$$

$$Q^s_{k+1} = p \tag{7}$$

$$Q^s_k = a + \frac{b}{d}(Q^s_{k+1} - c) \tag{8}$$

Terms A_k and B_k in Eq. (3) could be replaced by the terms P_{k+1} or P_k, similarly C_k and D_k in Eq. (4) by Q^s_{k+1} or Q^s_k. This yields 16 different system combinations defined by Eq. (3) and Eq. (4) that should be studied. The system combination further examined will have the following terms: $A_k = P_{k+1}, B_k = P_k, C_k = Q^s_{k+1}$ and $D_k = Q^s_k$. Proposed replacement changes the difference ΔP and ΔQ to state equations for P and Q. This yields the following set of equations:

$$P_{k+2} = \frac{d}{b}(P_{k+1} - \frac{bP_k - c + a}{d}) \tag{9}$$

$$Q^s_{k+2} = \frac{d}{b}(Q^s_{k+1} - (a + \frac{b}{d}(Q^s_k - c))) \tag{10}$$

In the performed modification we are considering the dynamics which is dependant on two different time state values in Eq. (9) rather than one. If we reformulate Eq. (9) and Eq. (10), the dependency of the *future-present-past* events could be observed:

$$P_k = \frac{bP_{k-1} + a - c}{d} + \frac{b}{d}P_{k+1} \tag{11}$$

$$Q^s_k = \frac{b}{d}Q^s_{k+1} + \frac{b}{d}Q^s_{k-1} + a - \frac{bc}{d} \tag{12}$$

Eq. (11) and Eq. (12) state that the value of the present is dependent on the past as well as on the future. In this case we anticipate future supply and demand.

2. Periodicity of the anticipative cobweb model

Different modes of cyclic behaviour response could be observed as the system's dynamic response, when parameter d is varied. Synchronization patterns are named by the shape of the Poincaré first-return map representing the values of P_k, P_{k+1}. The vertices of the solution converge to the edge point of the pentagram. Points on the vertices form the line at the periodic condition values for parameter d. The system is in transition to the next full polygon synchronization, which is estimated as the quad synchronization, where parameter d is near 0 (or hexagon, depending on the direction of parameter d variation).

The analytical determination of periodicity conditions is provided by the application of z-transform, which is the basis of an effective method for the solution of linear constant-coefficient difference equations. It essentially automates the process of determining the coefficients of the various geometric sequences that comprise a solution [16]. The application of z-transform on Eq. (9) and Eq. (10) with initial conditions stated by Eqs. (5)– (8) gives:

$$Y(z)=\frac{-y_1 z+y_0 dz-y_0 z^2}{-1+dz-z^2} \tag{13}$$

Inverse z-transform yields the following solution:

$$Y^{-1}(z)=2^{-1-n}y_0\left(d-\sqrt{-4+d^2}\right)^n-\frac{y_1\left(d-\sqrt{-4+d^2}\right)^n}{2^n\sqrt{-4+d^2}}+$$
$$+\frac{2^{-1-n}y_0 d\left(d-\sqrt{-4+d^2}\right)^n}{\sqrt{-4+d^2}}+2^{-1-n}y_0\left(d+\sqrt{-4+d^2}\right)^n+$$
$$+\frac{y_1\left(d+\sqrt{-4+d^2}\right)^n}{2^n\sqrt{-4+d^2}}-\frac{2^{-1-n}y_0 d\left(d+\sqrt{-4+d^2}\right)^n}{\sqrt{-4+d^2}} \tag{14}$$

In order to gain conditions for the periodic response of the system the following equation should be solved:

$$Y^{-1}(z)=y_0 \tag{15}$$

Let us compute a numerical example of periodic solution applying the z-transform. The period examined will be the period of *9*, i.e. $n=9$. In Eq. (15) one should put the condition $n=9$. One of the possible solutions for the initial condition worth examining is the following:

$$d=\frac{1}{(\frac{1}{2}(-1+i\sqrt{3}))^{\frac{1}{3}}}+(\frac{1}{2}(-1+i\sqrt{3}))^{\frac{1}{3}} \tag{16}$$

The term $(-1+i\sqrt{3})^{\frac{1}{3}}$ (let us denote the term as z^*) could be expressed in the following way by three different imaginary values in polar form:

$$z_1^*=\sqrt[3]{2}(\cos\frac{2\pi}{9}+i\sin\frac{2\pi}{9}) \tag{17}$$

$$z_2^*=\sqrt[3]{2}(\cos\frac{8\pi}{9}+i\sin\frac{8\pi}{9}) \tag{18}$$

$$z_3^*=\sqrt[3]{2}(\cos\frac{14\pi}{9}+i\sin\frac{14\pi}{9}) \tag{19}$$

By putting Eqs. (17), (18) and (19) into Eq. (16) and performing trigonometric reduction, one gets the following solutions:

$$d_1=2\cos\frac{2\pi}{9} \quad d_2=2\cos\frac{4\pi}{9} \quad d_3=2\cos\frac{8\pi}{9} \tag{20}$$

By inspecting Eq. (16) and considering the equation for the roots of complex numbers [13]:

IOP Conf. Series: Materials Science and Engineering **173** (2017) 012018 doi:10.1088/1757-899X/173/1/012018

$$\sqrt[n]{z}=\sqrt[n]{r}(\cos\frac{\theta+2k\pi}{n}+i\sin\frac{\theta+2k\pi}{n}) \tag{21}$$

the general form of the solution for parameter d could therefore be defined as:

$$d=2\cos\frac{2\pi m}{n} \tag{22}$$

where n is the period and $m=1,2,3,...,n-1$. A similar procedure could be performed for the arbitrary period n. A more general solution, which applies parameter b, which was fixed for the purpose of determining solutions, is:

$$d=2b\cos\frac{2\pi m}{n} \tag{23}$$

The solution could in some cases be expressed in an alternative algebraic or trigonometric form. Table 1 shows the solutions for parameter d up to the period $n=8$. Alternative solutions could be expressed as the roots of the polynomial. Table 1 incorporates the Shape symbols, which are important in the study of the response of dynamical systems. This is especially the case in the examination of complex nonlinear dynamical systems [27]. One of the important conditions gained by the proposed inspection is the value of the period $n=10$, which correlates closely with the period $n=5$. The value of parameter d is $d=\frac{1}{2}(1+\sqrt{5})$ with numerical value 1.61803... This solution represents the *Golden Ratio* (usually denoted with Φ) [9]. Some of the different representations of the solution for the parameter d value at period $n=10$ are:

$$d_{10}=\Phi=2\cos\frac{\pi}{5}=\frac{1}{2}(1+\sqrt{5})=1.61803398874989484820... \tag{24}$$

The first solution of parameter d at period $n=10$ connects the considered discrete system with the Fibonacci numbers given by the infinite series:

$$d_{10}=\Phi=1+\sum_{u=1}^{\infty}\frac{(-1)^{u+1}}{F_u F_{u+1}} \tag{25}$$

The fact, that the periodicity conditions of the examined discrete system incorporates the Golden Ratio number Φ could be observed in other studies of complex nonlinear expansions of the basic cobweb systems, e.g. [3] *Almost Homoclinic Tangency Lemma*. One should expect that the symmetric response in n-mapping to follow the pattern with a match at a certain point to the solution of synchronization values, for example, Φ. The value that inevitably emerges with *Golden Ratio* Φ is in our case present at period $n=5$, where the value for parameter $d=\frac{\sqrt{5}-1}{2}=0.61803...$, often called the *Golden Ratio Conjugate* (usually denoted with ϕ). Table 1 is similar to the gained parameter values for the domain of *2-D* dynamic attraction by [24]. The mentioned set of parameters is augmented with the value for the periods 8, which is not stated in [24].

Table 1. Synchronization parameter d values of periodicity conditions up to period 8.

Period n	Shape symbol	Shape description	Argument $\Omega = 2\pi\frac{m}{n}$	Algebraic or alternative trigonometric d value representation	Num. value $d = 2\cos\frac{2\pi m}{n}$
2*	⟷	*digon*	π		-2.00000
3	△	*triangle*	$\frac{2\pi}{3}$		-1.00000
4**	◇	*tetragon*	$\frac{\pi}{2}$		0.00000
5	⬠	*pentagon*	$\frac{2\pi}{5}$	$\frac{1}{2}(-1+\sqrt{5})$	0.61803
	☆	*pentagram*	$\frac{4\pi}{5}$	$\frac{1}{2}(-1-\sqrt{5})$	-1.61803
6	⬡	*hexagon*	$\frac{\pi}{3}$		1.00000
7	○	*heptagon*	$\frac{2\pi}{7}$	$\frac{1}{3}(-1+2\sqrt{7}\cos(\frac{1}{3}\arctan(3\sqrt{3})))$	1.24698
	✡	*heptagram*	$\frac{4\pi}{7}$	$\frac{1}{3}(-1+2\sqrt{7}\cos(\frac{4\pi}{3}+\frac{1}{3}\arctan(3\sqrt{3})))$	-0.44504
	✴		$\frac{6\pi}{7}$	$\frac{1}{3}(-1+2\sqrt{7}\cos(\frac{2\pi}{3}+\frac{1}{3}\arctan(3\sqrt{3})))$	-1.80194
8	○	*octagon*	$\frac{\pi}{4}$	$\sqrt{2}$	1.41421
	✵	*octagram*	$\frac{3\pi}{4}$	$-\sqrt{2}$	-1.41421

In Table 1 (*) indicates that since period 2 is on the boundary of the solutions interval, the periodic response of the system depends on the initial conditions. An example of numerical values of the period-2 response is: $a = 1, b = 1, c = -1, d = -2, p = 1$. The value for the tetragon marked with (**) in Table 1 is taken in the limit since the system of equations returns the undefined value when $d = 0$. Therefore one should consider the tetragon period condition as the value approaching zero, i.e. $d \to 0$. In this case, the system response is undetermined $(\frac{0}{0})$ at its critical point.

Numerical values of the solutions for parameter d confirm the findings of [25, 24] characterized by the following Prop. 1 about the domain of attraction for *2-D* dynamics by n-dimensional linear bifurcation analysis. Determination of the periodicity condition for the anticipative system of cobweb linear difference equations leads to the following proposition, which provides the interconnection between the periodicity cycles in the considered anticipative cobweb model and general bifurcation conditions for the nonlinear discrete dynamical systems:

Proposition 1 *Periodicity conditions of the anticipative cobweb model* $d = 2b\cos\frac{2\pi m}{n}$ *equals bifurcation condition on the flutter boundary determining* q *-periodic fixed points of the nonlinear discrete dynamical system* $trJ = 2\cos 2\pi\Omega$.

Here, Ω represents a rational fraction $\Omega = \frac{p}{q}; b = 1$. Prop. 1 provides dynamical interpretation of bifurcation periodicity conditions, which are important in the analysis of non-linear systems where manifestations of chaos and turbulence might occur [25]. There is a significant effort being made to further determine the conceptual framework of bifurcation analysis for which the central problem is bifurcation control. Prop. 1 provides a possible simplification of bifurcation condition representations. Geometrical visualization of such conditions is an important factor in the analysis of nonlinear discrete dynamical systems [27].

The emergence of system periodic stability in the shape of an n-sided polygon could be observed not only in economic systems [20]; the n-sided polygon and the Farey tree organization of the equilibriums could be observed in technical systems [26, 5] as, for example, in laser control as the paradigm of the chaotic system [11]. Data analysis in the field of economics by the Poincaré first-return map, which exercises the periodic character, is presented in [1].

According to Table 1 gained by the *z-transform*, the classification of the periodic solution could be drawn, as shown in Fig. 1.

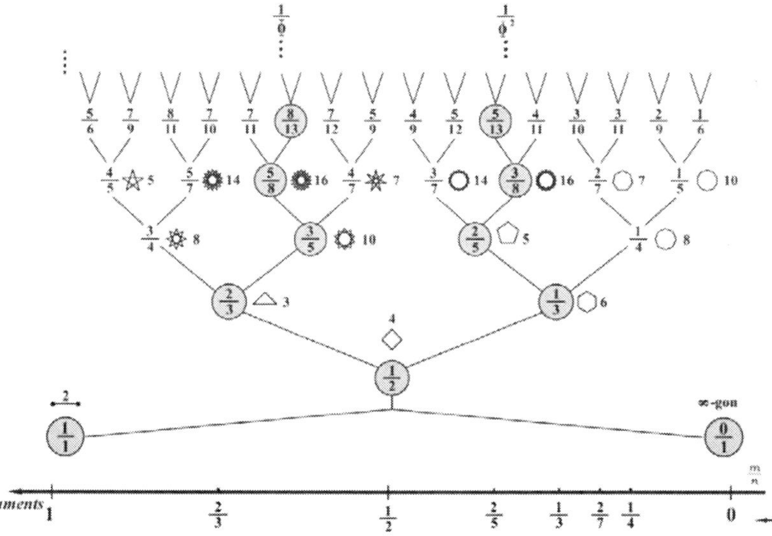

Figure 1. Periodicity of the incursive cobweb model in the form of an inverse Farey tree, where paths of Golden Ratio could be observed (greyed fractions)

The structure in Fig. 1 shows the inverse Farey tree which corresponds to the values of the periodic conditions gained by the *z-transform*. One of the questions that arose in the analysis of similar 2-D systems is the question concerning the rule and emergent organizational properties that determines periodicity [23]. In our case, the change of parameter d causes the system to switch between different periodic equilibriums. The ordering of the equilibriums is determined by the general Eq. (23). The rational fraction $\frac{m}{n}$, which is in our case transformed by Eq. (23) to the value of parameter d, corresponds to the Farey sequence, which could be represented by the inverse Farey tree in Fig. 1. The greyed fractions in the Farey tree point to the sequences of Fibonacci numbers [10] $\{F_i/F_{i+1}, i \in \aleph\}$ and $\{F_i/F_{i+2}, i \in \aleph\}$. A periodic parameter space region of system response is determined by the condition $det > 0$ and periodicity by $det < 0$. The classification at $d < 0$ specifies the angles, which are determined by the three points in the 2-D map in our case, $\alpha_n < \frac{\pi}{2}$; in the case where $d > 0$, the angles of the map are $\alpha_n > \frac{\pi}{2}$. The strongest periodicity points are determined by the polygon structures in 2-D mapping starting with *digon*, *triangle*, etc. Other periodicity is the subset of the main sections, which is determined by the $\sum \alpha_n$ and the Farey tree. The region of non-periodicity determines the response of the model that is equal to the conventional cobweb model.

One of the properties of the *hyperincursive*, nonlinear and chaotic systems, which are of special interest in the analysis of complex economic systems [2, 6, 17, 12], is that they have *cycles of every length*. It is important to be aware of the organizational structure of the periodic solutions provided by the emergent Farey tree incorporating the anticipative cobweb model and according to Prop. 1 also for the bifurcation conditions of nonlinear systems.

3. Equilibrium analysis

Equilibrium conditions of the studied systems corresponds to the periodic solutions and are therefore of primary importance for the description of system response. The equilibrium condition for the P

segment of the system provides an interesting response of the Q^s segment, which exhibits an hexagonal shape in the two dimensional Poincaré first-return map. Since hexagonal topology has a special meaning in the analysis of discrete dynamical systems [20], the equilibrium at $n = 3$ and $n = 6$ is examined. Other conditions are stated by the previously stated periodicity conditions, which are summarized in Table 1.

The equilibrium condition for the P segment of the anticipative cobweb system could be stated as:

$$\frac{d}{b}\left(\frac{p-a}{b}-\frac{b\frac{p-c}{d}-c+a}{d}\right)=\frac{p-a}{b}=\frac{p-c}{d} \tag{26}$$

The equilibrium values of the parameters for the P segment of the system are: $a = c = p$ and $b = d \neq 0$ and, in this case $P = 0$.

The equilibrium condition for the Q^s segment of the anticipative cobweb model could be stated as:

$$\frac{d}{b}(p-a-\frac{b}{d}(a+\frac{b}{d}(p-c)-c)) = p = a+\frac{b}{d}(p-c) \tag{27}$$

The Q^s segment of the system has no solution. When the equilibrium conditions for the P segment of the system are considered in fact in all the cases, the Q^s segment of the system could not be in a stable state. A graphic presentation of the equilibrium conditions $a = c = p$ and $b = d$ is shown in Fig. 2. The hexagonal shape is known as the possible optimum shape in the context of spatial economics [20]. The existence of hexagonal shapes is known in space economy and explained by the structural stability [20].

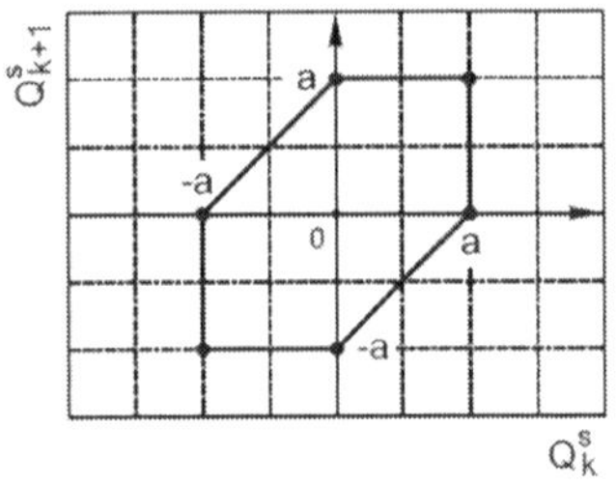

Figure 2: Response of the Q^s segment of the system while the P segment is in equilibrium

Proposition 2 *The equilibrium condition for the P segment of the anticipative cobweb system defined by the equations from Eq. (5) to Eq. (10) is:* $a = c = p$ *and* $b = d$. *Under these conditions, the* Q^s *segment of the system has the response of a hexagonal shape with vertices*

$$\{(a,0),(a,a),(0,a),(-a,0),(-a,-a),(0,-a)\}$$

in Q^s_k, Q^s_{k+1} *mapping.*

While the response of the system for the P segment is in equilibrium, the Q^s segment of the system has a hexagonal-like shape with a significant dimension of parameter a value and edge dimensions of a and $a\sqrt{2}$.

Proposition 3 *Triangular (△), i.e. three-period response in 2-D mapping is determined by the condition* $b = -d$.

In order to gain the term for the period in the P values, one should apply periodic condition. The values for times 1,...,4 should be symbolically expressed. By inserting Eq. (5) and Eq. (6) into Eq. (9) the following term is gained:

$$P_{k+2} = \frac{d}{b}\left(\frac{p-a}{b} - \frac{a-c+\frac{b(p-c)}{d}}{d}\right) \tag{28}$$

By repetition of a similar procedure for the equation for P_{k+3}, considering the period 4 condition, i.e. $P_{k+3} = P_k$, one should get the following equation:

$$\frac{d}{b}\left(\frac{d}{b}\left(\frac{p-a}{b} - \frac{a-c+\frac{b(p-c)}{d}}{d}\right) - \frac{p-c}{d}\right) - \frac{p-c}{d} = 0 \tag{29}$$

with solution $b = -d \in \Re/\{0\}$.

Proposition 4 *Hexagon (⬡), i.e. six-period response in 2-D mapping is determined by the condition* $b = d$.

The values for times 1,...,6 should be symbolically expressed. For example, determination of P_{k+2} is based on expressions for P_{k+1} and P_k, determination of P_{k+3} is based on expressions for P_{k+2} and P_{k+1}, etc. Periodic condition is expressed as $P_{k+6} = P_k$ with solution $b = d \in \Re/\{0\}$.

4. Agent-based anticipative cobweb model

This section describes the interactions between several economic entities modelled as agents. Agent interaction represents the alternative control mechanism, which should provide standing oscillations and global equilibrium-seeking behaviour found in real world cases. The initial *feedback-anticipative* cobweb model oscillates only for a certain parameter set, which represents a thin borderline in parameter space. If the system parameters are not on this border, the model either decays or produces explosive behaviour [19]. There are several approaches to fixing the mentioned problem. One of the methods for solving this problem is application of the *floor-roof* principle, which should limit capital stock values as well as raw materials, etc. [19]. In a developed *agent-based* system we move from simple linear dynamics to incorporation of the nonlinear rules as the delayed and anticipated information concept. The anticipated response of the system is therefore complex in behaviour due to the coevolving nature of the system [7]. In the proposed agent-based formulation one should assume that each agent has an individual, local view of the problem posed, which is addressed by the *feedback-anticipative* principle.

Real economic systems are determined by their periodic response. Let us assume that there are many different economic systems in our environment which interact. Here the economic system will be represented as agent A interacting with other agents with the goal of reaching general systemic equilibrium by differentiating parameter d, which determines the frequency response of the system. Here, the question arises: "Is it possible to control the economic system by changing the control parameter d, which actually alters the frequency response of the system?" The deviation in prices could be understood as the change of the system frequency response. In other words, if we change the price, the frequency of the system alters. Therefore, one should try to control the system by changing its frequency response.

Consider the following agent-based anticipative cobweb model of price P dynamics derived from equation Eq. (11):

$$\mathbf{P}_k = \frac{b\mathbf{P}_{k-1} + a - c}{\mathbf{d}_k} + \frac{b}{\mathbf{d}_k}\mathbf{P}_{k+1} \tag{30}$$

Initial conditions for Eq. (30) should be stated in matrix form by Eqs. (5, 6). In the above equation matrix annotation represents column vectors, which have the same arbitrary dimension n determined by the number of agents. For the computation of the new values of price P, shift operator ρ on sequence P is applied, which shift sequence $p \in P$ one step to the left:

$$\rho(\langle \mathbf{P} \rangle) = \langle \mathbf{p}_{n+1} \rangle \tag{31}$$

providing the forward shifted values for $\mathbf{P}_{k-1}$ and $\mathbf{P}_k$ in Eq. (30). The decision of change in parameter d will be dependant on the sum of two price values at time $k+1$ and time $k-1$. Here, the relative value of the price by taking the range of system response in the denominator will be considered:

$$\mathbf{e} = \frac{\xi_{k+1} + \xi_{k-1}}{|\lceil \xi_k \rceil - \lfloor \xi_k \rfloor|} \tag{32}$$

In Eq. (32) ξ represents the estimation chain for r time steps computed in a similar manner to $\mathbf{P}$ in Eq. (30) except for the initial conditions, which are stated in matrix form as in Eqs. (5, 6) for time 0, while for $\xi(0)$ shift operator ρ is applied forcing the anticipation principle as $\xi_k = f(\mathbf{P}_{k+1})$. Besides the notation for absolute value in the denominator, the *roof* and *floor* operators are applied. In order to perform the control by variation of parameter d, where n agents are present, the following state equation with the adaptive rule for $\Delta\mathbf{d}_k$ is introduced:

$$\mathbf{d}_{k+1} = \mathbf{d}_k + \Delta\mathbf{d}_k \tag{33}$$

where $\Delta\mathbf{d}$ determines the change in control parameter d:

$$\Delta\mathbf{d}_k = \begin{cases} \beta & if \quad e = \lceil \mathbf{e} \rceil \\ -\beta & if \quad e = \lfloor \mathbf{e} \rfloor \end{cases} \tag{34}$$

In the above definition of the agent's rule, the *floor* and *ceiling* functions over a vector of relative prices $\mathbf{e}$ considers only a finite number of lags. One should notice that the mentioned *floor-roof* operators are applied on vector $\mathbf{d}$ rather than on vector $\mathbf{P}$, which would mean the strict, conventional implementation of the *floor-roof* principle [19]. Parameter β is the *intensity* of agents reaction to the market disequilibrium; $\beta \in (0,1)$. Initialization of vector $\mathbf{d}$ is determined by random value $r_i \in [-2,2]$, which falls within the interval of periodic solutions for the anticipative cobweb system. Certainly, one could also assign an arbitrary value for $\mathbf{d}$ as this will also be considered.

The idea captured in the above definition considers a situation where an economic system, of which the market price in the past and estimated future is the highest, should be controlled by increasing the value of control parameter d, thus changing the frequency response of the system. The case at the lower end of the market price is inverse.

IOP Conf. Series: Materials Science and Engineering **173** (2017) 012018 doi:10.1088/1757-899X/173/1/012018

5. A note on application

Possible application of the developed agent-based model is in the field of stock exchange analysis and control where several economic sytems interact. Here it is our intention to show the manifestation of periodicity and control rule as described in previous sections on the real case. Ten top ranked stocks of the *Ljubljana Stock Exchange* [14] will be considered, namely: AELG, GRVG, HDOG, IEKG, ITBG, KRKG, LKPG, MELR, PETG and SAVA (Fig. 3).

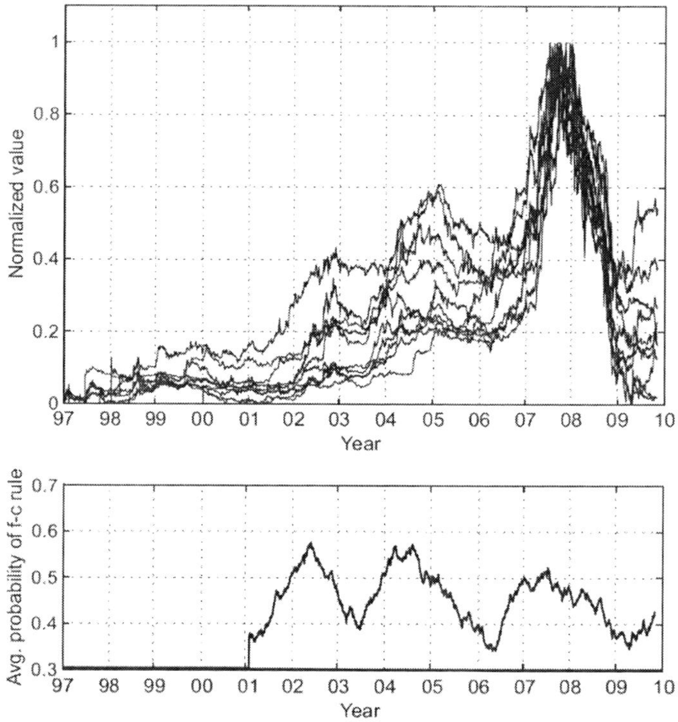

Figure 3: Normalized stock values (upper) and average probability of floor-ceiling rule (lower)

The period of the observation will be from 6.1.1997 to 5.11.2009 (3213 trading days) incorporating the critical fall in values from 2008 onward; several stocks came in to market after the initial date i.e. the last stock observed entered the marked on 6.1.2000 all of which present 30,059 data points. The normalized values of mentioned stocks are shown in the upper part of the Fig. 3; here equity rates are considered. The proposed floor-ceiling rule stated by Eqs. (32-34) provides a meaningful interaction between several economic systems which occurence in Slovenian top ten stocks is shown by Average probability in the bottom of Fig. 3; here the moving average of one year was considered. It is surprising, that the rule stated by Eqs. (32-34) for agent based model occurs with $p = 0.46$; $\sigma = 0.06$. The values are shown from 2001 onward since several stocks were not yet listed on the stock exchange market. Regarding the periodicity and the model stated by Eq. (11) one could ask, why is the determination of price at time t dependant on price at $t-1$ and $t+1$? In the stock exchange market this would be the case, when the price on the current day (t) could not be determined. The current price is not known before the closing on the end of the same day. Therefore for time t, only estimated price could exist, which is made on previous price ($t-1$) and expected price ($t+1$). There are methods to enhance the estimation of current price to extend the number of time-lags considered [8] which is the topic of rational expectations and intertemporal equilibrium. As the example regarding the periodicity of agents, the manifestation of period six in Slovenian stock exchange market is shown in Fig. 4; here the moving average of one year is taken. On y-axis the number of period six occurences are shown on average in one year for ten considered stocks, with $\bar{x} = 34$; $\sigma = 11$. That would mean, that the period six occurs on average every two weeks in one of the ten stock values. Period six is shown here as one possible manifestation of equilibrium as shown by Prop. 2. From Fig. 4 one could observe, that the Number of period six occurences is falling from the year 2004 onward which might be one of the indicators of ongoing instability in the stock exchange market.

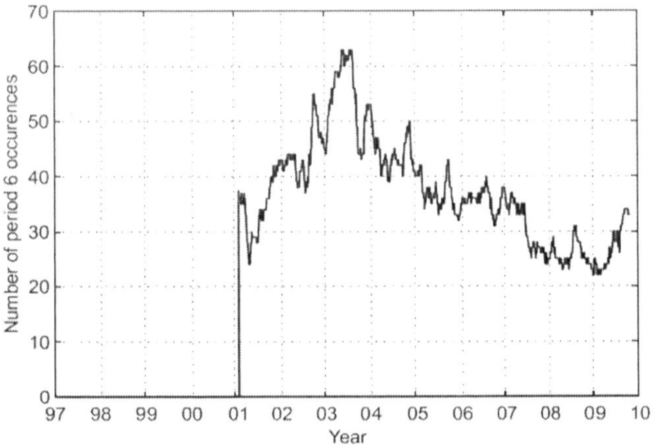

Figure 4: Number of period six occurences in blue chips

The importance of the question, how to control several interacting agents is apparent in the case of the global crisis which is indicated in 2008. It is a matter of further research to investigate the control strategies on the real cases which would provide global equilibrium or at least compensate for the global disturbances.

6. Discussion

Understanding the emergence of periodicity is a prerequisite for economic system control. The basic theory developed in this field considers the analysis of eigen vectors and eigen values [24]. The equality of the gained solutions determined by Prop. 1 for the anticipative cobweb model and general periodic solution for discrete nonlinear systems [24] provides a view of the Farey tree structure, which is [24] universal in nonlinear expansions of the considered systems.

Several anticipative cobweb systems were interconnected in order to form an agent-based model where interaction was determined by the discrete rule. By developing an agent-based model, the methodological platform is provided for a further examination of the interaction of several *feedback-anticipative* systems. The developed system provides promising results obeying the rules of interaction and self-synchronization found in real world economics systems. Another important fact is that agents with more interaction provide better conditions for general system stability. A detailed analysis of the agent-based model provides proof of system stability, which is one of the key conditions that should be meet by agent-based models simulating complex economic systems. The experimental examination of the two agent model indicates, that the entire model could be set in the global equilibrium mode. All the stated characteristics of the agent-based model as well as the response of the system for eight agents provides a promising methodological platform for studying the interaction between several economic entities such as agents. The proposed model provides the means for analyzing interaction, feedback, anticipation, frequency response, synchronization, standing oscillations and system equilibrium. An introduction of *feedback-anticipative* systems interconnection and control by varying the parameter, which influences system frequency response, represents a new perspective for the analysis of complex systems. The key values in the Farey tree hold for the agent-based model and have provided a possible starting-point for the search of analytical solutions in the parameter space.

The present study provides interconnection of Kaldor's cobweb model paradigm with an anticipative approach and *agent-based* paradigm. The common denominator as a thread of the paper is the evolutionary emergence principle hidden in the form of the Farey tree incorporating the *Golden Ratio* path.

The developed system confirms the applicability of the control rule where system response is controlled by the variation of parameter $\mathbf{d}$, which influences the frequency response of a particular agent.

By examining the developed agent-based system, it could be concluded that interaction of discrete oscillatory agents by itself results in stable working conditions. In the present case the rule of interaction provided system stability and a synchronization property which is found in real world systems [22] and significantly determines interacting economic systems.

Acknowledgment
This research was supported by the Slovenian Research Agency (Programme No. P5-0018 & Proj. No. RU/16-18-040).

References
[1] Berry B and Kim H 2000 Long waves 1790–1990 *Chaos Theory in the Social Sciences: Foundations and Applications* pp 215–236
[2] Bischi G, Dieci R, Rodano G and Saltari E 2001 *Journal of Evolutionary Economics* **11** 527–554
[3] Brock W and Hommes C 1997 *Econometrica* **65** 1059–1095
[4] Cavalli F, Naimzada A and Parisio L 2015 A cobweb model for electricity markets *12th International Conference on the European Energy Market (EEM)* pp 1–5
[5] Chechin G and Ryabov D 2004 *Physical Review E* **69** 036202.1–036202.9
[6] Gallas J and Nusse H 1996 *Journal Of Economic Behavior & Organization* **29** 447–464
[7] Hogg T and Huberman B 1991 *IEEE Transactions on Systems Man and Cybernetics* **21** 1325–1332
[8] Hommes C 1998 *Journal Of Economic Behavior & Organization* **33** 333–362
[9] Knott R 2006 "Fibonacci numbers and the golden section" http://www.mcs.surrey.ac.uk/Personal/R.Knott/Fibonacci/ (22/11/2006)
[10] Kocic L and Stefanovska L 2004 *Facta Universitatis, Series: Mechanics, Automatic Control and Robotics* **4** 491–496
[11] Kociuba G, Heckenberg N and White A 2001 *Physical Review* **E 64** 056220–1–056220–8
[12] Kopel M 1997 *Journal of Evolutionary Economics* **7** 269–289
[13] Kreyszig E 1993 *Advanced Engineering Mathematics* (Hoboken, NJ: John Wiley & Sons)
[14] LSE 2009 "Ljubljana stock exchange" http://www.ljse.si (10/11/2009)
[15] Gori L, Guerrini L and Sodini M 2014 *Discrete Dynamics in Nature and Society*
[16] Luenberger D 1979 *ntroduction to Dynamics Systems: Theory, Models and Applications*
[17] Matsumoto A 1997 *Discrete Dynamics in Nature and Society* **1** 135–146
[18] Naimzada A and Pecora N 2016 *Journal of Difference Equations and Applications* **22** 766–779
[19] Puu T and Sushko I 2004 *Chaos, Solitons and Fractals* **19** 597—612
[20] Puu T 2005 *Networks and Spatial Economics* **5** 5–20
[21] Rosen R 1985 *Anticipatory Systems: Philosophical, Mathematical and Methodological Foundations*
[22] Rosenblum M and Pikovsky A 2003 *Contemporary Physics* **44** 401—416
[23] Schenk K E 2005 *Journal of Evolutionary Economics* **16** 231–253
[24] Sonis M 1999 *Discrete Dynamics in Nature and Society* **4** 333–343
[25] Sonis M 1996 *Discrete Dynamics in Nature and Society* **1** 45–56
[26] Sprott J 1994 *Physical Review E* **50** R647–R650
[27] Strogatz S 1994 *Nonlinear dynamics and chaos: with applications to physics, biology, chemistry and engineering*

IOP Conf. Series: Materials Science and Engineering **173** (2017) 012019 doi:10.1088/1757-899X/173/1/012019

Hybrid binary GA-EDA algorithms for complex "black-box" optimization problems

E Sopov

Reshetnev Siberian State Aerospace University, 31 Krasnoyarskiy Rabochiy Ave., 660037, Krasnoyarsk, Russia

E-mail: evgenysopov@gmail.com

Abstract. Genetic Algorithms (GAs) have proved their efficiency solving many complex optimization problems. GAs can be also applied for "black-box" problems, because they realize the "blind" search and do not require any specific information about features of search space and objectives. It is clear that a GA uses the "Trial-and-Error" strategy to explorer search space, and collects some statistical information that is stored in the form of genes in the population. Estimation of Distribution Algorithms (EDA) have very similar realization as GAs, but use an explicit representation of search experience in the form of the statistical probabilities distribution. In this study we discus some approaches for improving the standard GA performance by combining the binary GA with EDA. Finally, a novel approach for the large-scale global optimization is proposed. The experimental results and comparison with some well-studied techniques are presented and discussed.

1. Introduction

Many real-world problems can be stated as a problem of finding an optimal alternative from a list of options. The problems are decision making, design, structure and parameters optimization, control and many other. Many real-world design and decision making support problems are complex and bad-formalized, thus the quality criterion is usually considered as the "black-box" model [1]. Moreover, alternatives cam be represented by complex structures that contain variable parameters of many different types, including categorical, integer, rank, binary and others. There exist many efficient and well-studied techniques that can deal with the global black-box (BB) optimization problems.

One of the main difficulties for designing better BB optimization techniques is the limitation of the number of objective evaluations. There exist the so-called "cheap" BB problems, where the objective can be evaluated sufficiently many times (many thousands), and the "expensive" or "costly" BB problems, where the objective evaluations are strongly limited (a few hundreds). In the case of costly problems, we cannot provide adequate analysis of the search space and the objective landscape. Thus the "blind" search that implying more advanced optimization techniques like genetic algorithms (GAs) is more preferable [2]. At the same time GAs are also computational costly methods even they apply parallel stochastic search. Moreover, the GA's performance usually decreases when the dimensionality of the search space increases. There exists the challenge of designing and investigating approaches that are able to better exploit an experience of the past search trials and to improve the convergence to the optimal solution.

In this study we will combine the Estimation of Distribution Algorithm (EDA) and the GA for extracting statistical data about the GA's past search experience (using the EDA conception), and for using such statistical distribution in the GA.

The rest of the paper is organized as follows. Section 2 describes related work. Section 3 describes the proposed approaches. In Section 4 the results of numerical experiments are discussed. In the Conclusion the results and further research are discussed.

2. Related work

The BB optimization refers to a problem setup in which an optimization algorithm is supposed to optimize (minimize or maximize) an objective function and the algorithm may only query the value of the objective for a candidate-solution, but it does not obtain gradient information, and in particular it cannot make any assumptions on the analytic form of the objective. The goal of optimization is to find an as good as possible value the objective within a predefined time, often defined by the number of available queries to the black box (the number of the objective evaluations). Problems of this type regularly appear in practice, when optimizing parameters of a model is in fact hidden in a black box (there is no possibility to analyze the objective), or the problem is too complex to be modeled explicitly.

A large variety of optimization algorithms for the BB optimization has been proposed. The majority of techniques that demonstrates high performance with arbitrary BB problems are based on evolutionary (including genetic) and bio-inspired algorithms. Nevertheless, the BB optimization is still a challenge for the algorithms. In recent years many BB optimization competitions have been held to compare performances of different traditional and new approaches (the most known are the BBOB/COCO as a part of the GECCO conference, and the BBComp as a part of the IEEE CEC conference). The given results show that BB problems are still hard for existing optimizers.

Another essential difficulty for optimization algorithms is related with an effect that is called the "curse of dimensionality". Performance of algorithms usually decreases when the dimensionality of the search space increases. Nowadays, optimization problems with many hundreds or thousands of objective variables are called large-scale global optimization (LSGO) problems. BB LSGO problems have become a great challenge even for optimizers especially with non-separable problems that excludes a straightforward variable-based decomposition. Nevertheless, some assumption can be done, and there exist many efficient LSGO techniques for the continuous search space [3].

Many real-world optimization problems encode different complex structures and contain variables of many different types, which cannot be represented only by real values. In this case binary representation of candidate solutions can be used. In this study we will use the binary GAs. As we can see from papers, the majority of approaches for BB and BB LSGO are designed for the continuous search space, and there is a lack of approaches using the binary GAs as the core technique.

As previously mentioned, population-based stochastic algorithms collect and exploit some statistical information about search space that is not presented in explicit way. At the same time there exist a family of algorithms that are also operate with a set of candidate-solutions at each trial search step, and estimate and use an explicit probability distribution to generate new solutions. Such approaches are called Estimation of Distribution Algorithms [4]. There also exist binary EDAs.

There are some straight-forward attempts to combine a GA with the binary EDA. The main idea to estimate a distribution of genes with one-values in chromosomes of a current population. The distribution then is applied to generated next generation instead of applying the standard crossover. There exist some well-studied approaches like Probabilistic Incremental Learning Algorithm (PBIL) [5], hybrid EDA-EA [6,7] and other.

In [8] similar approach has been proposed, in [9] the approach has been applied for multi-objective optimization problems.

3. Proposed approaches

A binary GA is conventional evolutionary algorithm (EA) with a population of solutions, which are represented by fix-length vector containing zero- or one-value in each position. The binary GA and the EDA have very similar schemes. As we can from Table 1, the only differences in algorithms the way they collect and exploit search experience. GAs collect and update statistics inside population, EDAs do that using external explicit distribution.

Table 1. Comparison of general steps of GA and EDA algorithms.

GA	EDA
1. Initialize population with random. 2. Evaluate current population. 3. Form a new population using genetic operators. 4. Update current population using the given strategy. 5. Until a stop criterion is satisfied, repeat steps 2-4.	1. Initialize population with random. 2. Evaluate current population. 3. Form a new population according to the probability distribution. 4. Update the distribution using the given strategy. 5. Until a stop criterion is satisfied, repeat steps 2-4.

We can estimate a distribution of one-values in population of the GA by evaluating a probabilities vector for the population using the binary EDA way (see example on Figure 1):

$$P^k = (p_1^k, p_2^k, \ldots, p_n^k),\ p_i^k = P\left(\left(x^i\right)^k = 1\right) = \frac{1}{N}\sum_{j=1}^{N}\left(x_j^i\right)^k, \forall i, k \quad (1)$$

where p_i^k is the probability of the one-value in the i-th gene in chromosome, n is a chromosome length, x_j^i is the value of the j-th gene of i-th chromosome, k is the current generation number.

Gene position	1	2	3	4	5	6	7	8
Individual 1	1	1	1	1	0	0	0	0
Individual 2	1	1	0	0	0	0	0	1
Individual 3	0	1	0	1	1	0	0	0
Individual 4	0	1	1	0	0	0	1	1
Individual 5	0	1	0	0	0	0	0	0
Individual 6	0	1	1	0	1	0	0	1
Individual 7	1	1	1	0	0	0	0	0
Individual 8	1	1	1	1	0	0	0	1
Individual 9	0	1	0	1	0	0	0	0
Individual 10	1	1	1	0	1	0	1	1
Distribution	0,5	1	0,6	0,4	0,3	0	0,2	0,5

Figure 1. Example of distribution estimation in a GA.

3.1. *Optimal solution prediction for the BB problems using EDA-based GA*

We can investigate the performance of binary GAs using a visual representation of the distribution changes over the GA run. In [10] the following approach have been proposed. After the given GA finishes the run, we create a diagram for each component of the probabilities vector. The vertical axis contains probability values; the horizontal axis contains numbers of generation. An example of a component changes inside a GA run is presented on Figure 2.

Analyzing many diagrams for different BB optimization problems we can find that the components of the probabilities vector frequently converge to one-value if optimal solution contains one in corresponding position and or to zero-value if the optimal solution contains zero. Our hypothesis is that probability converges to one (or zero) if corresponding gene of the best-found solution is equal to one (or zero). The higher algorithm performance the more often probabilities converge to the correct value.

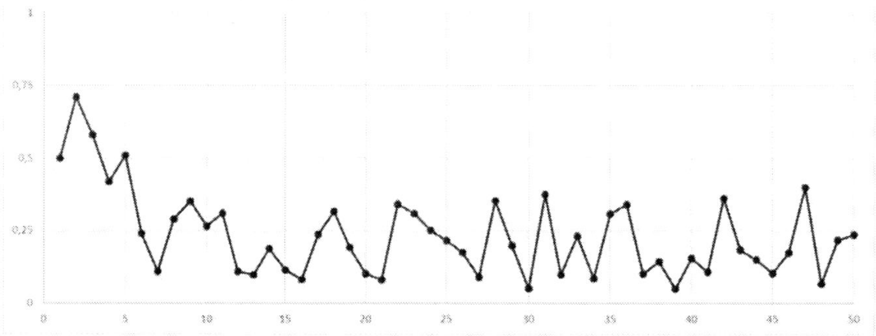

Figure 2. Example of a probabilities vector component changes.

We can propose the following strategies of implementing the convergence feature in GAs:

- *Straight-forward optimal solution prediction.*

When applying a GA to an optimization problem, the distribution of one-values in populations are estimated and collected. Then we visually analyze the convergence of each component of the probabilities vector and define genes of the optimal solution as:

$$x_i^{optimal} = \begin{cases} 1, if\ p_i \to 1 \\ 0, if\ p_i \to 0 \end{cases} \quad (2)$$

The predicted value can precise or match the best-found individual.

- *Integral criterion for the prediction.*
-

$$x_i^{optimal} = \begin{cases} 1, if\ \sum_{g=1}^{MaxGen}(p_i^j - 0.5) > 0 \\ 0, otherwise \end{cases}, \quad (3)$$

where MaxGen is maximum number of generations.

The predicted value can precise or match the best-found individual.

We can also evaluate prediction on early generations and include the predicted solution into current population (for example, replacing the worst individual) to enhance search process.

- *Weighted integral criterion for the prediction.*

The main idea of the given modification is that the probability values on the later iterations have the greater weights as algorithm collects more information about search space.

We have estimated the performance of the standard conventional GA, the GA using the distribution for updating next population instead of using a crossover operator and the GA with the optimal solution prediction using the weighted integral criterion at the final generation. We have included the following well-studied benchmark problems, that are known as complex for a GA optimizer: Rastrigin and Shifted Rotated Rastrigin, Rosenbrock, Griewank, De Jong 2, "Sombrero" ("Mexican Hat"), Schekel, Katkovnik and other. We have also included some deceptive "trap" problems like Ackley Trap, Whitley Trap, Goldberg Trap, Trap-4, Trap-5, DEC2TRAP, Liepins and Vose.

The numerical experiments have shown that the GA with prediction outperforms the standard approach. More detailed information on the experimental results can be found in [10].

3.2. Problem decomposition for the LSGO problems using EDA-based GA

As increasing dimensionality of the BB optimization problems decrease the performance of a GA, we will focus on the LSGO problem.

There exist a great variety of different LSGO techniques that can be combined in two main groups: non-decomposition methods and cooperative coevolution (CC) algorithms. The first group of methods are mostly based on improving standard evolutionary and genetic operations. But the best results and the majority of approaches are presented by the second group. The CC methods decompose LSGO problems into low dimensional sub-problems by grouping the problem subcomponents. CC consists of three general steps: problem decomposition, subcomponent optimization and subcomponent coadaptation (merging solutions of all subcomponents to construct the complete solution).

The problem decomposition is a critical step. There are many subcomponent grouping methods, including: static grouping, random dynamic grouping and learning dynamic grouping. A good survey on LSGO and methods is proposed in [3]. As we can observe in papers, almost all studies are focused on continuous LSGO, and there is a lack of techniques for binary (or other discrete) representations.

The main idea of the LSGO problem decomposition methods is based on the divide-and-conquer approach which decomposes the problem into single-variable or multiple-variable low dimensional problems. In this case, only part of the variables are used in the search process; the rest are fixed and their values are defined using some strategy (for example, values from the best-found solution are used).

The finding of an appropriate decomposition is part of the general search process. It is obvious and has been presented in many studies that the best performance is achieved with separable LSGO problems. In the case of non-separable problems, the performance strongly depends on the decomposition strategy.

In this work, we will formulate the following requirements for the proposed decomposition method:

- The grouping should be dynamic to realize the "exploration and exploitation" strategy.
- The grouping should be random to avoid the greedy search and the local convergence.
- The grouping should be based on the past search experience of the whole population (to provide the global search options).
- The grouping should be adaptively scalable to provide efficient decomposition at every stage of the search process.

We will use previously discussed convergence property to define the values for fixing genes in chromosomes at the grouping stage. If the i-th position in a chromosome at the t-th generation is fixed, its value is defined by the corresponding value of the probability vector:

$$x_i\ (t) = \begin{cases} 0,\ if\ p_i(t) < (0.5 - \delta) \\ random,\ if\ (0.5 - \delta) \leq p_i(t) \leq (0.5 + \delta) \\ 1,\ if\ p_i(t) > (0.5 + \delta) \end{cases} \quad (4)$$

where δ is a threshold (a confidence level), $\delta \in (0, 0.5)$.

We will explain the proposed approach using Figure 3. The diagram visualizes an arbitrary component of the probability vector for an arbitrary run of a GA on a benchmark BB problem. For the chosen gene the corresponding value of the optimal solution is equal to zero. As we can see from Figure 3, the GA starts with random initialization, thus the value of the probability vector is equal to 0.5. At the first generations the GA actively explores the search space and number of 1's and 0's genes are almost equal, thus the value of the probability vector is still about 0.5. After that, the GA locates a promising region in the search space and increases the number of 0's in this position, thus the value of the probability vector decreases towards zero.

The confidence level δ is a parameter that defines a threshold for the probability value around 0.5, where we cannot make a decision about the gene value.

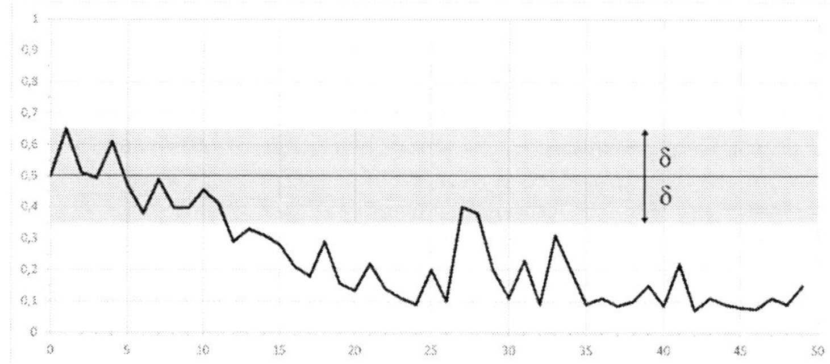

Figure 3. Making decision about a gene's values fixation using the analysis of the dynamic of the probability vector component.

Although a decision about fixed variables is made by stand-alone components, the estimated distribution contains information about the problem solving in general. Thus the method is not focused only on separable LSGO problems.

Next we need to define the number of variables that will be fixed. There exist many strategies. For example, the splitting-in-half method divides an n-dimensional problem into two $n/2$ subcomponents. In general, we will define the number of fixed variables as a percentage of the chromosome length and will denote it as α. The value of α can be constant or can change during the run of the algorithm. The variables and corresponding components of the probability vector are fixed for some predefined number of generations, which is called an adaptation period (denoted as t_{adapt}). The list of fixed components is randomly defined.

In this paper, the straight-forward approach is used, α and t_{adapt} are predefined and constant.

The main advantage of such EDA-based decomposition is that we do not lose the previously collected statistic as we fix components of the probability vector. The GA solves the problem of reduced dimensionality and updates the probability only for active components. After each adaptation period we will randomly fix other components, and the previously fixed components will continue updating their saved values.

We will describe the proposed LSGO algorithm in detail.

First, we need to encode the initial problem into a binary representation. The standard binary or Grey code can be used. A chromosome length n is defined. Next, specific parameters of the EDA-based decomposition and the chosen GA, maximum number of fitness evaluations (*MaxFE*) or maximum number of generations (*MaxGEN*) are defined. The maximum number of generations can be substituted with any other stop condition (for example, time-based condition).

Finally, the following algorithm is used (see Figure 4).

As is known, the island model GA can outperform the standard single-population GA for many complex optimization problems [11]. We can also decrease the computational time by implementing the island GA with a parallel multi-core or multi-processor computer.

```
   Input: n, N, α, δ, t_adapt, MaxFE, the GA
operators' parameters.

   Initialization:
   Randomly generate a population of N individuals
of the length n.
   Calculate P(0) using formula (1).

   Main loop.
   Until MaxFE is reached:
   1. Problem decomposition stage: Start new
adaptation period. Fix random α components in
chromosomes and in the probability vector.
   2. Subcomponent optimization stage: Run the GA
for t_adapt generations:
         a. Fitness evaluation. Set values in fixed
positions of chromosomes according to P(t) using
formula (2).
         b. Perform selection, crossover and mutation
operations.
         c. Create next generation, update the
probability vector P(t) for active components.

   Output: the best-found solution.
```

Figure 4. EDA-based decomposition GA for LSGO.

4. Numerical experiments

We have used 15 large-scale benchmark problems from the CEC'2013 Special Session and Competition on Large-Scale Global Optimization [12]. These problems represent a wider range of real-world BB LSGO problems. There are 3 fully-separable problems (denoted as f1-f3), 8 partially separable problems (f4-f7 with a separable subcomponent and f8-f11 with no separable subcomponents), 3 problems with overlapping subcomponents (f12-f14), and 1 non-separable problem (f15).

The experiment settings are:

- Dimensions for all problem are D=1000;
- The standard binary encoding is used with accuracies: ε=0.1 for f1, f4, f7, f8 and f11-15, ε=0.05 for f3, f6 and f10, and ε=0.01 for f2, f5 and f9;
- For each problem the best, mean, and standard deviation of the 25 independent runs are evaluated;
- Maximum number of fitness evaluations is MaxFE=3.0e+6;
- The performance estimation is performed for the number of fitness evaluations equal to 1.2e+5, 6.0e+5 and 3.0e+6.

The EDA-based decomposition GA settings are:

- Population sizes are N=1000 for the single-population version, N=500 for the island version with 3 islands, and N=400 for 5 islands;
- The adaptation period is t_{adapt}=100;
- The probability threshold is δ=0.05, 0.15 and 0.25;
- Numbers of fixed components are α=25%, 50% and 75% of the chromosome length.

All algorithms have been implemented in Visual Studio C++ using the OpenMP for parallel computing with multi-core PC. Free C++ source codes of the benchmark problems are taken from (http://goanna.cs.rmit.edu.au/~xiaodong/cec13-lsgo/, 2013).

We have carried out the above-mentioned experiments and have established the following. In the case of single population, the best performance on average is achieved with 50% fixed components and δ=0.15. In the case of the island model, the best results are obtained by the 5 island model with 75% fixed components and δ=0.15. Almost for every considered value of parameters, the island model outperforms the single population version of the algorithm.

The experimental results for the best found settings are presented in Table 2.

Table 2. Experimental results for the EDA-based decomposition GA with 5 islands and α=75%.

		f1	f2	f3	f4	f5	f6	f7	f8
1.2e5	Best	1.42E+07	9.63E+03	1.08E+02	1.39E+11	6.11E+14	2.90E+05	7.05E+08	3.97E+15
	Mean	5.50E+07	1.06E+04	4.52E+01	9.15E+11	7.17E+14	7.78E+05	2.76E+09	2.71E+16
	StDev	2.98E+07	1.53E+03	1.44E+01	5.63E+11	7.45E+08	2.61E+05	1.44E+09	6.77E+15
6.0e5	Best	6.89E+03	9.11E+03	3.04E+00	1.95E+10	3.07E+14	5.21E+05	2.01E+08	2.09E+14
	Mean	1.99E+04	1.25E+04	1.30E+01	9.07E+10	5.03E+14	6.05E+05	9.41E+08	2.18E+15
	StDev	1.68E+03	1.18E+03	6.32E-01	6.03E+10	2.50E+07	2.60E+05	7.56E+08	1.52E+15
3.0e6	Best	4.59E-05	1.82E+03	2.94E-05	6.60E+09	7.59E+14	6.25E+04	7.65E+07	4.49E+13
	Mean	5.68E-04	3.34E+03	4.81E-01	2.32E+10	9.75E+14	4.75E+05	2.53E+08	3.64E+14
	StDev	4.29E-04	2.54E+02	2.28E-01	1.14E+10	2.18E+06	3.35E+05	8.35E+07	5.21E+14
		f9	f10	f11	f12	f13	f14	f15	Average
1.2e5	Best	1.08E+09	8.87E+06	1.46E+11	3.87E+06	2.66E+10	1.88E+11	3.61E+07	3.05E+14
	Mean	1.80E+09	7.14E+07	3.47E+11	4.36E+08	2.98E+10	5.78E+11	2.69E+08	1.85E+15
	StDev	4.27E+08	1.57E+07	2.25E+11	7.89E+08	1.12E+10	3.67E+11	9.91E+07	4.52E+14
6.0e5	Best	6.42E+08	7.91E+06	1.34E+10	2.40E+03	6.28E+09	5.68E+10	1.80E+07	3.44E+13
	Mean	1.25E+09	1.38E+07	9.84E+10	6.66E+03	1.47E+10	1.03E+11	2.43E+07	1.79E+14
	StDev	5.21E+08	1.65E+07	1.18E+11	5.53E+03	4.83E+09	6.70E+10	8.68E+06	1.02E+14
3.0e6	Best	4.15E+08	6.18E+06	2.60E+10	7.72E+02	8.02E+09	1.42E+10	2.40E+07	5.36E+13
	Mean	8.06E+08	1.61E+07	7.01E+10	2.30E+03	1.27E+10	1.69E+11	3.05E+07	8.93E+13
	StDev	1.72E+08	7.89E+06	4.29E+10	2.41E+03	2.96E+09	4.81E+10	5.13E+06	3.47E+13

The summary results are compared with other techniques presented at the CEC'13 competition. The algorithms are DECC-G (differential evolution (DE) based cooperative coevolution (CC) with random dynamic grouping), VMO-DE (variable mesh optimization using differential evolution), CC-CMA-ES (Covariance Matrix Adaptation Evolution Strategy using Cooperative Coevolution), MOS (Multiple Offspring Sampling (MOS) based hybrid algorithm), and SACC (smoothing and auxiliary function based cooperative coevolution) [13]. We have averaged the performance estimates of all algorithms over all problems and have ranked algorithms by the Best and the Mean values. The results are in Table 3.

As we can see from Table 3, the proposed approach has taken 4th place by the Best criterion and 5th place by the Mean value. We should note that all algorithms except the proposed are specially designed for continuous LSGO problems. The EDA-based decomposition GA does not use any knowledge about search space. Moreover, the chromosome length in the binary algorithm is greater than in the case of the continuous space. Nevertheless, the EDA-based decomposition GA outperforms the CC-CMA-ES by two measures and the DECC-G by the Best value on average.

Table 3. LSGO approaches comparison.

Algorithm	SACC	MOS	VMO-DE	DECC-G	CC-CMA-ES	EDA-GA
Best	9.80E+12	2.17E+11	4.90E+13	5.80E+13	6.25E+13	5.36E+13
Ranking by Best	2	1	3	5	6	4
Mean/ StDev	8.0E+13/ 5.08E+13	5.33E+11/ 2.04E+11	5.32E+13/ 4.81E+12	7.7E+13/ 1.02E+13	8.58E+13/ 2.39E+13	8.93E+13/ 3.47E+13
Ranking by Mean	4	1	2	3	6	5

Although the proposed approach yields to some of LSGO algorithms, our hypothesis is that the proposed approach will be a good tool for solving complex real-world LSGO problems, which usually contain not only continuous variables, but mixed-type variables and can represent arbitrary complex structures. Further investigations of the algorithm structure and parameters can probably improve its performance. In particular, the α value can be adjusted adaptively during the algorithm run using information about the probability vector convergence.

5. Conclusions and further work

In this study we have presented and discussed an approach of combining the binary GA and the EDA algorithms to improve the GA performance via estimating and exploiting explicit statistics about one-values distribution in the GA's population. The estimated distribution can be used for many purposes. We have proposed the optimal solution prediction method which is based on the analysis of the distribution components' convergence. The results of numerical experiments have shown that the GA with such prediction outperform the standard GA. We have also proposed a novel LSGO algorithms with EDA-based decomposition which is also based on the distribution analysis for fixing genes values in chromosomes. The approach performs well on complex BB LSGO benchmark problems and demonstrates results comparable with the state-of-art techniques.

In further work, more detailed analysis of the EDA-based decomposition GA parameters will be provided. A self-configuration will be introduced into the algorithm. We also will try to apply the distribution analysis for other classes of optimization problems.

6. Acknowledgements

The research was supported by the President of the Russian Federation grant (MK-3285.2015.9).

References

[1] Audet Ch 2014 A Survey on Direct Search Methods for Blackbox Optimization and Their Applications *Mathematics Without Boundaries Springer New York* pp 31–56

[2] Srinivas M and Patnail L M 1994 Genetic algorithms: A Survey *Computer* **27 (6)** pp 17-26

[3] Mahdavi S, Shiri M E and Rahnamayan Sh 2015 Metaheuristics in large-scale global continues optimization: A survey *Information Sciences* **295** pp 407–428

[4] Hauschild M and Pelikan M 2011 An introduction and survey of estimation of distribution algorithms *Swarm and Evolutionary Computation* **1** pp 111-128

[5] Baluja Sh 1994 Population-Based Incremental Learning: A Method for Integrating Genetic Search Based Function Optimization and Competitive Learning *Technical Report Carnegie Mellon University* 41 p.

[6] Dong W, Chen T, Tino P and Yao X 2013 Scaling up estimation of distribution algorithms for continuous optimization *IEEE Trans. Evol. Comput.* **17 (6)** pp 797–822

[7] Wang Y and Li B 2008 A restart univariate estimation of distribution algorithm: sampling under mixed gaussian and lévy probability distribution *IEEE Congress on Evolutionary Computation CEC* pp 3917–3924

[8] Sopov E, et al. 2009 A Modified probabilistic genetic algorithm for the solution of complex constrained optimization problems *Vestnik of Siberian State Aerospace University* **5(26)** pp 31-36

[9] Sopov E and Sopov S 2011 On probabilistic genetic algorithm with adaptive mutation and Pareto-solutions prediction for complex multi-objective optimization problems *Vestnik of Samara State Aerospace University* **6(30)** pp 273-283 (in Russian).

[10] Sopov E and Semenkina O 2015 The optimal solution prediction for genetic and distribution building algorithms with binary representation *IOP Conf. Ser. Mater. Sci. Eng..* **70** 10 p

[11] Gonga Y-J, et al. 2015 Distributed evolutionary algorithms and their models: A survey of the state-of-the-art *Applied Soft Computing* **34** pp 286–300

[12] Li X, et al. 2013 Benchmark functions for the CEC 2013 special session and competition on large-scale global optimization *Technical Report Evolutionary Computation and Machine Learning Group RMIT University*.

[13] Li X, et al. 2013 Technical report on 2013 IEEE Congress on Evolutionary Computation Competition on Large Scale Global Optimization *Available at: http://goanna.cs.rmit.edu.au/~xiaodong/cec13-lsgo/competition/lsgo-competition-sumary-2013.pdf.*

IOP Conf. Series: Materials Science and Engineering **173** (2017) 012020 doi:10.1088/1757-899X/173/1/012020

Why don't you use Evolutionary Algorithms in Big Data?

[1]Vladimir Stanovov, [1]Christina Brester, [2]Mikko Kolehmainen, [1]Olga Semenkina

[1]Siberian State Aerospace University, Krasnoyarsk, Russian Federation, 660037, Krasnoyarsky Rabochy Av., 31
[2]University of Eastern Finland, Kuopio, Finland, 70211, Yliopistonranta, 1 E

Email: vladimirstanovov@yandex.ru

Abstract. In this paper we raise the question of using evolutionary algorithms in the area of Big Data processing. We show that evolutionary algorithms provide evident advantages due to their high scalability and flexibility, their ability to solve global optimization problems and optimize several criteria at the same time for feature selection, instance selection and other data reduction problems. In particular, we consider the usage of evolutionary algorithms with all kinds of machine learning tools, such as neural networks and fuzzy systems. All our examples prove that Evolutionary Machine Learning is becoming more and more important in data analysis and we expect to see the further development of this field especially in respect to Big Data.

1. Introduction

Tons of data is measured and collected every day by specialists from many areas. All this information is supposed to be analysed to discover new non-trivial knowledge and used to train effective models in Machine Learning. However, a huge amount of data does not necessarily equate to the high quality of its processing. Firstly, many algorithms are stuck with the curse of dimensionality. It has been shown that Big Data might be a crucial obstacle for conventional methods of data analysis [1]. Moreover, measurements are often gathered from different sources and after the fusion might duplicate each other. A vector of variables is usually heterogeneous: it may comprise numerical and categorical values, as well as unstructured textual information. Many datasets contain missing values and abnormal measurements which require experts to design robust models [2]. Then, in [3] authors emphasize that the model performance might deteriorate with the inclusion of more and more irrelevant variables. Although some models (such as Naïve Bayes) are robust to irrelevant data, their effectiveness decreases if there are correlated variables in the dataset [4]. Therefore, based on all these facts we highlight the necessity of thorough data pre-processing and developing advanced methods which can work with Big Data.

In this paper we focus on three aspects of the Big Data problem which are feature selection, instance selection and handling missing values. The first two aspects relate to the large number of variables and instances in the dataset which is typical for Big Data. One of the widely applied extensive methods for high dimensionality is the use of parallel computations and modern parallel architectures. However, this approach is not always applicable due to resource limitations. In this case, Data Reduction (DR) methods should be used, which represent a more advanced approach for dealing with a large number of features and instances. At the same time, while solving real problems researchers or analysts either do not use the DR methods or apply the simplest ones: as a feature selection method they may use correlation analysis to eliminate highly-correlated variables and for instance selection they apply the "k-nearest neighbour"-based methods. The third aspect is also highly important: on the one hand, many effective methods cannot handle missing values, which causes restrictions on algorithmic tools, but, on the other hand, exclusion of features and instances containing missing values from the dataset may lead to the loss of useful information.

The open questions in this area forced us to find a new effective paradigm which could be used in order to solve the above-mentioned problems. After testing different methods and reviewing modern approaches we have found that there is a group of algorithms which are underestimated in the sense of Big Data, namely, are evolutionary algorithms (EAs) [5]. Compared to other optimization algorithms, they have a number of undoubted advantages: evolutionary algorithms (in particular, genetic algorithms (GAs)) tackle the curse of dimensionality successfully; they are used in various search spaces (for example, it might be the space of relevant feature vectors or informative subsets of instances); they can optimize several criteria of different types at once. These and some other qualities of EAs are beneficial in data analysis, and therefore, we expect to see further development of evolutionary machine learning tools. In this study we illustrate the positive effects of using EAs in feature selection, instance selection and handling missing values. We describe a number of applications of EAs in real-world and benchmark high dimensional problems. Our main goal is to observe a relatively large area of Big Data including a number of issues and show that EA-based approaches should be used as an advanced alternative to conventional methods.

The rest of the paper is organized as follows: Section 2 contains a brief description of EAs and their useful qualities for Big Data. In Section 3.1 we introduce GA-based feature selection methods and illustrate their effectiveness for several real-world problems. Then, instance selection approaches, which use an evolutionary search, are presented in Section 3.2. In Section 3.3 we discuss a missing values problem and show how to solve it with the usage of EAs. The last section is devoted to conclusions.

2. Evolutionary Algorithms and their useful qualities for Big Data

EAs are stochastic population-based optimization algorithms which always include three principal components: a set of candidate-solutions, a rule for comparing candidates and choosing the best ones, and operators to generate new solutions [6]. For several decades researchers have been developing various modifications of EAs: many self-adaptive [7] and co-operative [8, 9] algorithms have been proposed, many effective operators have been implemented [10], and a large number of successful applications of EAs in real-world problems have been demonstrated so far [11, 12]. All these have become possible due to the unique properties of EAs which quite often allow us to tackle complex problems that could not be solved with any other algorithms (Table 1).

Firstly, EAs can work in high dimensional search spaces, whereas other algorithms are just stuck with optimizing [13]. In terms of feature selection this property is very useful: there are datasets with hundreds and thousands of attributes ("features" or "variables" mean the same as "attributes"). A vector of features might be coded with a binary string: 1 means relevant features and 0 denotes irrelevant ones. We only need to introduce an optimized criterion and run the algorithm. As opposed to conventional feature selection methods such as stepwise selection, EAs allow us to compare various subsets of features, whereas many other methods work iteratively, adding or removing one variable at a time (forward selection or backward elimination). Moreover, EAs can optimize not only one, but several criteria at once, which signifies the wide flexibility of this algorithmic tool for feature selection.

Secondly, EAs are often used to train models. In this process the search space is a set of different models which is quite unusual for traditional optimization algorithms. An important benefit here is the possibility to combine training a model and instance selection: we assess the performance of each model using a particular subset of training instances. However, in addition to this, the so-called *adaptive* instance selection, there is another way to decrease the number of training examples. EAs are easily parallelized due to their population-based nature and this quality is effectively used in training data rotation [14]. The main idea is to divide the dataset into subsamples and at one round train models not on the whole sample but on the smaller set of instances. At the next round, subsamples are "rotated" and models are trained on other instances. Another advantage is related to online learning and the possibility to adjust the trained model after receiving a piece of new data. In this case we do not need to start the learning process from scratch and use a lot of resources in vain. So, EAs could be applied in the dynamic environment: the population of candidate-solutions (for example, models) adapts for new data quickly.

Recently EAs have been widely used for data imputation, and the reason is that EAs are global optimization techniques, capable of creating models describing the dependency between tens of variables, such as neural networks or fuzzy logic systems. Several examples of such applications can be found in [7, 15, 16]. Finally, EAs could be used to generate fuzzy rule bases, i.e. models which can work with missing data without imputation.

Table 1. Main advantages of EAs in different Big Data issues.

Big Data issue		EA useful quality
Feature selection (3.1)		▪ The possibility to compare subsystems of features due to the binary representation of a feature vector; ▪ EAs work effectively in the high dimensional search space of feature subsystems; ▪ EAs can optimize a number of criteria at once.
Instance selection (3.2)	Training data rotation (3.2.1)	▪ The possibility to parallelize computations due to the population-based nature of EAs; ▪ Cooperative modifications with several parallel populations may be used with data rotation.
	Adaptive instance selection (3.2.2)	▪ Instance selection has been developed for population-based methods; ▪ Instance selection might be easily embedded in training a model.
Handling missing values (3.3)		▪ Several models that can handle missing values can only be generated by EAs.

In the next section we illustrate all these positive effects of using EAs in real Big Data problems.

3. The use of Evolutionary Algorithms in Big Data

3.1 Feature selection methods

Feature selection is an important part of data pre-processing: attributes might have a low variation level, correlate with each other or be measured with mistakes. For high dimensional problems conventional feature selection methods (forward selection and backward elimination) do not work effectively. Their abilities are limited by the iterative nature of these methods: it is impossible to estimate the cumulative contribution of several variables because only one feature is added or removed at a time. Therefore, a number of alternative evolutionary feature selection methods have been proposed.

Generally, if we train a model to assess the quality of a reduced feature set (based on the model performance) we are talking about a wrapper approach. However, for Big Data it requires a lot of computational resources and it is not reasonable to use it. The other approach is a filter. It involves statistical criteria to compare various feature subsystems and looks more preferable for Big Data [17].

In terms of the filter approach EAs are used as optimizers. The feature vector is coded with a binary string where 1 and 0 denote relevant and irrelevant variables respectively. In [18] the authors combine various statistical criteria and investigate multi-objective models for feature selection. They apply the evolutionary filter approach to cardiovascular predictive modelling. It is important to remember that in this area dimensionality reduction leads to a decrease in the medical costs of clinical trials. As a result, for this cardiovascular dataset it became possible to reduce the number of features significantly: from 433 down to 38 without any loss in the performance of the model.

The similar evolutionary filter approach was used for speech-based speaker state identification. In this problem the total number of variables, which are acoustic characteristics, is 384. Based on this data it is possible to recognize speakers, their genders and even emotions. In [19] a self-adaptive GA is incorporated into the feature selection as an optimizer. In the series of experiments it has been shown that: we can find a feature vector which is relevant to a particular problem, the number of variables might be decreased by a factor of two and this leads to an increase in the model performance. This approach was tested on datasets in English, German and Japanese and in many trials it outperformed alternative feature selection methods such as PCA and IGR.

3.2 Instance selection methods

In this subsection we will concentrate on Instance Selection (IS) methods [20], including Training Set Selection (TSS) and training data rotation. IS focuses on creating a subsample from the set of available training instances, i.e. preparing the data for the training algorithm. IS methods also include active learning, boosting and sampling methods.

Using GAs and EAs when learning models such as neural networks or fuzzy logic systems on Big Data requires a lot of iterations, during which the training sample is used. Because of this, applying instance selection methods, which create a smaller sample decreasing the computation time, combined with evolutionary approaches looks promising. However, in this case the problem of relevance of the subsample to the original sample arises. Creating a subsample may result in losing the information, but at the same time, it may allow better results to be obtained by filtering out non-informative cases and concentrating on instances that represent the main dependency between variables. This is why creating an IS method that would effectively use the original sample to create subsamples is an important task in Big Data analysis.

3.2.1 Training data rotation

The idea of training data rotation was introduced in [14]. The population of individuals is divided into N subpopulations and the same number N of threads is initialized. Thus, the fitness function evaluation for different subpopulations is implemented in parallel threads. To achieve an additional reduction in computational time, Ishibuchi et al. (2013) proposed dividing the whole training data set into N subsamples and the use of each part by different subpopulations at certain rounds of optimization. The term "rotation" denotes the exchange of training examples among subpopulations.

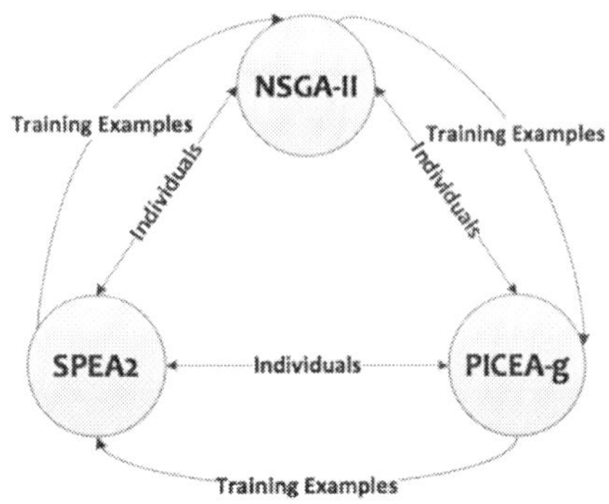

Figure 1. Data rotation in the cooperative EA [19].

The work of cooperative EAs is based on the usage of several populations, and therefore, we can easily incorporate data rotation into this type of EAs. In [8] the cooperative algorithm was implemented as a set of "islands": three different heuristics (SPEA2, NSGA-II and PICEA-g) were used as parallel working islands which exchanged the best individuals after a particular number of generations (a process called "migration"). Then they combined this "island" model with data rotation [21]: the sample was divided into three parts randomly and, in addition to the migration of individuals, they fulfilled the exchange of subsamples among "islands". At each third migration stage (after all parts of a sample were used by all populations) the training data was divided into three new parts (Figure 1). This instance selection approach was successfully applied in the speech-based emotion recognition problem while selecting informative features. In the experiment the databases contained from 480 to 4836 instances (four corpora in different languages were used). As a result, it was shown that this instance selection approach (data rotation) allowed the computational time spent on feature selection to be reduced (by a factor of 2.55 on average) without any detriment to the model accuracy. The most significant decrease in the amount of time was achieved for the largest databases.

3.2.2 Adaptive instance selection

In this subsection we will consider the adaptive IS approach applied to the Fuzzy Genetics Based Machine (GBML) learning method [22, 23]. The adaptive instance selection method was introduced in [24] and also described in [25]. We will give a brief description of this approach here.

The adaptive instance selection was developed for classification problems, although it can be extended for regression problems. This method does not require any data pre-processing, and the instances are selected into the training set based on how well they are classified.

In the first step, the subsample is created from the available set of instances by selecting random instances with uniform probabilities. Next, the classifier training procedure starts on this subsample. A counter U_i is set for every instance in the subsample, which signifies the number of times when an instance i was used and correctly classified. After the end of the adaptation period, which usually lasts tens or hundreds of iterations, the best current solution is used to update the counters. Only the counters of instances present in the subsample are updated, i.e. if an instance i was classified correctly, then $U_i = U_i + 1$, otherwise the counter is reset to $U_i = 1$.

In this manner, the instances that were classified correctly for several adaptation periods receive higher counter values. After the end of each adaptation period, a new subsample is created using the new counter values. The probability that an instance i will be chosen into the subsample is:

$$p_i = \frac{1/U_i}{\sum_{j=1,n} 1/U_j} \quad (1)$$

Increasing the counter value reduces the probability, while instances with counters that were recently reset to 1 receive higher probabilities.

The adaptive instance selection algorithm sets lower probabilities to instances which are easy to classify, filtering them out of the sample. At the same time, instances which are difficult to classify and unused instances get higher probabilities. This method follows two main strategies: the exploration of unknown areas of the feature space and the exploitation of information about classification accuracy to build better separation between classes.

In the case of this method being applied to a population-based algorithm, i.e. GAs, genetic programming or other EAs, the best solution on the subsample might not be the best on the whole sample, but the population may still contain a solution which has greater generalization abilities. This is why, after each adaptation period, all individuals in the population should be tested on the whole training sample, to find the best solution.

Another possibility, which arises from the adaptive instance selection method, is artificial balancing of the subsample, which becomes possible because the instances are selected with certain probabilities. Instance selection allows a balanced, if possible, or more balanced subsample to be created out of the original imbalanced sample by applying the presented selection algorithm separately to each class, and fixing the number of instances to be chosen for each class.

The described adaptive instance selection algorithm has been applied to the Hybrid Evolutionary Fuzzy Classification Algorithm, HEFCA, previously described in [24, 25]. The algorithm was tested on 9 datasets from the KEEL [26] and UCI [27] machine learning repositories.

The testing was performed with different sizes of the subsample, which varied from 5% to 30% with a 5% step, and different adaptation period length, 50, 100, 200 and 400 generations. The number of individuals in the population was 210, the number of generations was 50000, and the maximum number of fuzzy rules was set to 40. The balancing strategy, which creates balanced subsamples was applied, and the results were compared to the original HEFCA without the adaptive IS, as well as the HEFCA with a 10.5 times larger computational resource. A set of other similar fuzzy classification methods are also presented for comparison in Table 2. The values in the table are the relative error values in per cent.

Table 2. HEFCA with the adaptive IS compared to other methods.

Dataset	HEFCA _IS balanced	HEFCA original	HEFCA larger resource	Fuzzy GBML [14]	Parallel Fuzzy GBML [14]	GP-Coach [30]	IVFS-Coop [32]	IVFS-Amp [31]	FARC-HD [28]	Bio HEL [29]
Magic	15.08	15.73	15.87	15.42	**14.89**	20.18	19.82	20.82	15.49	-
Page-blocks	**3.25**	3.96	3.78	3.81	3.62	8.77	6.57	5.84	4.99	-
Penbased	3.81	7.46	5.02	**3.07**	3.30	17.80	17.00	21.73	3.96	6.00
Phoneme	16.88	16.48	**15.36**	15.43	15.96	-	-	-	17.86	-
Ring	**5.08**	5.82	5.52	6.73	5.25	-	12.57	16.89	5.92	-
Satimage	12.93	14.22	12.86	15.54	12.96	27.50	-	-	12.68	**11.60**
Segment	5.19	6.45	5.62	5.99	5.90	24.04	-	-	-	**2.90**
Texture	**4.45**	7.75	4.90	4.64	4.77	-	-	-	7.11	-
Twonorm	4.81	6.06	5.57	7.36	**3.39**	15.17	-	-	4.72	-

The classifier combined with the adaptive instance selection and balancing strategy appeared to be the best method for 3 out of 9 problems, overcoming not only the original algorithm, but also the algorithm with a 10.5 times larger computational resource on 7 problems out of 9. The acceleration ratio achieved with the adaptive instance selection method was from 3.09 to 21.5 times depending on the size of the subsample. The results which are not worse than those received by the original method without IS can already be obtained with a sample size of 10-15%, giving an average acceleration from 6 to 12 times.

3.3 Handling missing data

Today the methods for processing missing data are well developed, especially for the case of Missing Completely At Random (MCAR) [33]. Most methods focus on the imputation of missing data based on the other data available, which requires a significant number of computations. The reason for focusing on data imputation is that most data analysis methods cannot process missing values or else they show very poor results if all instances or features with missing values are deleted from the dataset.

In fact, even artificial neural networks use imputation to fill in the missing values: they may consider inputs for missing values as 0, but this also means imputation because for instances without missing values they have certain numerical values for these inputs, which means that they fill in gaps with 0.

For Big Data, imputation could be a challenging task due to the amount of data and the percentage of missing values. This is why in this section we shall consider a data analysis technique capable of building models on a dataset with missing values without any pre-processing.

The hybrid evolutionary fuzzy classification algorithm used in the previous section was modified to handle missing values. For the fuzzy logic rule base, the processing of missing data is similar to the processing of "Don't Care" conditions (DC), which are used to ignore the variable value in a particular rule. The difference for missing values is that even if in a rule q for a variable i the condition is not DC, during the fuzzy inference procedure, it is considered as a DC condition. One of the reasons why this method has not been used before is that it requires a specific EA and special operators for creating new rules from the sample instances and combining them in a crossover.

To see the difference in the classification quality depending on the number of missing values, 4 datasets from the UCI ML Repository [27] were chosen, namely: Page-blocks, Phoneme, Segment and Texture. For every dataset 5 tests were performed, one without missing values and 4 with missing values, generated randomly (uniformly) with probabilities 0.1, 0.2, 0.3 and 0.4. In Table 3, the relative error values in per cent are presented for the every test. The tests were performed with smaller computational resources: 100 individuals, 5000 generations and 40 rules maximum. The adaptive IS used 15% of the sample and 50 generations as an adaptation period.

Table 3. HEFCA with adaptive IS on missing data.

Missing percent	0	0.1	0.2	0.3	0.4
Page-blocks	5.12	6.92	8.19	8.52	8.60
Phoneme	18.30	20.91	22.65	23.82	24.49
Segment	7.45	14.35	18.72	23.03	28.74
Texture	7.27	15.25	24.20	36.68	46.74

The results differ very much depending on the dataset, for example for Page-blocks adding up to 40% of the missing values does not change the accuracy significantly, while for the Texture dataset the error rate follows the percentage of missing values.

5. Conclusions

The Big Data problem attracts many researchers nowadays. It is supposed to be one of the most promising directions in data analysis. Various attempts to process high-dimensional datasets in an effective way have been undertaken recently. However, it is still an open question.

In this paper we focus on EAs and their useful qualities to solve the Big Data problem. We believe that many researchers underestimate the effectiveness of EAs in Big Data. Therefore, we have described various applications of EAs in the area of data analysis and proved the usefulness of EAs in decreasing the time spent on analysis and the number of features in the dataset, and also showed the possibility to process the data without imputing missing values. The presented implications are supported by the experimental results. We have observed a number of benchmark and real-world Big Data problems that have been successfully solved with the use of EAs.

The constant development of EAs and evolutionary machine learning may provide researchers with an effective tool for Big Data. Therefore, in the near future we may also expect interest in this scientific area to grow.

Acknowledgements
This research is supported by the Russian Foundation for Basic Research, within project № 16-31-00349. The reported study was funded by RFBR and Government of Krasnoyarsk Territory according to the research project № 16-41-243036.

References

[1] Qiu J, Wu Q, Ding G, et al. 2016 EURASIP *J. Adv. Signal Process* **67**.
[2] Zhang Z 2015 Missing values in big data research: some basic skills *Annals of Translational Medicine* **3(21)** p. 323.
[3] Bagherzadeh-Khiabani F, Ramezankhani A, Azizi F, Hadaegh F, Steyerberg E W, Khalili D 2016 A tutorial on variable selection for clinical prediction models: feature selection methods in data mining could improve the results *J Clin Epidemiol* **71** pp 76–85
[4] Kohavi R, John G H 1997 Wrappers for feature subset selection *Artificial Intelligence* **97** pp 273–324
[5] Holland J H 1975 *Adaptation in natural and artificial systems*
[6] Goldberg D E 1989 *Genetic algorithms in search, optimization and machine learning*
[7] Khritonenko D, Semenkin E 2013 Distributed self-configuring evolutionary algorithms for artificial neural networks design *Vestnik SibSAU* **4(50)** pp 112–116
[8] Brester Ch, Semenkin E 2015 Cooperative Multiobjective Genetic Algorithm with Parallel Implementation *Advances in Swarm and Computational Intelligence, LNCS* **9140** pp 471–478
[9] Akhmedova S, Semenkin E 2013 Co-operation of biology related algorithms *IEEE Congress on Evolutionary Computation* pp 2207–2214
[10] Semenkin E, Semenkina M 2012 Self-configuring genetic algorithm with modified uniform crossover operator *Advances in Swarm Intelligence* pp 414–421
[11] Ryzhikov I, Semenkin E 2013 Evolutionary strategies algorithm based approaches for the linear dynamic system identification *Lecture Notes in Computer Science* **7824** pp 477–484
[12] Zaloga A, Yakimov I, Burakov S, Semenkin E, Akhmedova S, Semenkina M, Sopov E 2015 On the application of co-operative swarm optimization in the solution of crystal structures from x-ray diffraction data *Lecture Notes in Computer Science* **9140** pp 89–96
[13] Lin Q, Liu W, Peng H, Chen Y 2013 Efficient Genetic Algorithm for High-Dimensional Function Optimization *Proceedings of the 9th International Conference on Computational Intelligence and Security* pp 255–59
[14] Ishibuchi H, Mihara S, Nojima Y 2013 Parallel Distributed Hybrid Fuzzy GBML Models With Rule Set Migration and Training Data Rotation *IEEE Transactions on fuzzy systems* **21(2)**
[15] Mahajan R, Kaur G 2013 Neural Networks using Genetic Algorithms *International Journal of Computer Applications* **77(14)** pp 6–11
[16] Thounaojam D M, Khelchandra T, Singh K M, Roy S 2016 A Genetic Algorithm and Fuzzy Logic Approach for Video Shot Boundary Detection *Computational Intelligence and Neuroscience* p 11
[17] Venkatadri M, Srinivasa Rao K 2010 A multiobjective genetic algorithm for feature selection in data mining *International Journal of Computer Science and Information Technologies* **1(5)** pp 443–448
[18] Brester C, Kauhanen J, Tuomainen T P, Semenkin E, Kolehmainen M 2016 Comparison of two-criterion evolutionary filtering techniques in cardiovascular predictive modelling *Proceedings of the 13th International Conference on Informatics in Control, Automation and Robotics* **1** pp 140–145
[19] Sidorov M, Brester C, Semenkin E, Minker W 2014 Speaker State Recognition with Neural Network-based Classification and Self-adaptive Heuristic Feature Selection *Proceedings of the 11th International Conference on Informatics in Control, Automation and Robotics* **1** pp 699–703
[20] Cano J R, Herrera F, Lozano M 2007 Evolutionary Stratified Training Set Selection for Extracting Classification Rules with trade off Precision-Interpretability *Data & Knowledge Engineering archive* **60(1)** pp 90–108
[21] Brester C, Semenkin E, Sidorov M 2016 Multi-objective heuristic feature selection for speech-

based multilingual emotion recognition *Journal of Artificial Intelligence and Soft Computing Research* **6(4)** pp 243–253

[22] Fernandez A, Garcia S, Luengo J, Bernado-Mansilla E, Herrera F 2010 Genetics-Based Machine Learning for Rule Induction: State of the Art, Taxonomy, and Comparative Study *IEEE Transactions on Evolutionary Computation* 14(6) pp 913–941

[23] Booker L B, Goldberg D E, Holland J H 1989 Classifier systems and genetic algorithms *Artif. Intell.* **40(1–3)** pp 235–282

[24] Stanovov V, Semenkin E, Semenkina O 2015 Self-configuring hybrid evolutionary algorithm for fuzzy classification with active learning *IEEE Congress on evolutionary computation*

[25] Stanovov V, Semenkin E, Semenkina O 2015 Instance Selection Approach for Self-configuring Hybrid Fuzzy Evolutionary Algorithm for Imbalanced Datasets *Sixth International Conference on Swarm Intelligence* **9140(I)** pp 451–459

[26] Alcalá-Fdez J, Sánchez L, Garcia S, Jesus M J, Ventura S, Garrell J M, Otero J, Romero C, Bacardit J, Rivas V M, Fernández J C, Herrera F 2009 KEEL: A software tool to assess evolutionary algorithms for data mining problems *Soft Comput.* **13(3)** pp 307–318

[27] Asuncion A, Newman D 2007 UCI machine learning repository *University of California, Irvine, School of Information and Computer Sciences* http://www.ics.uci.edu/~mlearn/MLRepository.html.

[28] Alcala-Fdez J, Alcala R, Herrera F 2011 A fuzzy association rulebased classification model for high-dimensional problems with genetic rule selection and lateral tuning *IEEE Trans. Fuzzy Syst.* **19(5)** pp 857–872

[29] Bacardit J, Burke E K, Krasnogor N 2009 Improving the scalability of rule-based evolutionary learning *J. Memetic Comput.* **1(1)** pp 55–67

[30] Berlanga F J, Rivera A J, del Jesus M J, Herrera F 2010 GP-COACH: Genetic programming-based learning of compact and accurate fuzzy rulebased classification systems for high-dimensional problems *Inf. Sci.* **180(8)** pp 1183–1200

[31] Sanz J A, Fernandez A, Bustince H, Herrera F 2010 Improving the performance of fuzzy rule-based classification systems with interval-valued fuzzy sets and genetic amplitude tuning. *Inf. Sci.* **180(19)** pp 3674–3685

[32] Sanz J, Fernandez A, Bustince H, Herrera F 2011 A genetic tuning to improve the performance of fuzzy rule-based classification systems with interval-valued fuzzy sets: Degree of ignorance and lateral position *Int. J. Approx. Reason.* 52(6) pp 751–766

[33] Schafer J L, Graham J W 2002 Missing Data: Our View of the State of the Art. *Psychological Methods* **7(2)** pp 147–177

IOP Conf. Series: Materials Science and Engineering **173** (2017) 012021 doi:10.1088/1757-899X/173/1/012021

Measuring the global information society – explaining digital inequality by economic level and education standard

H Ünver

Ulm University, Institute Databases/Artificial Intelligence & Research Institute for Applied Knowledge Processing/n (FAW/n), Lise-Meitner-Str. 9, 89081 Ulm/Germany

E-mail: halit.uenver@uni-ulm.de

Abstract

A main focus of this research paper is to investigate on the explanation of the *'digital inequality'* or *'digital divide'* by economic level and education standard of about 150 countries worldwide. Inequality regarding GDP per capita, literacy and the so-called UN Education Index seem to be important factors affecting ICT usage, in particular Internet penetration, mobile phone usage and also mobile Internet services. Empirical methods and (multivariate) regression analysis with linear and non-linear functions are useful methods to measure some crucial factors of a country or culture towards becoming information and knowledge based society. Overall, the study concludes that the convergence regarding ICT usage proceeds worldwide faster than the convergence in terms of economic wealth and education in general. The results based on a large data analysis show that the digital divide is declining over more than a decade between 2000 and 2013, since more people worldwide use mobile phones and the Internet. But a high digital inequality explained to a significant extent by the functional relation between technology penetration rates, education level and average income still exists. Furthermore it supports the actions of countries at UN/G20/OECD level for providing ICT access to all people for a more balanced world in context of sustainable development by postulating that policymakers need to promote comprehensive education worldwide by means of using ICT.

1. Introduction

Indeed, one of the most important terms in the field of information society is 'digital divide'. Historically, it was high on the agenda of the European Union [26], a topic of the World Summit on the Information Society (WSIS) took place in Geneva in 2003 [33] and Tunis in 2005 [34] as well as at the World Summit on Sustainable Development (WSSD) in Johannesburg [32]. Nowadays, a lot of issues related to 'digital divide' are discussed at the UN Internet Governance Forum [11, 12, 13] by using the term 'digital inequality' as a more adequate and precise description of the technological situation of a country.

1.1 Terms

As we know, good infrastructure (traffic infrastructure, technological infrastructure, housing infrastructure, energy infrastructure etc.) is one of the most important components of wealth and economic power [27]. The economic strength of a country can be quantified by the (financial) value of all goods and services produced within a given time period in a country [20]. GDP per capita (*GDPpC*) is one of the main indicators of economic analysis, both in spatial and in temporal international comparisons. For international analysis, the purchasing power parity (PPP) is used. The social balance or the distribution of income is, in addition to GDP per capita, an important parameter for understanding the social situation in a country. It is empirically and theoretically well known that neither too much nor too little inequality is good for a country [17] [35].

A country needs to invest in education and research to maintain economic welfare. In a global perspective, it is still important to observe the number of people being able to read and write. The adult literacy rate is the percentage of the population age 15 and above who can read and write a short, simple statement concerning their everyday life [29]. Generally, literacy also encompasses numeracy, the ability to make simple arithmetic calculations like adding numbers. The UN Education Index is a more complex measure to compare between different educations standards in countries. It is a combination of expected and average school time of individuals in a country [465]. Statistically possible repetitions of classes are included. From a mathematical viewpoint, the expected years of schooling is of a predictive character.

Innovation and technology promote economy and education, enabling convenience in communication, which is the basic operation of a society with enormous effects on the performance of a country [19]. One index for the technological development of a country, in terms of technological deployment and communication, is to measure the percentage of population using any (digital) technology, while ignoring the type of access (private/ shared), connection (wired/wireless) or location (household, school, office, cafe). For example, the Internet penetration rate (IPR) is the estimated number of Internet users out of a total population. This includes those using the Internet from any device (including mobile phones) over the last 12 months, measured through household surveys. If household surveys are not available, an estimate can be derived based on the number of Internet subscriptions [15].

The mobile phone connection is becoming more and more important due to the fact of technological convergence which provides the use of the Internet from any mobile device. The mobile phone penetration rate (MPR) includes both the number of mobile phone subscriptions and the number of active prepaid accounts that have been used within the last three months. It does not include the connections via data cards or USB - modems, connections of public mobile data services, private trunked radio, telepoint, radio paging and telemetry services [14].

1.2 Data and Methods

The World Bank provides data regarding the economic development level (GDP, GDPpC, income distribution) of countries [31]. Technology penetration rates (MPR, IPR) are available (partly free of charge) on the International Telecommunication Union website [16], where also some other ICT key indicators can be downloaded. Data for measuring the education level of a country and its population can be downloaded from the statistical division of the UNESCO [30] and UN Development Programme websites [465]. On those websites data is available for national adult/youth literacy rates, enrolment ratios and education index by international standard classification of education level (ISCED). This study focuses on adult literacy rate and education index. Due to the lack of data regarding literacy rate for many developed and developing countries, polynomial interpolation is used to have enough data for pair-by-pair comparison concerning technology penetration rates.

The data management process in this work is essentially based on the authoritative architecture of data warehouse systems [4]. Figure 1 shows the ETL process (*extract, transformation, load*), which is applied for the data management. The extraction component (1) has the function to procure data from the different data sources. Among other things, there are questions regarding the exchange of data, like formats to be considered. Furthermore, certain data subsets need to be extracted from the sources, why separation is of importance. A decisive factor here is also the question of the time of extraction. This is done as needed, whereas a periodic, event-based or immediate extraction is generally possible. After extraction, the needed data for analysis is stored in the work area. Step (2a) in *figure 1* is the so-called *transformation* component. The data obtained from different sources must be standardized in a certain way in order to make a valid processing possible. The unifying and processing of the data is carried out in this component. The *loading* up component in step (3) passes on the prepared data from the workspace to the corresponding database (Access DB). Furthermore, this component is responsible for the historicization of the data in a certain regard. This means old data is not simply deleted or overwritten but

IOP Conf. Series: Materials Science and Engineering **173** (2017) 012021 doi:10.1088/1757-899X/173/1/012021

rather provided with a timestamp and stored. A change at a data set therefore has the consequence that two data sets exist with a different timestamp. In step (4a) the corresponding data are brought together for analysis purposes from different sources. For this use the *structured query language* (SQL) is used. Step 5 serves now to fetch the data for regression analysis. The corresponding data are exported from the database program. Afterwards the data are loaded in Step (6a) into the statistics program *SPSS* or in step (6b) into the numeric program *SciLab*. It comes in step (7a) or (7b) to the *correlation and regression analysis* by using the numeric program SciLab or SPSS, in which some queries (e.g. the categorization of the countries after economy performance) also take place in step (7c) directly at the Access DB.

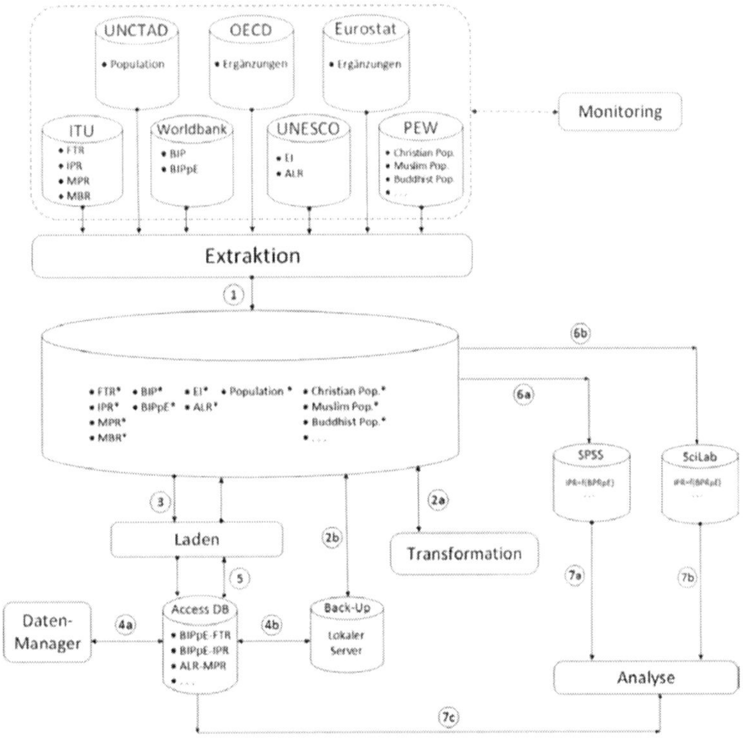

Figure 1. Adapted ETL-process.

Correlation quantifies the degree to which two variables are related, and this regards a linear form of mutual dependency [8] [23]. Regression finds the best line or function (method of least squares), which predict the dependent variable from the independent variable [117] [22]. The best fit function from the class of (test) functions is determined by using the least squares method. For this purpose, one minimizes the sum of the mean square of errors (*MSE*), wherein the coefficient of determination (*R2*) is a quality measure for the adaption of individual functions. The function which fits most to the data by means of method of least squares is then finally used for the relation between the dependent (e.g. mobile phone usage) and independent variable (e.g. literacy) in a certain year. A lot of known functions in the fields of economy, natural, physical and social science are used to identify the relation between technology penetration and different factors in a country. Interdisciplinary reasons for using some known functions are as follows:

1. Linear functions are easy and one of the first approach, if humans try to understand real-life phenomena with given data.
2. Exponential functions are often used to represent growth and decay, like population growth or depreciations

3. Logarithmic functions are commonplace in scientific formulae, and in measurements of the complexity of algorithms and in many other different applications like in the measurement of earthquakes and sound
4. Logistic functions are used to model real-life quantities whose growth levels off because the rate of growth changes, from an increasing growth rate to a decreasing growth rate.
5. Planck's law describes originally the electromagnetic radiation emitted by a black body in thermal equilibrium at a definite temperature. Anyway, in general, it describes an initial increase of the dependent variable at higher levels of the independent variable, and then a decrease of the dependent variable at higher levels of the independent variable.

Function	**Formular**
1. Linear	$y_i = a + b \cdot x_i + \varepsilon_i$
2. Exponential	$y_i = a + b \cdot e^{(c \cdot x_i - d)} + \varepsilon_i$
3. Logarithmic	$y_i = a + b \cdot \log(x_i - c) + \varepsilon_i$
4. Logistic	$y_i = \dfrac{a}{b + c \cdot e^{(d \cdot x_i)}} + \varepsilon_i$
5. Planck's Law	$y_i = \dfrac{a \cdot x_i^b}{c \cdot e^{(d \cdot x_i)} - 1} + \varepsilon_i$

Only bivariate correlations between two variables can be examined with a correlation coefficient and simple linear regression. If one would like to investigate on the relationship between several variables, there is the so-called multivariate and multiple regression analysis. There is a dependent variable Z (e.g. IPR) and two independent variables (e.g. GDPpC and Education Index). Furthermore, it is assumed that a simple regression analysis for the relationship between the internet user rate and the GDPpC provides a non-linear function *f(x)*. The function *g(y)* corresponds to the non-linear relationship between the IPR and the Education Index. The corresponding multiple regression model with the corresponding weights A and B is:

$$\boxed{Z = A \cdot f(x) + B \cdot g(y)}$$

The underlying functions *f(x)* and *g(y)* are of e.g. 3-parametric logarithmic type:

$$z.B: f(x) = a + b \cdot \log(x - c)$$
$$g(y) = d + e \cdot \log(y - h)$$
$$\Rightarrow Z = A \cdot (a + b \cdot \log(x \cdot c)) + B \cdot (d + e \cdot \log(y - h))$$

Now, based on the method of least squares it is to minimize the function $\varphi(A, B)$:

$$\sum_{i=1}^{n} (z_i - A \cdot f(x_i) - B \cdot g(y_i))^2 = \varphi(A, B) \to min$$

For this we calculate the partial derivatives of $\varphi(A, B)$ below. These equations are set to zero to find the appropriate minimum.

IOP Conf. Series: Materials Science and Engineering **173** (2017) 012021 doi:10.1088/1757-899X/173/1/012021

$$\frac{\partial \varphi}{\partial A} = 2 \cdot \sum_{i=1}^{n} (z_i - A \cdot f(x_i) - B \cdot g(y_i)) \cdot f(x_i) \stackrel{!}{=} 0$$

$$\Rightarrow \sum_{i=1}^{n} z_i \cdot f(x_i) = \left(\sum_{i=1}^{n} f^2(x_i), \sum_{i=1}^{n} g(y_i) \cdot f(x_i) \right) \cdot \begin{pmatrix} A \\ B \end{pmatrix}$$

$$\frac{\partial \varphi}{\partial B} \stackrel{!}{=} 0$$

$$\Rightarrow \sum_{i=1}^{n} z_i \cdot g(y_i) = \left(\sum_{i=1}^{n} g(y_i) \cdot f(x_i), \sum_{i=1}^{n} g^2(y_i) \right) \cdot \begin{pmatrix} A \\ B \end{pmatrix}$$

The result is a system of equations, which delivers the regression weights A and B after its resolution according to the following procedure.

$$\begin{pmatrix} \sum_{i=1}^{n} x_i^{*2} & \sum_{i=1}^{n} x_i^* y_i^* \\ \sum_{i=1}^{n} x_i^* y_i^* & \sum_{i=1}^{n} y_i^{*2} \end{pmatrix} \cdot \begin{pmatrix} A \\ B \end{pmatrix} = \underbrace{\begin{pmatrix} \sum f^2(x_i) & \sum f(x_i) g(y_i) \\ \sum f(x_i) g(y_i) & \sum g^2(y_i) \end{pmatrix}}_{=N} \cdot \begin{pmatrix} A \\ B \end{pmatrix}$$

$$= \underbrace{\begin{pmatrix} \sum z_i f(x_i) \\ \sum z_i g(y_i) \end{pmatrix}}_{=\beta} = \begin{pmatrix} \sum z_i x_i^* \\ \sum z_i y_i^* \end{pmatrix}$$

$$\boxed{\Rightarrow \begin{pmatrix} A \\ B \end{pmatrix} = \beta N^{-1}}$$

In doing so, we can find out the different importance (weight) of economic level and education standard for the deployment of the Internet in a country. Certainly there is also a relation between the economic power and the education level of a country. Based on the findings of this work and the different importance of economy and education for the deployment of ICT, in particular mobile phone usage and Internet penetration, it was able to suggest or advice important decision makers and international organisations, such as at the Internet Governance Forum in 2014 (Turkey) and 2015 (Brazil) or at UN/G20/OECD level.

3. Literature Overview and Distinction

In the 1990s, the term *'digital divide'* emerged to describe technology haves and have-nots. Current research regarding digital divide has a descriptive character. Those studies emphasize the digital divide by using demographical, economical or educational data, in general at an individual level of technology usage or digital skills of people in a country [2]. This study investigates the digital divide more from the perspective of *'digital inequality'*. There are mainly four successive kinds of access in the appropriation of digital technology [7]. These kinds of access are a) *motivation*, b) *physical* and *material access*, c) *digital skills* and d) *digital usage*. Here, the focus is on access types b) and c) by searching for functional relations between the economic power, education level and technology penetration rates of countries. It is assumed that the digital skills of a society are higher, if the education level of this society is higher. Physical and material access corresponds to mobile phone usage rate (MPR) and Internet penetration rate (IPR).

A number of empirical studies have been done on the topic of digital divide. A few of them highlighted that income level [1] [2], income distribution [9] [36], education level [5] [10] [21], size of population [28] and urbanization [3] have essential correlation with Internet penetration levels of countries. Andres et al. [2] stated that low-income countries have a steeper Internet diffusion curve than that of high-income countries. Although this result is rational, because low-income countries can *'leapfrog'* technological developments, it has to be mentioned, that the splitting to only two categories of income levels is questionable. Zhang [36] found out a positive contribution of GDP per capita (PPP, current int.$) to IPR and a negative influence of income distribution measured by the Gini-Index. Here it should be noted; a higher average income corresponds to a more equal income distribution in general [18] and that's not discussed anyway in the results of Zhang. Furthermore he did not explain the relation between GDP per capita and Internet penetration in form of a detailed function. Kiiski et. al. [21] showed that the average years of schooling is considerably a positive factor for the Internet hosts per capita in a country. Even so another study find out that the degree to which the difference in Internet rates depends upon education level is surprisingly small [6]. This is in contrast to our findings.

In the past, studies about using technology acceptance explain, how attitudes determine Internet penetration [3]. Such studies show, that some of the faster rates of growth in Internet use have been among individuals who are older, less educated, of minority status or with lower incomes. A similar research work to this study examines the relation of the Internet penetration rate with the human development level over the decade from 2000 through 2010 [25]. These results support the argument that a digital divide exists between developed and developing countries. The main outcomes of this study are that there is a positive correlation between human development level and Internet penetration rate and that the correlation has become slightly stronger from 2000 through 2010.

However, Internet Usage rates associated with these demographic groups are lower than that of the general population [24]. Another study points out, that more educated people use the Internet more actively and their use is more information oriented, whereas the less educated seem to be interested particularly in the entertainment functions of the Internet. If we look to the digital divide with respect to Internet penetration, there is still an inequality between developed and developing countries [25].

However, there seems to be no related scientific work analyzing the empirical relation between education index and Internet penetration, especially over the decade from 2000 through 2012. Therefore, this issue is tackled in this study. Furthermore, most of the related work describes the relation between economical development, education level and Internet penetration with linear regression functions and with logarithmic or exponential functions. In this study, more non-linear regression models like the logistic function or functions according to Planck's law are used to describe the relation between the different parameters in detail. Specifically the multivariate regression analysis for the relation between the ICT penetration rates as the dependent and education level and income situation as the different independent variables (described by non-linear regression functions), is a essential contribution to the scientific field related to digital divide or digital inequality.

4. Empirical Results and Functional Analysis

4.1 Relation between Economic Power and ICT usage

4.1.1 GDPpC and Mobile telephony (MPR)

Table 1 shows the number N of analyzed countries, the minimum (min), maximum (max), span (sp), mean (av), median (md) and standard deviation (sta) for the GDP per capita (GDPpC) and mobile penetration rate (MPR) in a worldwide perspective in the years 2000 and 2013.

Table 1. Descriptive statistics of mobile penetration rate (MPR) and GDP per capita (GDPpC)in year 2000 and 2013.

	2000 (N=177)						2013 (N=178)					
	sp	min	max	av	md	sta	sp	min	max	av	md	sta
GDPpc	90636	270	91011	10955	5702	14434	131154	603	131757	17898	11764	19760
MPR	79.68	0.01	79.69	15.89	4.46	22.63	233.07	5.61	238.68	101.48	106.09	38.77

If one describes the worldwide digital inequality in the context of *MPR* and the average *GDPpC* with the help of *sp* and *sta* as easy dispersion metrics, it turns out that the inequality in the use of mobile phones has increased between 2000 and 2013, whereas also the inequality regarding the *GDPpC* increased. Looking closer at the data for the year 2000 in *figure 1* one can recognize that states with a low *GDPpC* have low *MPR*. The greater the *GDPpC* values become, the greater also the *MPR* values become. Moving into the direction of higher *GDPpC* the *MPR* first reaches a maximum and afterwards drops again. In this initial situation the *Planck's function* (black) returns a very good adjustment (within the considered class of functions and according to the method of the squares) of the relationship between the *GDPpC* and the *MPR* in 2000.

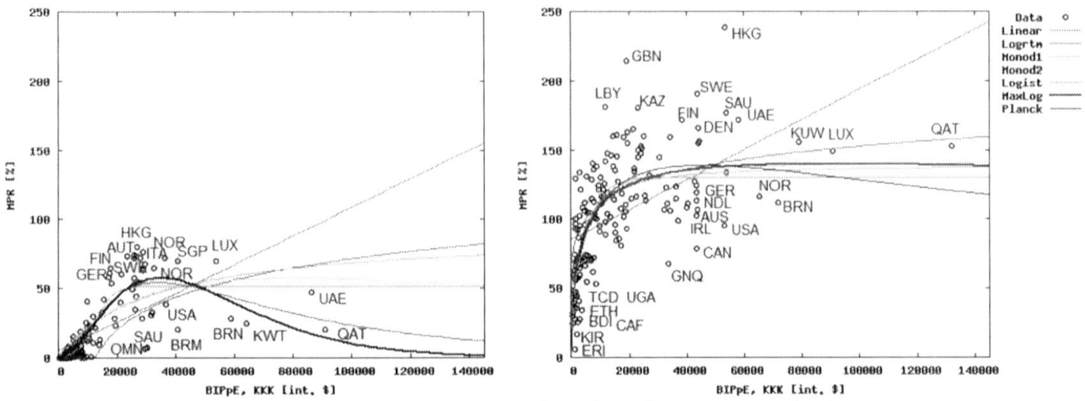

Figure 1. mobile phone penetration rate (MPR) as a function of GDP per Capita (GDPpC), year 2000 (left) und 2013 (right).

More than a decade later, a first observation is that the data concerning 2013 reach significantly higher values for MPR than the data in 2000. Furthermore the MPR is over 100% in many countries in 2013. This means that there is averaged more than one mobile device per person in a country. Overall, all states have increased their MPR, so that the catch-up in the mobile phone usage is very clear worldwide. In *figure 1* is also a list of selected functions such as *linear* (red), *logrtm* (blue), *monod1* (pink), *monod2* (gold), *logist* (green), *maxLog* (brown) und *Planck* (black). The 4-parameter function in equation 5 (see section 1.2) with the parameters a, b, c, and d (expression of the Planck function) delivers the best fit for 2000. After verification of the particular functions for each year between 2000 and 2013, it turns out that only in 2013 the *maxLog* function delivers a better adaptation than the *Planck* function. *Table 2* shows the different fitting functions and the associated values for the sum of squared errors (*MSE* or *MSE/N*) and the coefficient of determination (*R2*).

Table 2. Sum of mean square of errors (MSE) and coefficient of determination (R2) for regression MPR = f(BIPpK)

	2000 (N=177)			2013 (N=178)		
	MSE	MSE/N	R^2	MSE	MSE/N	R^2
Linear	48565	274.4	0.439	209077	1174.6	0.281
Logartm	32312	182.6	0.627	147529	828.8	0.439
Monod1	32674	184.6	0.622	148515	111.4	0.489
Monod2	30556	172.6	0.647	148515	105.3	0.489
Logist	21324	120.5	0.754	150714	102.8	0.482
MaxLog	21471	121.3	0.752	**148380**	**105.2**	**0.491**
Planck	**20788**	**117.4**	**0.760**	151586	101.2	0.479

In *figure 2* one can study the development of the best fit functions for the relationship between *GDPpC* and *MPR* from 2000 (blue) till 2013 (black). The result shows that the MPR has increased in all areas of *GDPpC* from year to year and has taken a similarly good development over all *GDPpC* ranges in more than a decade. This result is not only an artefact of the adaptation in the optimisation process; rather the investigation of the database laid for this optimisation delivers the same result. Based on *figure 2* one can recognise, that the *MPR* has raised over all *GDPpC* areas i.e. accordingly in all countries. This can be seen on the one hand in the higher course of the curve for 2013 (black) as the curve course for 2000 (blue). On the other hand, the data points for the year 2000 (blue) are not or barely overlaid with those data points for 2013. In relative terms, the shift of centre of data is stronger upwards (in the direction of *MPR*) than to the right (in the direction of *GDP*).

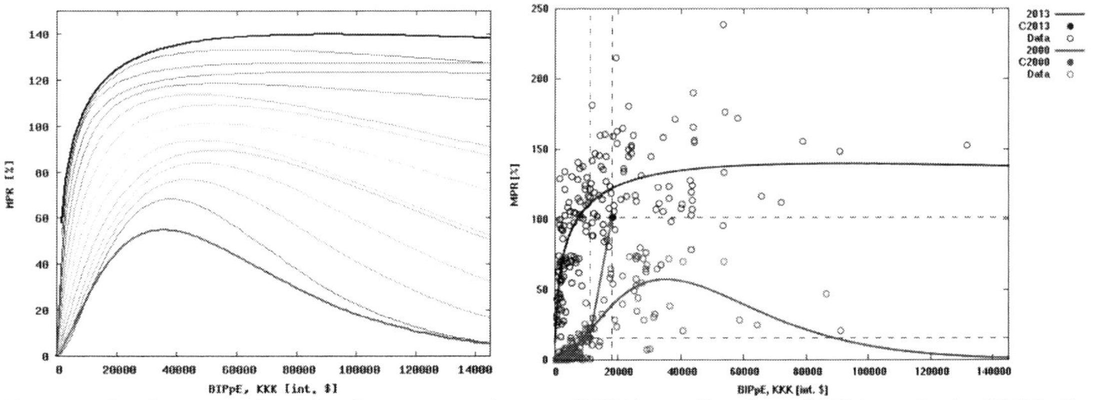

Figure 2. development of mobile phone penetration rate (MPR) as a function of GDP per Capita (GDPpC) between 2000 and 2013 (left); data points, best fit function, centre (right).

If one performs a growth analysis over the whole time from 2000 to 2013 for the *GDPpC* and the *MPR* in the suitable *GDPpC* categories A, B, C, D and E, where category A stands for the richest and category E for the poorest countries, one is able to examine, to what extent the *GDP* growth and the *MPR* growth are connected within and between these categories. The bar chart in figure 3 shows on the x-axis, the respective *GDPpC* category and on the y-axis the growth of *GDPpC* (blue) and *MPR* (burgundy). Hence the highest growth of *GDPpC* and MPR took place in category C (medium *GDPpC*). Both growth rates lie in category C with about 112 %. In the categoryA (highest *GDPpC*) and B (high *GDPpC*) is a similarly high *MPR* growth of approx. 94% and 94. 5%, whereby the *GDPpC* growth lies in category A with about 62 % and in category B with approx. 100 %. The *GDPpC* growth is in category D (low *BIPpE*) 109 % with a MPR-growth of 90 %. In category E (lowest *GDP*) was held the slightest *MPR* growth, namely by about 65 % with a *GDPpC* growth of nearly 77 %. The result indicates that the *GDP* and the *MPR* growth have a strong coherence (high correlation), especially if one does not consider the richest countries (category A). This does not surprise if one minds that the richest states can not show such a high performance as economically weaker states, because they start from a clearly higher level of *GDPpC*.

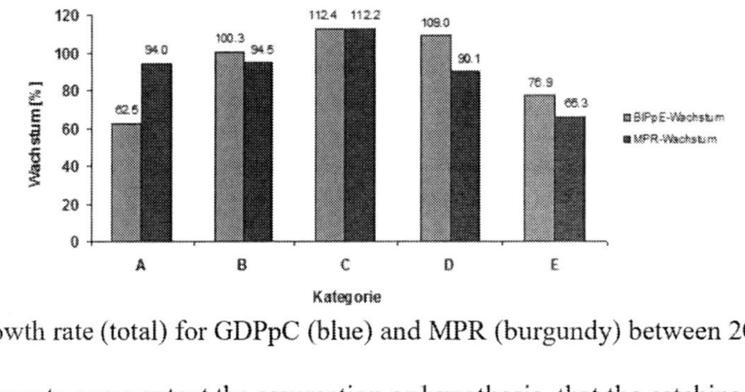

Figure 3. Growth rate (total) for GDPpC (blue) and MPR (burgundy) between 2000 and 2013.

This study confirms to some extent the assumption or hypothesis, that the catching up generally in the use of ICT and in particular in the mobile phone usage runs faster worldwide than catching up with respect to prosperity in general. The size and/or the growth/shrinking of the population have no significant influence on the *MPR*. A hypothesis represented in this work is that we will experience a convergence in the ICT use. This thesis is confirmed so far for the mobile phone usage, as that in 2013 the worldwide average *MPR* has reached about 100 % and that the *MPR* growth between 2000 and 2013 over all *GDPpC* areas took place at a similarly high level, also due to the fact that in many countries there is more than one mobile phone connection and that the availability of prepaid systems has favoured the mobile phone usage in the poorer regions of the world ('mobile miracle'). In this context today one can describe the situation in the following way: The world population calls up mobile and has carried out the jump of the local-engaged communication (fixed network) to the person engaged communication (mobile phone).

4.1.2 GDPpC and Internet Usage (IPR)

Aberrantly to the considerations in the previous section, the question gets answered, how big the amount of a country is in the world gross domestic product (world GDP) and which amount a country has in the worldwide Internet users (world IPR). The population size is factored out in the considerations of the economic achievement and the Internet usage. *Figure 4* shows on x-axis the economic achievement of a country as a portion in the world *GDP* and on the y- axis the Internet usage extent of a country as a portion in the world IPR in the year 2000 (left) and 2013 (right).

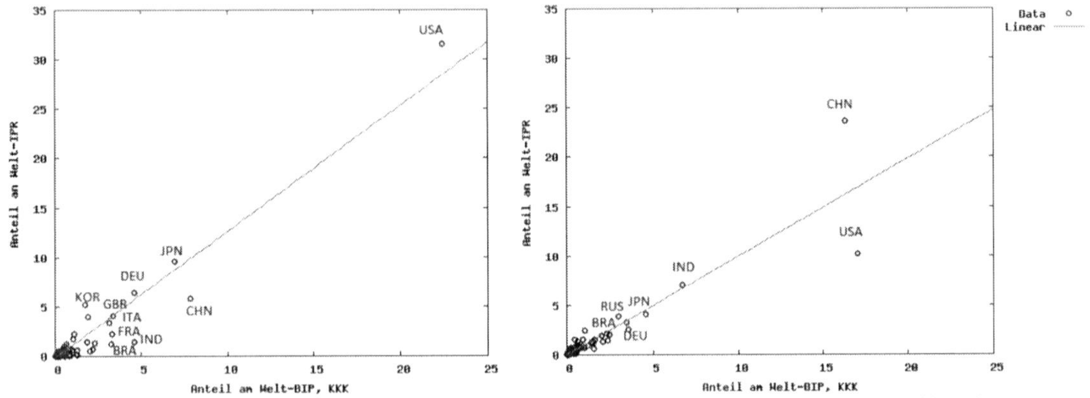

Figure 4. Proportion of world IPR as a function of the proportion of world GDP, 2000 (left) and 2013 (right).

The United States has accounted for about 23 % of world economic output and about 30 % of Internet users (worldwide). If one divides the states into one group above and one below the regression line, the USA, Japan, Britain and Korea are located above and states like China, India, Brazil, Russia and the so-called BRIC states, but also France and Italy below the line. As *figure 4* shows, there is a high correlation in the year 2000. Furthermore, it is in such a way, that the relations after 13-year development in 2013 still can be described well by a linear model, i.e. that still a high correlation is given. This easy analysis points to a narrow respect between the economic power of a country and the Internet use in this country. This is also valid, if one ignores the population number with regard to the economic power. In a per capita consideration the differences between the richer and poorer countries are still much greater. This is also due to the large population of countries such as China and India. It is interesting in this context that in 2013 all the BRIC countries are above the linear regression model and the USA, Japan and Germany lie under it. This development is a consequence of that the BRIC states, which have a relatively large population, increased their IPR between 2000 and 2013 significantly. Since the alignment of communication is still faster than the economic performance, the size of population of BRIC states leads to a correspondingly large Internet population.

Table 3 shows for the *GDPpC* and the *IPR* the values of the descriptive statistics. The *IPR* has in 2000 a maximum of 52% (Norway) and this value increases till 2013 to approximately 96.5% (Iceland). The *IPR* minimum value of 0.01% (Dem. Rep. of Congo) in 2000 increases slightly and is almost 1% (Eritrea) in 2013. There are again the African countries, as in the *MPR*, which have the lowest values for the *IPR*. The average increases over the same period from nearly 8% to about 41.6%. It is interesting, that the median of 39.2% in 2013 is nearly at the same level as the average value, although the median of nearly

2% in 2000 is on a four times lower level as the average. This is a consequence of the significantly higher balance of the proportions till 2013, as further noted in another context above.

Table 3. Descriptive statistics of Internet penetration rate (IPR) and GDP per capita (GDPpC)in year 2000 and 2013.

	2000 (N=177)						**2013 (N=181)**					
	sp	min	max	av	md	sta	sp	min	max	av	md	sta
GDPpC	90741	270	91011	10919	5732	14351	141961	603	142564	18484	11653	21728
IPR	51.99	0.01	52.00	7.97	1.98	12.91	95.65	0.90	96.54	41.57	39.20	29.02

We consider further in detail the empirical relationship between *GDPpC* and *IPR* per country. Here are significantly greater differences to be expected, as above in considering the economic output (*GDP*) of each state. The relationship is no longer linear. In *figure 4.12* (left), it is evident that states with a lower *GDPpC* have a lower *IPR*. The greater the value of the *GDPpC* becomes, the greater also becomes the IPR, indeed, for quite big values of the *GDPpC* it falls again. This is not surprising in this respect, as nearly every Internet access in year 2000 occurred via a fixed network connection by dial-up (modem). The mobile Internet access hardly existed worldwide. So the fixed network was a necessary condition for the Internet connection, which was (only) developed in richer countries. The Scandinavian countries Denmark, Sweden, Norway and Finland have worldwide one of the highest IPR in 2000. But also Australia, Germany, Great Britain, Japan, Canada, Korea, Switzerland, Singapore and the USA have high values for the *IPR*. There are countries such as Bahrain, Brunei, Kuwait, Oman, Saudi Arabia, and the United Arab Emirates, but also surprisingly Luxembourg, which have a high *GDPpC* and in relation to this huge *GDPpC* value a relatively low *IPR* value compared to the remaining rich states. These are primarily the oil states that enforce a form of Planck's function for the relationship between *GDPpC* and *IPR*. Reason is found in the in parts of nature 'borrowed' prosperity, mainly resulting from the sale of oil and gas, not from an internationally competitive, diversified productive economy, which itself would require massive dissemination of technology, Internet usage and innovation in this area.

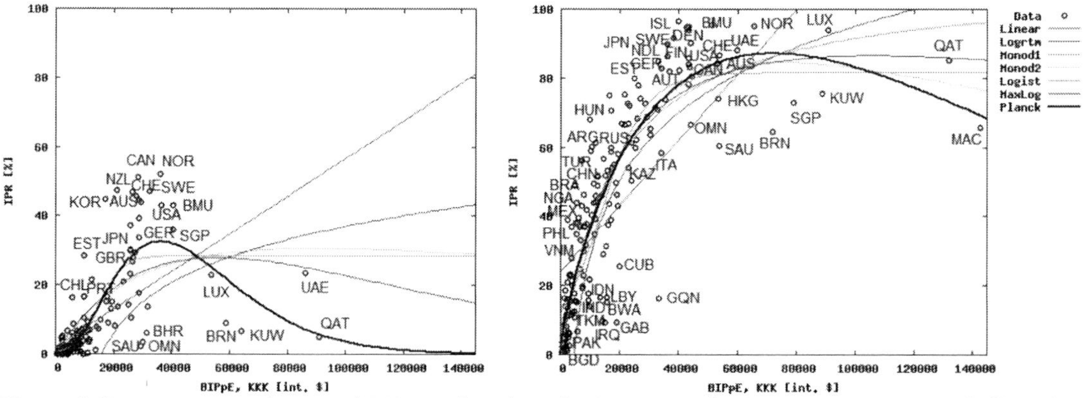

Figure 5. Internet penetration rate (IPR) as a function of GDP per Capita (GDPpC), year 2000 (left) und 2013 (right).

The values for the *IPR* have in 2013 increased significantly compared to 2000. The statement from 2000 that states with low *GDPpC* have a rather low *IPR* is also valid for 2013, in spite of occurred catching-up processes. Furthermore the obvious correlation is valid that higher *GDPpC* values cause higher *IPR* values. Table 4.12 lists the results of the regression analysis with the values for the sum of squared errors (*MSE* or *MSE/N*) and the coefficient of determination (R2) for all functions tested for the years 2000 and 2013. As already mentioned, the *Planck* function fits best, followed by the functions *maxlog* and *logist* with a similarly good adjustment in 2000. In 2013, the *Planck* function again has the best fit, but also the rest of the tested functions, except for the linear, describe the relationship similarly well. This points to the fact that the functions with an inhibition lose relatively in strength and functions with saturation win relatively in strength for this relationship.

Table 4. Sum of mean square of errors (MSE) and coefficient of determination (R2) for regression MPR = f(BIPpK).

	2000 (N=177)			2013 (N=178)		
	MSE	MSE/N	R^2	MSE	MSE/N	R^2
Linear	18440	104.2	0.372	74357	410.8	0.509
Logartm	14177	80.10	0.517	35274	194.9	0.767
Monod1	14292	80.75	0.513	33760	186.5	0.777
Monod2	13690	77.34	0.534	32466	179.4	0.786
Logist	10946	61.84	0.627	35212	194.5	0.768
MaxLog	10261	57.97	0.651	33032	182.5	0.782
Planck	**9760**	**55.14**	**0.668**	**32011**	**176.9**	**0.789**

Figure 6 shows the development of the best fit function for 2000 (blue) to 2013 (black). For all years between 2000 and 2013 the Planck function achieves the best fit. The results for the fit functions show that the *IPR* has expanded similar for all *GDPpC* areas between 2000 and 2007. The (Planck) fitting functions for the years 2008 and 2009 run in this respect slightly differently, because states with the biggest *GDPpC* have lower *IPR* values, than during the years before. However, this is not appropriate as the data show. So this is an artefact of the regression optimization. This results from the fact that the countries in the medium *GDPpC* area between 2008 and 2009 tend to have a larger (relative) increase in their *IPR* than the countries in the top *GDPpC* area. The reasons for this are to be found in the beginning of the mobile Internet usage in countries with a medium *GDPpC*. One can observe the same phenomenon in 2013.

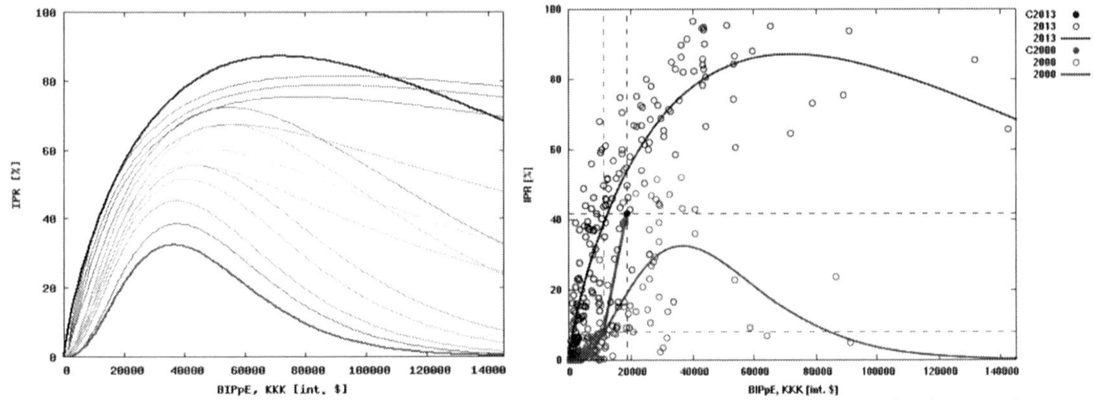

Figure 6. Development of Internet penetration rate (IPR) as a function of GDP per Capita (GDPpC) between 2000 and 2013 (left); data points, best fit function, centre (right)

In *figure 6* is to be recognised once more that the *IPR* increased over all *GDPpC* areas in all countries. The curve for 2000 (blue) runs clearly below the curve for 2013 (black) and there is no intersection of the curves. Furthermore, there is hardly a superposition of the two point clouds. The shift of focus is more upwards (towards *IPR*) than to the right (towards *GDPpC*) relatively regarded.

If one carries out a growth analysis from 2000 to 2013 for the *GDP* and the *IPR* in the corresponding *GDPpC* categories A, B, C, D and E (division, see appendix A.1) the relationship between *GDP* growth and *IPR* growth can be specified. Figure 4.15 shows on the x-axis the respective *GDPpC* category and on the y-axis the growth rates for the *GDPpC* (blue) and the *IPR* (red). The respective number refers to the overall growth between 2000 and 2013. Here category A stands for the richest and category E for the poorest countries.

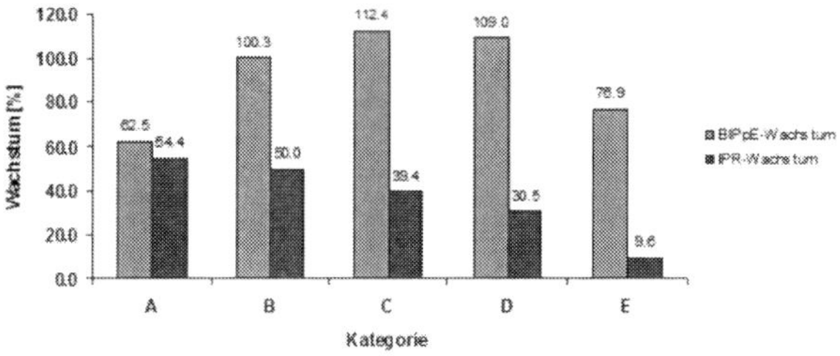

Figure 7. Growth rate (total) for GDPpC (blue) and IPR (burgundy) between 2000 and 2013.

The highest average IPR growth of 54.4% took place in the highest GDPpC category A, but in this category also the slightest GDPpC growth of 62.5% is given what, nevertheless, absolutely means a high growth, because the richest countries have a higher starting level than poorer countries. This finding is interesting that the average (absolute) IPR growth is lower, the poorer the group of states is. This means that in category B, the second strongest IPR growth takes place with 50%, followed by the category C with 39.4%, category D with 30.5% and category E with 9.6%. On the one hand the thesis that the catching-up in the usage of Internet gets confirmed generally. But on the other hand this finding shows that regarding to the global spread of Internet access, especially in the developing countries still much is to be done. In summary, the empirical relationship between the GDP and the IPR both in 2000 and in 2013 is positive. The Planck function modelled this context (within the considered class of functions) best for all years from 2000 to 2013. The course of the Planck function in 2000 is similar to the the one of the relationship between the GDPpC and FTR, suggesting also the then existing physical connection of Internet and Fixed Telephony (see section 4.4.2). Up to 2013 almost all states have increased their IPR and on average the richest states show the highest rise and the poorest states the lowest. With the relative increase it is exactly the reverse. This observation underlines the thesis that the catching-up in Internet usage around the world runs (clearly) faster than catching up with regard to prosperity in general.

4.2 Relation between Education Level and ICT usage

In a society and culture that is dependent on the written language in addition to oral communication, the literacy of people is a distinct advantage compared to illiteracy. The use of the Internet for example is restricted significantly for illiterate people, because it works on a text-based nature until today mainly. As a comparison, the use of fixed or mobile phones requires (almost) no literacy skills.

4.2.1 Literacy and Mobile Telephony

To use the mobile phone, you have to be able to identify at least the digits 0 to 9. This requires no skills for reading and writing of letters or words. The identification of numbers and corresponding basic operations such as adding numbers is part of skills are indicated in the adult literacy rate (*ALR*). It is expected that the empirical relationship between the *ALR* and the mobile phone rate (MPR) is low, but should be lower in 2012 or functionally run significantly different than in 2000. The reason for this is the rapid spread of mobile telephony in this period across all countries in the world, also within illiterate people. Table 5 shows the descriptive statistics for the intersection of the data for ALR and MPR for the years 2000 and 2012.

Table 5. Descriptive statistics of adult literacy rate (ALR) and mobile phone penetration rate, year 2000 and 2013.

	2000 (N=148)						2012 (N=153)					
	sp	min	max	av	md	sta	sp	min	max	av	md	sta
ALR	90.3	9.54	99.79	81.99	89.79	20.87	68.48	31.37	99.85	86.05	93.80	16.91
MPR	76.39	0.02	76.41	17.52	5.64	23.37	182.43	4.98	187.41	102.55	106.17	40.46

In 2000, the global average *ALR* is about 81.2%, while this rate increased to about 86.2% in 2012, although in the same period, the world population has increased by almost 16% from 6 to 7 billion people. Worldwide development programs relating to the MDGs and similar campaigns contribute to this progress. The *sp* of the *ALR* has decreased worldwide from about 90% to almost 70%, which is also a positive signal in the direction of improving global levels of education. The data confirms that most countries with a low *ALR* are located on the African continent. In Africa, the high population growth is a major challenge when it comes to increase the *ALR* in total. The Baltic countries belong to the countries with the highest *ALR*. These high values for the *ALR* are also consequences of the culturally high importance of education in the old Soviet Union. In both years the median value is with 89.1% in 2000 and 93.8% in 2012 slightly higher than the average. Most times it is the other way round, for example, at income distributions, i.e. the average is (often significantly) higher than the median.

What is it the reasoning? In many distribution situations there are great outliers upwards as very high incomes that amount a multiple of the average income. In the present education data it is different. Almost all states are with regard to the *ALR* anyway at a level greater than 50%, many greater than 80%. An *ALR* = 100% is the upper limit. There are no outliers upwards, only those down. The median is about 90% in 2000. The mean value is therefore determined by smaller ALR close to 50% and lower. This drives the mean value downwards in the direction of about 80%, what is less than the median. In 2000 the data points are strewn along the ALR and MPR axis, which is why the exponential model and the power function describe the relation well, which means that only for high values of the ALR also accordingly high values of the MPR are reached. It is probably the case that in 2000 the mobile telephony was closely linked for financial reasons with the fixed telephony, although the respective physical infrastructure was decoupled.

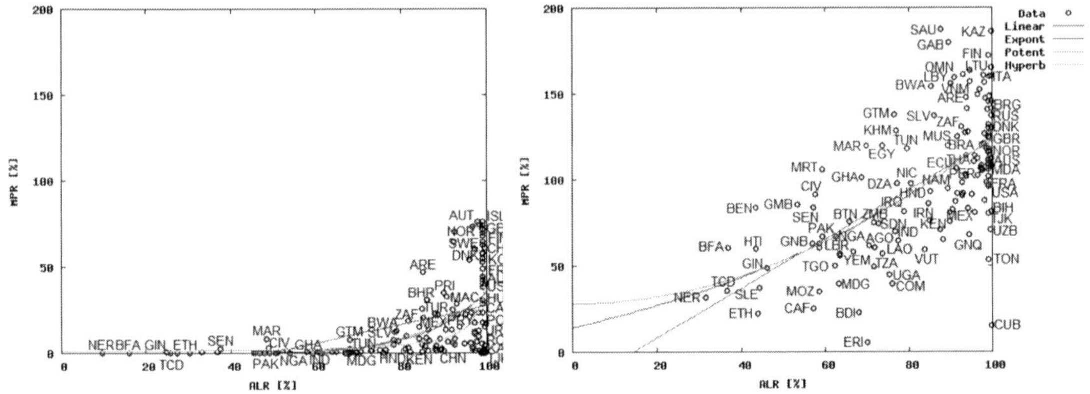

Figure 8. Mobile phone penetration rate (MPR) as a function of GDP per capita (GDPpC), year 2000 (left) and 2013 (right).

The relationship between the ALR and MPR changes up to 2012 so far that also for relatively low values of the ALR high values of the MPR exists. This change is clearly visible in the functional modelling of the relationship between ALR and MPR, because in 2012 the linear, exponential and power function have almost identical values for the sum of squared errors (MSE or MSE / n) and the coefficient of determination (R2).

Table 6. Sum of mean square of errors (MSE) and coefficient of determination (R2) for regression MPR = f(ALR).

	2000 (N=148)			2012 (N=153)		
	MSE	MSE/N	R^2	MSE	MSE/N	R^2
Linear	61603	416.2	0.232	160139	1046.6	0.357
Expont	54214	366.3	0.324	159100	1039.8	0.361
Power	54311	366.9	0.323	158923	1038.7	0.361

Catching up in the mobile telephony use in the developing and emerging countries in which worldwide most illiterates live, decisively contributes to this positive development. The 100% mark of the MPR is achieved from an ALR of about 60%.

4.2.1 Education Index and Internet Usage

The use of the Internet still requires at least the ability to read and to write. But what is the functional relationship between the Internet penetration rate (*IPR*) and the level of education as measured by the Education Index (*EI*)? A basic thesis is as follows: The higher the *EI* of countries is, the higher the *IPR* should be. Another theory, which is described on the basic thesis with the importance of the educational level of countries for the *IPR* decreases continuously between 2000 and 2012 and will most likely decrease in the future. Table 4.31 shows the descriptive statistics for the intersection of the data for *EI* and *IPR* for the years 2000 and 2012.

Table 7. Descriptive statistics of Internet penetration rate (IPR) and Education Index (EI), year 2000 and 2012.

	2000 (N=161)						2012 (N=174)					
	sp	min	max	av	md	sta	sp	min	max	av	md	sta
EI	0.863	0.111	0.974	0.579	0.605	0.213	0.812	0.177	0.989	0.6487	0.685	0.202
IPR	51.98	0.02	52.00	8.14	1.98	13.12	95.41	0.8	96.21	39.02	36.38	28.42

A first sight at the point cloud in figure xx shows that in both years a positive correlation exists between the EI and the IPR. If one looks at the previous section 4.2.5 about the relationship between the educational standard, measured by the Alphabetization Rate (ALR), and the IPR, the _rst di_erence is to be observed here. While indeed a high ALR is a necessary but not a su_cient condition for a high IPR, you have at the top of EI also correspondingly high, but no low values for the IPR in 2000. One can see this also in the fact that many data points are spread along the EI axis up to a value of about 0.6 and up from this EI value the IPR increases. A compression of the data along the IPR axis is here, as opposed to the correlation between ALR and IPR, in 2012 not available. One also notes this di_erence with the comparison of the single regression models. The hyperbolic function (yellow green) delivers the worst adaptation for the connection between EI and IPR.22 The relationship between EI and the IPR is modelled with the help of a power function (green) best of all, followed by the exponential function (blue) which is almost as good as the power function. The linear regression model delivers (red) only partly a good adaptation (see table 8).

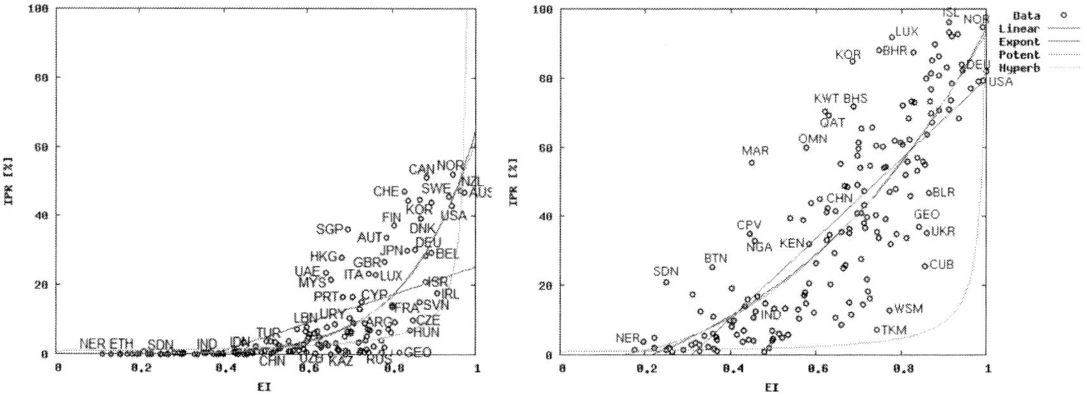

Figure 9. Internet penetration rate (IPR) as a Education Index (EI), year 2000 (left) and 2012 (right).

Table 8. Sum of mean square of errors (MSE) and coefficient of determination (R2) for regression IPR = f(EI)

	2000 (N=161)			2012 (N=174)		
	MSE	MSE/N	R^2	MSE	MSE/N	R^2
Linear	15623	97.04	0.433	46135	265.14	0.670
Expont	8859	55.02	0.678	42011	241.44	0.699
Potent	**8818**	**54.77**	**0.680**	**41817**	**240.33**	**0.701**
Hyperb	19429	120.68	0.294	312670	1796.95	-0.168

Up to 2012 the dispersion of the point cloud changes what is to be seen good optically. The power function gives in 2012, as well as in 2000, the best fit within the considered class of functions. The exponential function delivers the second-best adapting. The linear model approximates the data in 2012 significantly better than in 2000. Basically all three functions deliver a similarly good modelling of the connection between the EI and the IPR. A high EI delivers a high value of the IPR and, however, already a middle EI level is sufficient to reach rather high values of the IPR. Countries with a low EI also tend to have a low IPR, although there are also countries that, despite a relatively low EI have rather high values for their IPR. Catching up in Internet use has progressed worldwide for all countries faster than the development in the area of education. This fact suggests that global convergence processes in Internet use run significantly faster than the convergence in the context of education. Interesting is the negative value of the coefficient of determination (R2), which is obtained for the hyperbolic function in 2012 (see *table 8*) and occurs in this work for the first time in this form. This is first contra intuitive, because one would think the R2 to be a squared size, because it's symbolic description. This would then always have values greater than or equal to zero. However, this is not the case here. It does not concern a size in the square, but from the definition follows that in the case in which the adaptation of a function to the data is worse than a horizontal line at the level of the average value, the size R2 is negative. In this example one can interpret this as follows: The negative value of the R2 says that the carried out modelling of the data with certain (here as hyperbolic assumed) function is worse than the modelling of the data with a steady function which runs at the level of the average of the IPR.

4.3 Multivariate Regression

By univariate regression (only) relationships between two variables can be examined. If one uses instead multiple variables for predicting a size, one goes to the field of multivariate and multiple regression analysis (see section xx). When considering several independent variables in general, the question arises, which of them affects the dependent variable at most or at least. The effect of each explanatory variable on the dependent variable is measured (indirectly via the associated estimators) by the respective regression coefficient. If the independent variables own different units, the dimensions of the regression coefficients cannot be compared (directly). Standardised regression coefficients compensate the effect of different scales. Thus all variables of the regression model receive a statistically uniform unit. The contact with the negative weights which can appear in the optimisation process just like that is interesting also. Comparing the dimension of the influence of the respective independent variables one has to use the absolute value and compare them with each other. In this work the regression coefficients (A, B, C and D) are used as weight variables that specify which importance the respective predictor enters the forecast.

4.3.1 Combination of the factors MPR, GDPpC, EI, ALR

Looking at the mobile phone rate (MPR) as the dependent variable and the GDP per capita (GDPpC), the Education Index (EI) and the adult literacy rate (ALR) as the (three) independent variables, the question is to which extent the individual independent variables affect the MPR. Besides, the interaction between three explanatory variables is to be considered once more implicitly with. The approach is the following:

$$MPR = A \cdot f_1(GDPpC) + B \cdot f_2(EI) + C \cdot f_3(ALR) + D$$

Based on the results in previous sections the presumption is that the MPR is related to the GDPpC most, the EI second most and the ALR lowest. The relationship between the GDPpC and the MPR was modelled in 2000 using a Planck function, which will be used now as class of underlying transformation functions f1 for the GDPpC data. In 2012, the transformation function f1 is a MaxLog function. The relationship between MPR and EI has been described both in 2000 and 2012 using a power function, which is now used as a transformation function f2 for the EI data. Furthermore, the relationship between MPR and ALR in 2000 and 2012, was modelled by an exponential function f3. For 2000, the following regression weights A, B, C and D result:

$$A = 0.737, B = 0.485, C = 0.291, D = -3.563$$

The result shows that the MPR is related most strongly in 2000 to the GDPpC. It further shows that the EI affects the MPR the second strongest. Besides, the MPR is determined differently strongly by GDPpC and EI. In comparison to these both explanatory variables the ALR has a lower influence on the MPR. Considering the proportional influence of all independent variables, the following relationship is found: The GDPpC affects the MPR to 48.7%, the EI to 32.1% and the ALR to 19.2%. How can this result be interpreted?

In 2000, the mobile phone has been used almost exclusively in the more richer states. Thus, the GDPpC was the determining factor for the MPR worldwide. In these countries, furthermore, the ALR is consistently very high, i.e. in this group of states the ALR has no great importance for the height of the MPR for the purposes of a di_erentiation characteristic. The lower meaning of the EI concerning the MPR is determined above all by states like Greece, Hong Kong, Italy, Korea, Portugal, Singapore, Spain etc., that have a relatively high MPR, but a low EI have in comparison to states like Belgium, Denmark, Germany, Finland, the Netherlands, Norway or Sweden. It can be assumed that the ALR in 2012 will have an even lower impact on the MPR, because firstly, the ALR has increased worldwide and on the other, the mobile phone is now used almost all over the world by many people. Moreover, in this context is to be expected that the GDPpC and the EI will determine the MPR again similarly strongly.
For 2012, the following regression weights A, B, C, D result:

$$A = 0.206, B = 0.184, C = -0.085, D = -2.412$$

The MPR is still influenced strongest by the GDPpC, followed by the EI, indeed, with the difference that now both explanatory variables exert an according to portion similarly strong influence on the MPR. This result corresponds to the expectations. The GDPpC has the strongest influence on the MPR in 2000 and 2012, but decreases proportionally in this period. The EI increases it's proportionally influence during the considered period, while the connection with the ALR sinks a little bit. But what could be the reasons for this?

Worldwide, the MPR has massively increased between 2000 and 2012, so that almost the entire world population nowadays uses the mobile phone. One could logically conclude from this that there might be no more recognizable connection between the MPR and the independent variables, because the mobile phone is used by everybody no matter whether somebody is rich or poor, educated or uneducated, able to and write or not. In spite of this fact the analyses deliver a differentiated relationship between MPR, GDPpC, EI and ALR. This is due to the fact that the MPR is not limited upwards. There are many countries with a MPR significantly over 100%, because one person can have and has more than one mobile phone access. A MPR value over 100% is recognizable over all GDPpC areas. So this is in fact not only in rich states the case. Furthermore one can not observe any compensatory effects by the growth of the population in a country or in the whole world with regard to the increase of the MPR. Another reason could lie in the fact that the MPR shows a saturation effect in 2012 worldwide indeed, but, however, there are still some countries in Africa with a low prosperity level and educational standard (Ethiopia, Burundi, Eritrea, Kiribati, Congo, Mozambique, Niger, Rwanda, Sierra Leone, Togo, Chad, Uganda, Central Africa etc.) in which still less than half of the population has a mobile phone.

4.3.2 Combination of the factors IPR, GDPpC, EI, ALR
If one looks at the Internet penetration rate (IPR) as the dependent variable and the GDP per capita (GDPpC), the Education index (EI) and the adult literacy rate (ALR) as the independent variables, is the question again with which weight the single independent variables in_uence the IPR. Here are implicitly the interactions between the three explanatory variables to consider with. It is placed the following approach:

$$IPR = A \cdot f_1(GDPpC) + B \cdot f_2(EI) + C \cdot f_3(ALR) + D$$

For 2000 the following regression weights A, B, C and D result:

$$A = 0.671, B = 0.584, C = -0.239, D = -0.222$$

The result shows that the IPR is most related to the GDPpC, second most to the EI and lowest to the ALR. Here the GDPpC and the EI have a similar influence on the IPR. This indicates once more to a (possible) high correlation of economic power and education level of a country. Compared to GDP and EI the ALR has (significantly) less impact on the IPR. A negative regression weight, as it is the case for C, is an 'artefact' of the optimization. Concerning the contents the negative regression coefficient C concerning the ALR means that with rising ALR values there is a smaller probability for the MPR to accept high values. In comparison to that positive regression coefficients, as it is the case for the GDPpC and the EI, with rising values for the GDPpC and the EI lead to rising probabilities for high MPR values. Considering the proportional influence of all independent variables, one finds the following relationship: The GDPpC affects the IPR to 44.9%, the EI to 39.1% and the ALR to 16.0%.

Though the reading and writing ability is one of the basic competence of people to be able to use the Internet, indeed, the access to the Internet is first a question of cost. A wide literacy of the population is further a necessary condition for a correspondingly high level of prosperity in a country, but not a sufficient condition. A high level of prosperity in turn allows more likely to be able to afford Internet access. In 2000 an Internet connection is to be found almost only in rich countries and in these countries nearly the whole population is alphabetised. From this results the low correlation between IPR and ALR. Within this group of rich countries there are differences regarding the education level in form of the EI, what explains the strong meaning of the EI.

What is the situation after 12 years of development in 2012? The following regression proportions A, B, C and D result:

$$A = 0.624, B = 0.441, C = -0.023, D = -3.762$$

It becomes clear first of all that all the regression coefficients have reduced (according to amount). This is related to the used transformation functions. It also turns out that the IPR is still most related to the GDPpC followed by the EI. The IPR is influenced by the GDPpC to 57.3%, by the EI to 40.5% and the ALR to 2.2%.

The IPR has risen between 2000 and 2012 worldwide on just 40%. In this time frame the biggest (absolute) growth concerning the IPR has taken place in the rich states, what explains the increase of the meaning of the GDPpC as explanatory variable. In comparison with that the slightest IPR growth took place in the poorest states. Within the group of rich states an accordingly high EI is given. Furthermore, these countries already have an ALR very close to the 100% limit, that consequently has almost no significance for the explanation of IPR level. Now the IPR has increased worldwide, also as a result of access to the mobile Internet. How is it to be explained then that the weight of the ALR is so low in 2012? The reason could lie in the fact that within the countries in which the Internet is widespread, i.e. a high IPR, there are (clearly) higher differences with regard to the EI than with regard to the ALR, because most countries have a very similar ALR. The result is clear, because the ALR within the states, which use the Internet, is either already very high before 2000 or have risen until 2012 on a very high level. By technological means, as already described on top, the narrow coupling of the Internet access with the fixed network infrastructure is responsible for this in 2000, while up to 2012 the Internet access has partially decoupled from the fixed network infrastructure, namely by the success of the mobile Internet.

5. Conclusion

The investigations concerning the empiric or functional relationship between the GDP per capita (GDPpC) of a country and the different access rates of technology, e.g. Internet and mobile penetration rate, show a positive correlation in the years 2000 and 2012/2013, as well as also for all years in between. The relationship between the GDP per capita and the mobile phone access rate (MPR) is best modelled in 2000 with the Planck function and in 2013 with a maxLog function. For all GDPpC categories a nearly similar MPR growth is to be observed. In 2013 many countries have a MPR of 100%, what leads also to a worldwide MPR of about 100%. Thus, one can say that most people have a mobile phone access now. It is convincingly represented that the catching up in mobile phone usage runs much faster than the convergence in prosperity. The result for the relation between the absolute GDP produced and the Internet penetration rate (IPR) shows first that a strong linearly positive correlation exists between the portion of a country in the world GDP and its portion in the worldwide Internet users. Concerning this a few rich states were represented very over-proportional in the Internet world. Reason for this is the largely physical coupling of an Internet access to the fixed network in the year 2000. However, between 2000 and 2013 the IPR growth was greatest in the richest and richer countries and smallest in the poorer and poorest countries. The hopes in this area are based on the mobile Internet access on the basis of technological convergence of mobile telephony and Internet. Indeed, up to 2013, it came to better balanced relations and in fact the mobile Internet usage increased significantly during the previous years.

A (very) high ALR is generally a necessary condition on a high technology penetration in a country, but by far still no sufficient one. There is no country which shows a relatively high deployment of ICT with a low ALR. This statement is valid in a weaker form for the EI as a stronger measure of the educational standard. There are isolated states which show moderately high values of the respective access rate with a relatively low EI. This situation is different a little bit for the mobile telephony in this respect, as that also for relatively low values of the ALR and the EI high values of the MPR are given, because the mobile phone is used by everybody, no matter whether educated or uneducated.

The presented multivariate approach to determine the importance of the factors considered covers regressions as form of a linear combination. It becomes clear that a correlation between the level of wealth and the educational level of a country exists, both in 2000 and in 2012, and the fact that it is difficult to achieve sustainable progress in these two areas. The multiple regression analysis have confirmed that the access rates to ICT are most influenced by the GDPpC, the second most by the EI and the lowest by the ALR. Here, the GDPpC and EI have a similarly strong impact on the respective access rates. This also points to the implicitly existing (high) interaction between economic performance and level of education in a country. In comparison, the ALR has a significantly lower impact on the economic level. Overall, the studies conclude that the convergence regarding ICT usage proceeds worldwide faster than the convergence in terms of wealth and education in general, so the initial formulated thesis is confirmed. Based on this research, the digital divide is declining over the decade between 2000 and 2013, since more people worldwide use mobile phones and the Internet. But a high digital inequality explained to a significant extent by the functional relation between technology penetration rates, education level and average income still exists. Furthermore it supports the actions of countries at UN/G20/OECD level for providing ICT access to all people for a more balanced world in context of sustainable development by postulating that policymakers need to promote comprehensive education worldwide by means of using ICT.

Reference

[1] Amiri S 2013 Internet Penetration and its Correlation to Gross Domestic Product: An Analysis of the Nordic Countries *International Journal of Business, Humanities and Technology* **3(2)** pp 50-60

[2] Andres L, Cuberes D, Diouf M A, Serebrisky T 2010 The di_usion of the Internet: A cross-country analysis *Telecommunication Policy* **34** pp 323-340

[3] Bagchi K 2005 Factors contribution to global digital divide: Some empirical results *Journal of Global Information Technology Management* **8** pp 47-65
[4] Bauer A, Günzel H 2001 Data Warehouse Systeme - Architektur, Entwicklung, Anwendung
[5] Bonfadelli H 2002 The Internet and Knowledge Gaps: A Theoretical and Empirical Investigation *European Journal of Communication* **17(1)** pp 65-84
[6] Chinn M D, Fairlie R W 2007 The determinants of the global digital devide: A cross-country analysis of computer and Internet penetration *Oxford Economic Papers* **59** pp 16-44
[7] van Dijk J 2012 The Evolution of the Digital Devide. The Digital Devide turns to Inequality of Skills and Usage *Digital Enlightenment Yearbook*
[8] Freedman D, Pisani R, Purves R 2007 *Statistics* ISBN: 978-0393929720.
[9] Fuchs C 2009 The Role of Income Inequality in a Multivariate Cross-National Analysis of the Digital Divide. Social S Income Disparity and Digital Divide: The Internet Consumption Model and cross-country science *Computer Review* **27(1)** pp 41-58
[10] Hargittai E 1999 Weaving the Western Web: Explaining differences in Internet connectivity among OECD countries *Telecommunication Policy* **23** pp 701-718
[11] IGF 2009 Internet Governance Forum - Transcripts. Sharm-El-Sheikh http://www.intgovforum.org/cms/2009/sharm_el_Sheikh/Transcripts/Sharm%20El%20Sheikh%2018%20November%20Stock%2Taking%20Part%20I.pdf.
[12] IGF 2014 Internet Governance Forum. Istanbul. http://www.intgovforum.org/cms/igf-2014.
[13] IGF 2015 Internet Governance Forum. Jeao Pessoa. Brasilien. http://www.intgovforum.org/cms/igf-2015-website.
[14] ITU 2010 Definitions of World Telecommunication/ICT Indicators. https://www.itu.int/ITU-D/ict/material/TelecomICT_Indicators_Definition_March2010_for_web.pdf.
[15] ITU 2014 Measuring the Information Society. ISBN: 978-9261152918.
[16] International Telecommunication Union. Statistics. http://www.itu.int/en/ITU-D/Statistics/Pages/stat /default.aspx. 2013.
[17] Kämpke T 2012 Income distribution and majority patterns *International Journal of Computational Economics and Econometrics* **2** pp 155-178
[18] Kämpke T, Radermacher F J 2014 Income Modeling and Balancing - a rigorous treatment of distribution patterns *Lecture Notes in Economics and Mathematical Systems*
[19] Luhmann N 1995 *Social Systems*
[20] Mankiw G N 2004 Grundzüge der Volkswirtschaftslehre. 3. Auflage, Stuttgart, Schäfer-Poeschel.
[21] Kiiski S, Pohjola M 2002 Cross-country Diffusion of the Internet *Information Economics and Policy* **14** pp 297-310
[22] Pearson K 1896 Mathematical contributions to the theory of evolution. III. Regressions, heredity, and panmixia *Philosophical Transactions of the Royal Society Ser* **A187** pp 253-318
[23] Pearson K 1920 Notes on history of correlation *Biometrika* **13** pp 25-45
[24] Porter C E, Donthu N 2006 Using the technology acceptance model to explain how attitudes determine Internet usage: The role of perceived access barriers and demographics *Journal of Business Research* **59**
[25] Pratama A R, Al-Shaikh M 2012 Relation and Growth of Internet Penetration Rate with Human Development Level from 2000 to 2010. *International Business Information Management Association (IBIMA). Communications of IBIMA* **2012**
[26] Radermacher F J, van Dijk J A G M, Pestel R 1999 The European Way to the Global Information Society *IPTS Report* **32** pp 10-16
[27] Radermacher F J 2005 Die Zukunft der Infrastrukturen - Intelligente Netzwerke für eine Nachhaltige Entwicklung *Metropolis-Verl.* pp 97-112
[28] Quibria M G, Shamsun N A, Tschanh T, Reyes-Macasaquit M L 2003 Digital Divide: Determinants and policies with special reference to Asia *Journal of Asian Economics* **13** pp 811-825
[28a] UNDP 2014 Education Index. Human Development Reports. http://hdr.undp.org/en/content/education-index

[29] UNESCO 2012 Adult and Youth Literacy, 1990-2015. Analysis of data for 41 selected countries. UNESCO. ISBN: 978-92-9189-117-7.
[30] UNESCO 2013 Technology, broadband and education. Advancing the education for all agenda. A Report by the broadband commission working group on education.
[31] World Bank 2015 World Data Bank http://databank.worldbank.org/data/home.aspx
[32] World Summit on the Information Society 2003 World Summit on Sustainable Development. United Nations http://www.un.org/events/wssd/.
[33] World Summit on the Information Society 2003 World Summit on the Information Society. United Nations and International Telecommunication Union http://www.itu.int/dms_pub/itu-s/md/03/wsis/doc/S03-WSIS-DOC-0004!!PDF-E.pdf.
[34] World Summit on the Information Society 2005 World Summit on the Information Society. United Nations and International Telecommunication Union. http://www.itu.int/wsis/docs2/tunis/off/6rev1.html.
[35] Wilkinson R, Pickett K 2011 The Spirit Level: Why More Equal Societies Almost Always Do Better *Bloomsbury Press* ISBN: 978-1608193417.
[36] Zhang X 2013 Income Disparity and Digital Divide: The Internet Consumption Model and cross-country empirical research *Telecommunication Policy* **37** pp 515-529

IOP Conf. Series: Materials Science and Engineering **173** (2017) 012022 doi:10.1088/1757-899X/173/1/012022

Application of iterative hill climbing to the sound speed profile inversion in underwater acoustics

O Zaikin[1] and P Petrov[2,3]

[1] Matrosov Institute for System Dynamics and Control Theory SB RAS, Irkutsk, Russia
[2] V.I. Il'ichev Pacific Oceanological Institute FEB RAS, Vladivostok, Russia
[3] Far Eastern Federal University, Vladivostok, Russia

E-mail: zaikin.icc@gmail.com

Abstract. An application of the iterative hill climbing algorithm to the solution of inverse problems of underwater acoustics is discussed. Modal dispersion data extracted from a recording of a pulse acoustical signal is used as the input for the geoacoustic inversion procedure. The mismatch of the dispersion curves extracted from experimental data and computed from a given set of waveguide parameters is minimized over the parameters space. The solution of two test inversion problems is considered. It is shown that the iterative launches of the hill climbing algorithm allow to locate either the global minimum or a local minimum which is reasonably close to the former. In the latter case a reasonable estimate of the waveguide parameters is obtained.

1. Introduction

Global optimization problems emerge in many areas of research in present-day science including, for instance, electrical engineering, operations research and molecular modelling. In many practical cases the cost function of such problems exhibits very complicated behaviour and has many local extrema that render classical local optimization algorithms (e.g. Hessian-based and gradient-based minimization) inapplicable. Usually the solution of those problems requires certain metaheuristics such as tabu search, simulated annealing and iterated local search [1]. Iterated local search was shown to be highly efficient in the solution of the vehicle routing problems [2], the problem of solvers parameters tuning [3], etc. At the same time, this algorithm is quite easy to implement.

In this study we apply the iterated local search based on hill climbing [4] to the problem of the acoustical parameters estimation in a shallow-sea geoacoustic waveguide and show that it can be successfully used for this purpose. The paper is organized as follows. In Section 2 we outline the idea of dispersion-based geoacoustic inversion, in Section 3 we describe the iterative hill climbing algorithm and its implementation in the case of the inversion problem. The study is concluded by Section 4 where the computational examples are presented.

2. Geoacoustic inversion based on the dispersion data matching

The term geoacoustic inversion in the field of underwater acoustics designates a specific class of problems related to the reconstruction of the media properties using some data on the sound

IOP Conf. Series: Materials Science and Engineering **173** (2017) 012022 doi:10.1088/1757-899X/173/1/012022

propagation through these media (see e.g. [5, 6] and references therein). A typical geoacoustic waveguide in a shallow sea is comprised by a water column and several layers of bottom sediments where the acoustical parameters, the sound speed $c=c(z)$ and the density $\rho=\rho(z)$, vary with the depth z. By contrast to the direct problems where the media parameters are known and acoustic pressure field has to be computed, in the inverse problems some acoustic data (e.g. the acoustic field at several points) is known, while the media parameters are inverted for.

Recently it was found that dispersion data could be successfully used as the input for the inversion of the environment parameters [7, 8]. An important advantage of this technique is that the dispersion data can be obtained from a single-hydrophone measurements [7]. By contrast, other inversion methods usually rely on the measurements performed using arrays of receivers that are very expensive and complicated in deployment [5, 6].

By the dispersion data we mean hereafter the dispersion curves (DCs), i.e. the functions $\tau=\tau(f,m)$, where f is the sound frequency and m is the mode number [6]. In shallow-water waveguides acoustic modes with different numbers m and at different frequencies f arrive to the receiver (which is located sufficiently far from the source) at different moments τ [6, 9]. The delays of one mode with respect to another depend on the properties of the media, i.e. the sound speeds and densities in the water column and the bottom [9]. Measuring such delays and using the DCs as the input for the inversion algorithms we may therefore recover the media parameters [7, 10, 11, 12, 13].

Usually an inversion algorithm is based on the matching of the DCs obtained from the experimental data and DCs computed using certain model of the waveguide. The experimental DCs (EDCs) are obtained from a recording of a pulse signal by a single hydrophone (the receiver) which is located several kilometres away from the source that emits the signal. The extraction of the DCs proposed in [7] is based on the time-frequency analysis of the received signal and the use of the so-called warping operators that greatly enhance the possibility of the modes separation. The theoretical DCs (TDCs) are obtained by the formula (see e.g. [11])

$$\tau(f,m)=\frac{R}{g_v(\bar{A},f,m)}+\tau_0\,, \tag{1}$$

where R is the source-receiver range, g_v is the modal group velocity [9], τ_0 is the time offset correction. The waveguide is described by the set of parameters $\bar{A}$ (usually $\bar{A}$ contains the values of sound speed and density in various layers of the sea water and the bottom), and from these parameters one can compute the group velocities $g_v(\bar{A},f,m)$ by solving acoustical spectral problem [8, 9, 11].

Provided that EDCs $\tau_e(f,m)$ and TDCs $\tau_t(f,m)$ are obtained for certain vector of frequencies $f_1,f_2,\ldots,f_N$, we can compute the root mean square mismatch $E(\bar{A},\tau_0)$ of the experimental and theoretical dispersion data using by formula

$$E(\bar{A},\tau_0)=\sqrt{\frac{\sum_{i=1}^{N}\sum_{m=1}^{M}\left|\tau_t(f_i,m)-\tau_t(f_i,m)\right|^2}{NM}}\,, \tag{2}$$

where M is the number of modes detected on the spectrogram of the received signal. We can therefore estimate the waveguide parameters $\bar{A}$ and the scalar τ_0 by the minimization of error function $E(\bar{A},\tau_0)$:

$$[\bar{A}^*,\tau_0^*]=\min_{\bar{A},\tau_0}E(\bar{A},\tau_0) \tag{3}$$

where the values $\overline{A}^*$ and τ_0^* constitute the global minimum point for the function $E(A,\tau_0)$. Note that parameter space in this problem can have large number of dimensions [10], while the cost function usually has many local minima that deny the use of the simple local search algorithms.

In the previous work a variety of global minimization algorithms was used to carry out the solution of (3). In the papers [8, 14] a brute-force global search was used, in [7] the genetic algorithm was employed, and in the recent work [10] a trans-D Bayesian inversion method was developed.

In the present study we restricted our attention to the discrete (finite) search space with linearly-spaced trial values for each parameter in vector $\overline{A}$ and for τ_0.

3. The application of iterative hill climbing to minimization of the error function

According to the previous section, our goal was to minimize the objective function $E(A,\tau_0)$ in a finite search space. Following the authors of ParamILS (Iterated Local Search in Parameter Configuration Space [3]), for this purpose the simple hill climbing algorithm [4] was chosen. ParamILS is a state-of-the-art tool for tuning the parameters of solvers (SAT, SMT, etc.). With the help of ParamILS good speedups were achieved in the different scientific areas (protein folding, AI planning, etc.). The relative simplicity of the hill climbing algorithm was another reason why we chose it. Hill climbing algorithm was also successfully used for the environmentally adaptive optimization of the sonar settings [15].

According to the basic idea of simple hill climbing we developed the following algorithm:

1. start with some randomly chosen point $E(A,\tau_0)$ from a finite search space;
2. iteratively, adjust a single parameter of the point (here by parameters we mean τ_0 and the elements of $\overline{A}$), keeping the modification if the value of $E(A,\tau_0)$ improves and undoing it otherwise;
3. terminate when no single element adjustment yields an improvement of $E(A,\tau_0)$, or when the time limit was exceeded.

Note that this algorithm will typically terminate in a local minimum, in which changing value of any single element will not achieve any improvement. In order to jump from a local minimum an algorithm based on hill climbing is usually expanded by some metaheuristic: tabu search, simulated annealing or iterated local search. Following the authors of ParamILS algorithm [3] we chose iterated local search (ILS) [1] for this purpose.

The workflow of the iterative hill climbing algorithm can be described as follows. Starting from some initial point, ILS first performs simple hill climbing search until a local minimum $E(A,\tau_0)$ is reached, and then it cycles through the following steps:

1. apply perturbation to (A,τ_0) (in the form of multiple random changes of the parameters values);
2. perform simple hill climbing search until a new local minimum $(A',\tau_{0'})$ is reached;
3. accept the better (with respect to the value of E) of the two points (A,τ_0) and $(A',\tau_{0'})$ as the starting point of the next cycle.

4. Computational experiments

In previous work (see [14]) we developed a simple normal mode code capable of computing the modal group velocities $v_g(f,A)$ from a given waveguide parameters A. This routine was used as a part of brute-force search algorithm SSPEMDD [16] for the minimization of the error function E implemented using the Message Passing Interface (MPI). For the numerical experiments described below the iterative hill climbing algorithm suggested in the previous section was implemented using C++ and added to SSPEMDD toolbox. It employs the same normal mode evaluation routine based on the finite differences (see [14] for the details).

4.1. *Test case 1: inversion of the sound speed profile in the water*

In the first series of experiments we used the same synthetic dispersion data as in [14]. We estimated the sound speed profile (SSP) in the water using the dispersion curves in the frequency range from 150

Hz to 250 Hz at $R = 3500\ m$ (see [14] for the detailed description). We used five nodes at fixed depths $z_0,\ldots,z_4$ for the approximation of SSP by a vector $c_0,\ldots,c_4$ ($c(z_i) = c_i$ is the sound speed c at depth z_i) with the fixed (known) value of the sound speed c_0 at the sea surface $z_0 = 0$. The search space for this case is presented in table 1.

Table 1. Search space for the test case 1.

parameter	min. value	max. value	step
R	3400 m	3600 m	5 m
$c_j (j = 1,2,3,4)$	1450 m/s	1500 m/s	2.5 m/s

Note that in this test we set τ_0 to zero and estimate the correction to the source-receiver distance R instead. Also note that due to the processing errors the DCs extracted from a pulse signal do not match their theoretical counterparts exactly. More details on the DCs extraction technique can be found in [7] and references therein.

In [14] the brute-force search for this test case was carried out on 15 nodes of the "Academician V.M. Matrosov" computing cluster (Irkutsk Supercomputing Center of SB RAS). Each computing node of this cluster consists of two 16-core CPU AMD Opteron 6276, and therefore 480 CPU cores in total were used. All 7973721 points of the search space (see table 1) were processed in 8 hours and 35 minutes, as a result the global minimum point with $E = 0.0069710$ s was found.

We performed 20 separate launches of the iterative hill climbing algorithm from different randomly chosen start points, including 10 launches with 100 iterations and 10 launches with 1000 iterations. Each of these runs was executed on a single core of CPU AMD Opteron 6276. The characteristics of the obtained 20 local minima are shown in table 2 and in figure 1. One launch took 44 minutes on average in the case of 100 iterations and 8 hours 2 minutes on average in the case of 1000 iterations.

Table 2. The local minima found in the test case 1. The local minima that coincide with the global one are marked in bold.

serial number	local minimum, 10 iterations	local minimum, 100 iterations
1	0.0105368	0.0071304
2	0.0105368	0.0071304
3	0.0090778	0.0099037
4	0.0071304	0.0069710
5	0.0071304	0.0071304
6	0.0100120	**0.0069710**
7	0.0099037	**0.0069710**
8	0.0103995	0.0099037
9	0.0071304	0.0071304
10	0.0112283	**0.0069710**

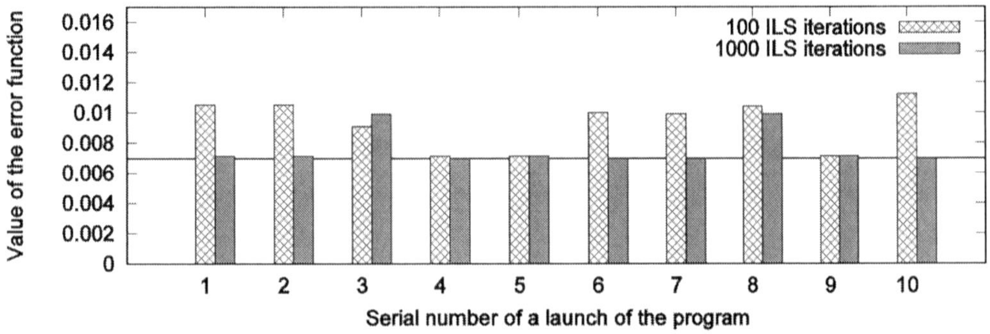

Figure 1. Results of 10 separate launches (with different random starting points) in the test case 1. The global minimum level $E = 0.0069710$ s is shown by a horizontal line.

4.2. Test Case 2: bottom parameters inversion

In this example we estimated the bottom parameters in shallow sea (sound speed c_b and density ρ_b) together with the arrivals offset τ_0 (see equation (1)) using the dispersion data extracted from a simulated pulse signal. We used the waveform from [17] for the emitted signal. The DCs were extracted in a frequency range from 20 Hz to 200 Hz. The sound speed in the water column was considered constant $c_w = 1500\ m/s$, and the bottom depth was set to $h = 90\ m$. The forward propagation problem was solved using our normal mode code [16] with $c_b = 1800\ m/s$, $\rho_b = 1.8\ g/cm^3$, and the signal in the receiver located at $R = 8000\ m$ from the source was recovered using the standard Fourier synthesis technique [9]. The computed signal is shown in figure 2, while the DCs are plotted in figure 3 against the spectrogram of the computed signal (see [7, 11] for the detailed explanation of the extraction technique).

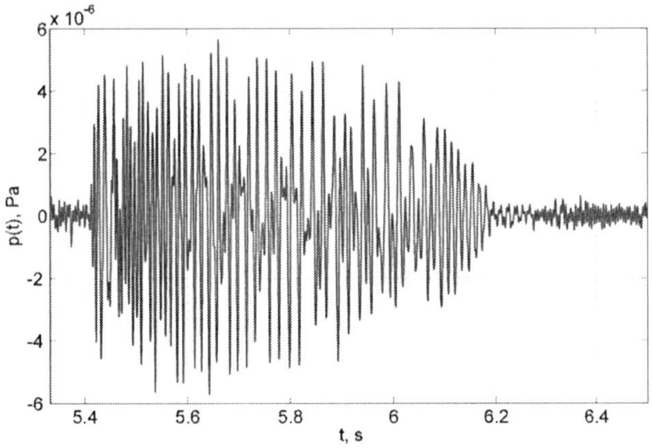

Figure 2. Simulated pulse signal at the range $R = 8000\ m$. This signal was used for the extraction of DCs in the test case 2.

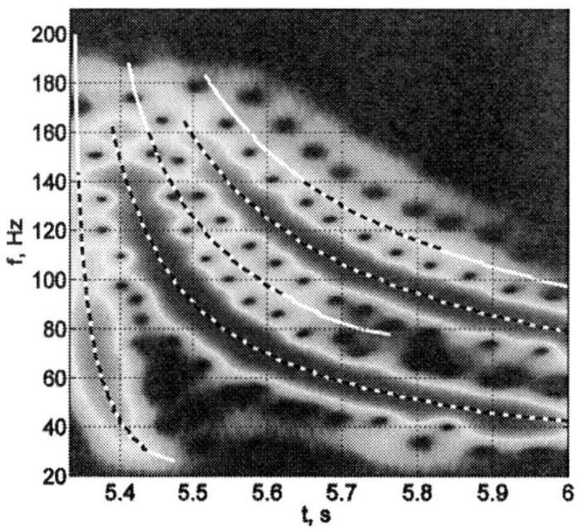

Figure 3. Spectrogram of the signal shown in figure 2 and the dispersion curves extracted from this spectrogram (plotted as black dashed lines). The DCs were used as the input for the inversion algorithm in the test case 2.

The search space for this example is described in table 3.

Table 3. Search space for the test case 2.

parameter	min. value	max. value	step
c_b	1600 m/s	1900 m/s	5 m/s
ρ_b	1.4 g/cm^3	2.0 g/cm^3	0.05 g/cm^3
τ_0	-0.015 s	0.015 s	0.005 s

For this case SSPEMDD was launched on 10 computing nodes of the computing cluster, and therefore 320 CPU cores in total were used. All 48373 points of the search space (see Table 3) were processed in 52 minutes, and the global minimum point with $E = 0.00267443\ s$ was found. We also performed 20 separate launches of the iterative hill climbing algorithm with different randomly chosen start points: 10 launches with 10 iterations and 10 launches with 100 iterations. The Error function values for the obtained 20 local minima are shown in Table 4 and in figure 4. One launch took 48 minutes on average in the case of 10 iterations and 7 hours 48 minutes on average in the case of 100 iterations (again one CPU core was used for each launch).

Table 4. The local minima found in the test case 2. The local minima that coincide with the global one are marked in bold.

serial number	local minimum, 10 iterations	local minimum, 100 iterations
1	0.00299985	**0.00267443**
2	0.00291565	0.00269999
3	0.00277798	0.00268122
4	0.00291565	0.00269999
5	0.00269999	0.00269999
6	0.00271570	**0.00267443**
7	0.00269999	**0.00267443**
8	0.00274744	0.00268122
9	**0.00267443**	0.00269999
10	0.00279155	0.00269999

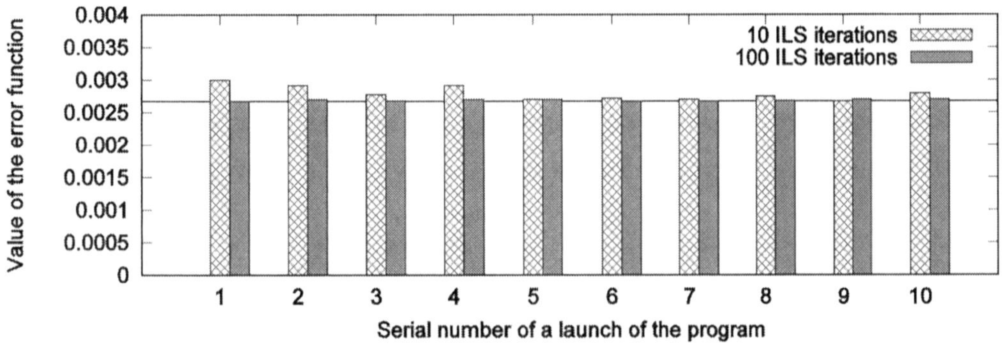

Figure 4. Results of 10 separate launches (with different random starting points) in the test case 2. The global minimum level $E = 0.00267443\ s$ is shown by a horizontal line.

It turned out that iterative hill climbing requires significantly less computational resources (in comparison with brute-force) to achieve a local minimum with the acceptable characteristics. In the test case 1 one launch of iterative hill climbing with 100 iterations took on average 5597 times less resources, and 513 times less resources for one launch with 1000 iterations. In the test case 2 one launch with 10 iterations took on average 346 times less resources, and 36 times less resources for one launch with 100 iterations.

5. Conclusion

In this study an application of the iterative hill climbing algorithm to the problem of geoacoustic inversion was considered. The results of the numerical experiments presented in the Section 4 show that it allows to obtain a reasonable estimate for the SSP or bottom parameters much faster than an exhaustive search algorithm even for a relatively low-dimensional search space (when the use of the latter is still feasible). Not that relatively small number of iterations is sufficient to achieve very good approximation for the set of waveguide parameters $\bar{A}$.

In future work we will apply our algorithm to the full-scale inversion problems that require higher dimensions of the search space (up to 20-30) and compare it with other methods in terms of accuracy and efficiency. We also plan to expand our hill climbing algorithm by another metaheuristics (e.g. simulated annealing and tabu search).

Acknowledgments

Authors express their gratitude to Dr Julien Bonnel (ENSTA Bretagne) for sharing his code and stimulating discussions. This work was partially supported by Russian Foundation for Basic Research (grants No. 15-35-20105-mol_a_ved, 16-05-01074-a and 16-07-00155-a) and by Council for Grants of the President of the Russian Federation (grant No. MK-4323.2015.5).

References

[1] Hoos H and Stützle T 2004 *Stochastic Local Search: Foundations & Applications*
[2] Penna P H V, Subramanian A and Ochi L S 2011 An iterated local search heuristic for the heterogeneous fleet vehicle routing problem *Journal of Heuristics* **19(2)** pp 201–232
[3] Hutter F, Hoos H H and Stützle T 2007 Automatic algorithm configuration based on local search *Proc. of the Twenty-Second Conference on Artifical Intelligence (Vancouver)* pp 1152–1157
[4] Russell S and Norvig P 2009 *Artificial Intelligence: A Modern Approach*
[5] Taroudakis M I and Makrakis G N (eds) 2001 *Inverse problems in underwater acoustics*
[6] Katsnelson B, Petnikov V and Lynch J 2012 *Fundamentals of shallow water acoustics*
[7] Bonnel J and Chapman N 2001 Geoacoustic inversion in a dispersive waveguide using warping

operators *Journal of the Acoustical Society of America* **130** EL101–EL107

[8] Zeng J, Bonnel J and Chapman N 2013 Inversion of seabed attenuation using timewarping of close range *Journal of the Acoustical Society of America* **134** EL394–EL399

[9] Jensen F, Kuperman W, Porter M and Schmidt H 2011 *Computational ocean acoustics*

[10] Warner G, Dosso S, Dettmer J and Hannay D 2015 Bayesian environmental inversion of airgun modal dispersion using a single hydrophone in the Chukchi sea *Journal of the Acoustical Society of America* **137** pp 3009–3023

[11] Petrov P 2014 A method for single-hydrophone geoacoustic inversion based on the modal group velocities estimation: application to a waveguide with inhomogeneous bottom relief *Proc. of the International Conference Day on Diffraction* pp 186–191

[12] Bonnel J, Nicolas B, Mars J and Walker S 2010 Estimation of modal group velocities with a single receiver for geoacoustic inversion in shallow water *Journal of the Acoustical Society of America* **128** pp 719–727

[13] Ballard M S, Frisk G V and Becker K M 2014 Estimates of the temporal and spatial variability of ocean sound speed on the New Jersey shelf *Journal of the Acoustical Society of America* **135** pp 3316–3326

[14] Zaikin O S and Petrov P S 2016 Algorithm of reconstruction of the sound speed profile in a shallow-water geoacoustic waveguide from modal dispersion data *Optoelectronics, Instrumentation and Data Processing* **52(3)** pp 259–265

[15] Lorentzen O J, Synnes S A V, Wiig M S and Glette M 2013 Environmentally adaptive sonar *Proc. of the First International Conference and Exhibition on Underwater Acoustics* pp 583–590

[16] SSPEMDD program. https://github.com/Nauchnik/Acoustics-at-home

[17] Petrov P, Zakharenko A and Trofimov M 2012 The wave equation with viscoelastic attenuation and its application in problems of shallow sea acoustics *Acoustical Physics* **58** pp 700–707

Complex of automated equipment and technologies for waveguides soldering using induction heating [1]

A V Murygin, V S Tynchenko, V D Laptenok, O A Emilova, A N Bocharov

Reshetnev Siberian State Aerospace University 31, Krasnoyarsky Rabochy Av., Krasnoyarsk, 660037, Russian Federation

E-mail: vadimond@mail.ru

Abstract. The article deals with the problem of designing complex automated equipment for soldering waveguides based on induction heating technology. A theoretical analysis of the problem, allowing to form a model of the «inductor-waveguide» system and to carry out studies to determine the form of inducing wire, creating a narrow and concentrated heat zone in the area of the solder joint. Also solves the problem of the choice of the temperature control means, the information from which is used later to generate the effective management of induction soldering process. Designed hardware complex in conjunction with the developed software system is a system of automatic control, allowing to manage the process of induction heating, to prevent overheating and destruction of the soldered products, improve the stability of induction soldering process, to improve the quality of products, thereby reducing time and material costs for the production.

Introduction

The waveguide systems are widely used in communication systems of spacecraft, ensuring the functioning of the apparatus and its on-Board equipment [1, 2].

However, induction soldering of thin-walled waveguides is a complicated technical process with a number of features [3].

Firstly, applied silumin solder has a melting temperature close to the melting temperature of the basic material. At speed induction heating 10 – 15 °C/sec an exact observance of the technological mode of the soldering process is required. Secondly, the component parts of a waveguide system such as a waveguide tube, flange or coupling in induction heating process are independent systems up to the point of melting solder between them [4]. That is, between the elements the heat exchange is practically absent, and their heating occurs independently from each other. Thirdly, it was found that during induction heating of the waveguide tracts of the maximum heat generation in the sections of elements zone (maximum density of induced eddy currents) do not coincide with the zones of the soldering [5].

In addition, a large range of brazed products with different mass-dimensional parameters leads to a difference in weight of similar products up to 25%. Also, during the heating elements of the waveguide tracts do not change color, eliminating the visual control of local overheating of solder products [6].

These factors significantly complicate the reproduction of the technological process of induction soldering of waveguide tracts, make demands on the requirements for accuracy in positioning the

[1] The reported study was funded by Russian Foundation for Basic Research, Government of Krasnoyarsk Territory, Krasnoyarsk Region Science and Technology Support Fund to the research project № 16-48-242029

products, to form the inductor, the parameters of the generator equipment for induction soldering and require highly skilled personnel [7, 8].

Analysis of the above listed features of induction soldering revealed the need to automate the process, and led to the creation of a hardware-software complex, allowing to accurately reproducing the process.

1. The theoretical analysis of the problem

When considering the specific physical processes as an object subjected to the induction heating, a compound of straight section waveguide tube and the flange has been chosen (Fig. 1). Selection of an object is related to its prevalence. The dependences obtained during the carried out research with minimal assumptions are valid also for all other areas of the waveguide tract subjected to the induction heating. The obtained dependences are important for understanding the physical processes in real objects.

Electrothermal processes occurring in the subject depend mainly on the external magnetic field, so for their correct description the aggregate system which consists of the alternating-current magnetic field source and the object subjected to heat under the effect of the field, should be considered. In simplified form (without considering the power of the induction equipment) system is represented by the inductor with an assembly unit placed in it (Fig. 1). The considered system is called «inductor – waveguide».

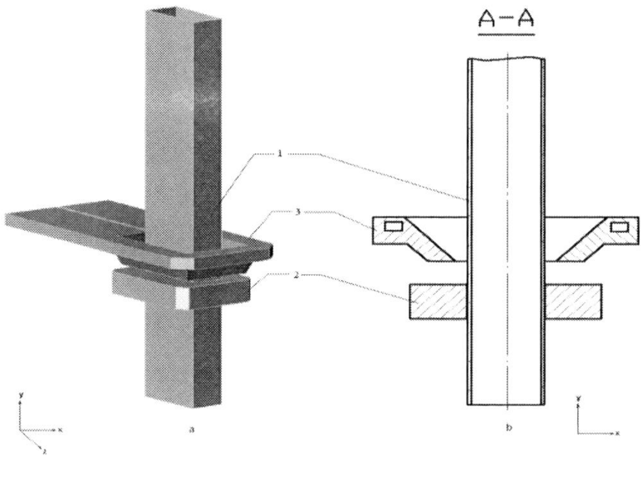

Figure 1. System of interest «inductor – waveguide»
1 - waveguide tube; 2 - flange; 3 - inductor

The mathematical model of this system is built without significant errors based on the following assumptions:

- the spatial configuration of the system allowed to restrict the consideration of the two-dimensional plane-parallel field (x, y), it was assumed that the geometry of the calculating areas, properties of environment and parameters characterizing the sources of the field remain unchanged in the direction of the z-axis (Figure 1 b);
- the field is considered to be quasi-stationary; the average frequency of the external current 65KHz allows to neglect the bias currents as compared with the currents in the conductive bodies;
- hysteresis losses are not accounted, because of their insignificance in comparison with the losses from eddy currents.

Analysis of alternating currents magnetic field consists of the calculation of the electric and magnetic field excited by an alternating current. The problem is formulated as a differential equation in partial derivatives relatively to the complex amplitude of the magnetic vector potential, recorded as:

$$\frac{\partial}{\partial x}\left(\frac{1}{\mu_y}\cdot\frac{\partial}{\partial x}A\right)+\frac{\partial}{\partial y}\left(\frac{1}{\mu_x}\cdot\frac{\partial}{\partial y}A\right)-i\omega\gamma A=-j \quad ,$$

The heat equation is used during solving the thermal problems:

$$\frac{\partial}{\partial x}\left(\lambda_x\frac{\partial T}{\partial x}\right)+\frac{\partial}{\partial y}\left(\lambda_y\frac{\partial T}{\partial y}\right)=-q-c\rho\frac{\partial T}{\partial t}.$$

System model «inductor – waveguide» built with the help of software package Elcut, is shown in Figure 2.

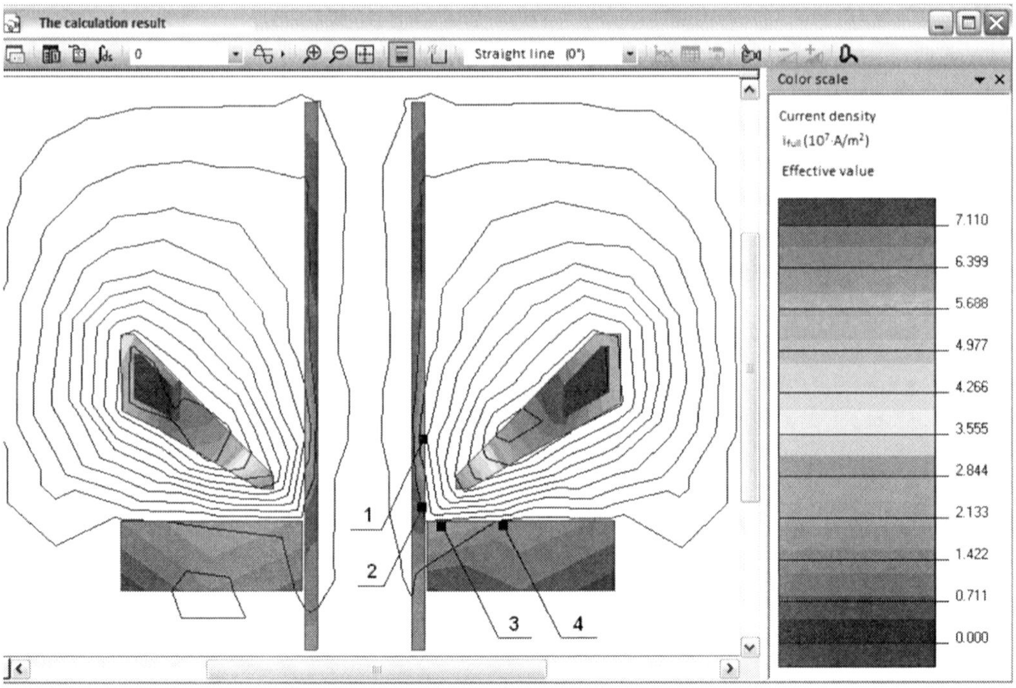

Figure 2. Model «inductor – waveguide» system
1,2,3,4 – the characteristic point limiting the soldering zone

On the basis of the obtained model research for the problem of the thin-walled rectangular pipe and flange heating in the process of heating during soldering was conducted. Local values of the various physical quantities, obtained in 4 characteristic points (Fig. 2) limiting the soldering zone are summarized in the Table 1.

As seen in the Figure 2 and Table 1, the maximum density of the eddy current, and hence the maximum heat radiation were observed at characteristic points 1 of the tube and 3 of the flange located in the soldering zone.

Table 1. Local values of physical quantities in the characteristic points of the research system.

Physical quantities	Point 1	Point 2	Point 3	Point 4
Induction B, 10^{-3} T	27,809	16,039	6,363	5,853
Strength H, 10^{3} ·A/m	9,6216	5,5494	2,2015	2,0253
Current density i_{full}, 10^{7}·A/m^{2}	2,1256	2,0732	1,9225	1,4784
Current density $i_{eddy..}$, 10^{7}·A/m^{2}	2,1256	2,0732	1,9225	1,4784
Heat radiation Q, 10^{7}·W/m^{3}	1,2211	0,14985	8,606	0,59069
Energy density W, J/m^{3}	133,78	44,504	4,2307	5,9277

As result of the conducted research the inducing wire shape was specified (Fig. 2), which allows to bring the inductor closer to the soldering zone, to create a narrow and concentrated heat zone eliminating overheating and destruction of the products outside the soldering zone, that provides high energy efficiency of the «inductor – waveguide» system.

2. Technological solutions

Let's consider the basic requirements for a technological process of assembling the elements of the waveguide paths by inductive soldering and induction equipment with high energy efficiency development and modernization by using high-tech means of pyrometry for implementing the robust feedback in control systems.

2.1 Inductor

Based on the thermal fields modeling a set of inductors for soldering lightweight waveguide tracts of different sections was developed. The proposed design of inductors provides equal distribution of temperature fields along the perimeter of the heated products in the soldering zone with the required accuracy (± 5 ° C), and also it provides the combination of maximum energy density zones with soldering area, which eliminates the possibility of overheating and destruction of soldered product outside of the soldering zone. Control of the temperature distribution along the contour of the heating carried out by means of thermocouples and pyrometers. General view of the inductor for soldering waveguide tubes with a cross section of the waveguide channel 19×9,5 mm is shown in Figure 3.

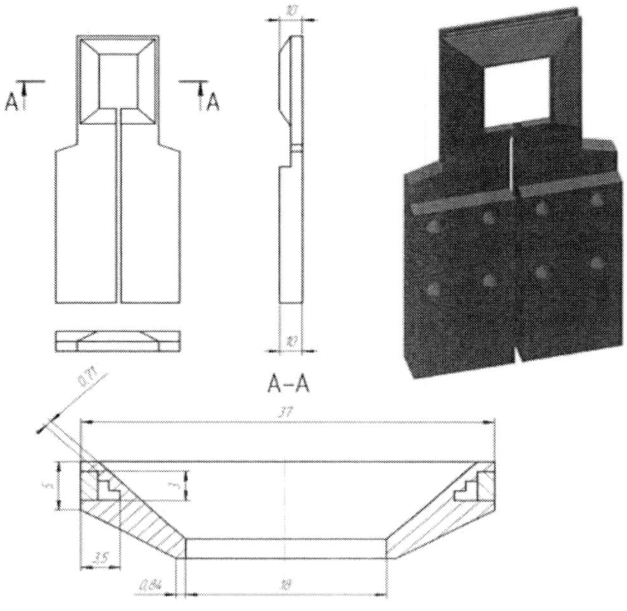

Figure 3. Inductor with optimal performance heating

2.2 Manipulator

To estimate the margin of error for the positioning of the waveguide in the inductor experimental research of the dependence of temperature gradient along the perimeter of the heated products in soldering zone from the displacement of the heated product relative to the inductor was conducted (Fig. 4). The data obtained in experiments show the need to position elements of waveguide tracts relative to inductor with an accuracy of at least ± 0.3 mm.

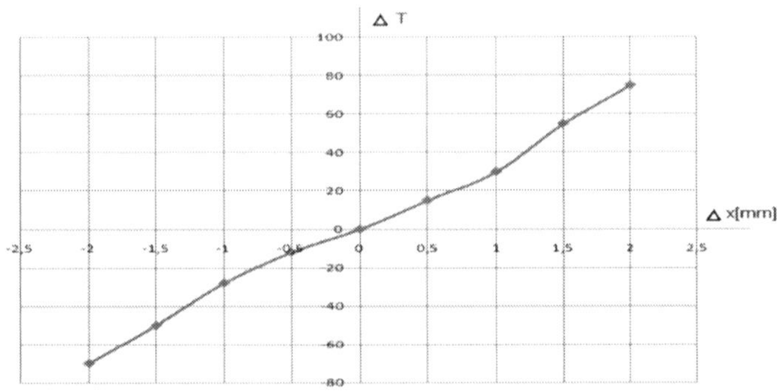

Figure 4. The dependence of the waveguide tube temperature gradient from the waveguide offset relative to inductor

Based on the experience of industrial applications of various devices and manipulators, which can be used for positioning elements of the waveguide tracts during soldering, as well as on the basis of the requirements of positioning accuracy (±0,3 mm), manipulator with enhanced functional characteristics was designed (Fig. 5).

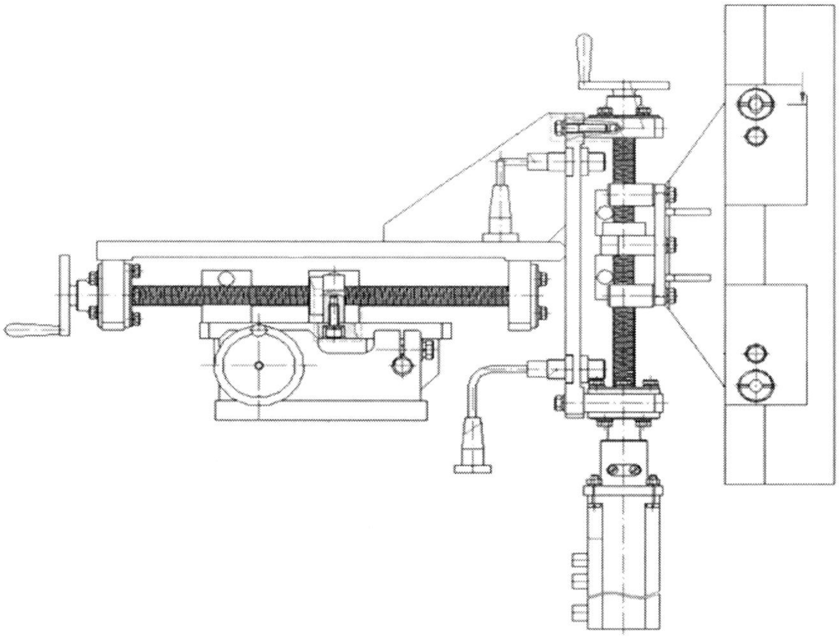

Figure 5. Manipulator with improved functional characteristics

The manipulator is equipped with displacement nodes with minimal backlash, which allows to position the soldered product with accuracy of ±0.1 mm relative to the inductor. Capture device of soldered parts and all the massive manipulator parts, that are located in close proximity to the inductor, are made of heat-resistant dielectric material. This allows to minimize the distortion of the inductor magnetic field, and, therefore, to provide a symmetrical distribution of temperature fields in soldered products. Electromechanical drive of the manipulator allows integrating it into the automated control system, with ability to the automated changing of position of the heated products directly in the soldering process.

2.3 Tools of temperature control

Decision about the use of contactless pyrometric sensors was made while development of the automated equipment complex. The main requirements for the choice was the high temperature control precision (not less than 0.5%), controlled spot's diameter not more than 1.5 mm. (to control the temperature in the gap of the inductor-waveguide), electromagnetic interference immunity and pair it with a computer.

Pyrometer AST 250 was selected as contactless temperature sensors because it satisfies the above listed requirements.

Four pyrometers were used in the study of thermal fields of the waveguides' elements, induced by the inductor (Fig. 3). As result of the conducted research it was determined that 2 pyrometers are enough for the automation of the induction soldering of waveguide tracts' elements process, one of which is used for temperature control on the waveguide flange and the other – on the waveguide tube.

2.4 Software of automated equipment complex

The soldering process programming consists of the specifying of the linear law of the waveguide tube temperature rise with subsequent stabilization of its temperature in the characteristic point above the melting point of the solder but below the melting temperature of the waveguide material.

A software product used to automate the process of induction soldering is a Windows-application, which work is possible in systems like Windows XP/7/8/8.1/10 [9].

It is programmed in C++ in the programming system Borland C++ Builder 6.0 using an object-oriented approach. Work with COM ports is implemented through WinApi, work with board PCI-1710 is implemented through third-party libraries bdaqctrl.h.

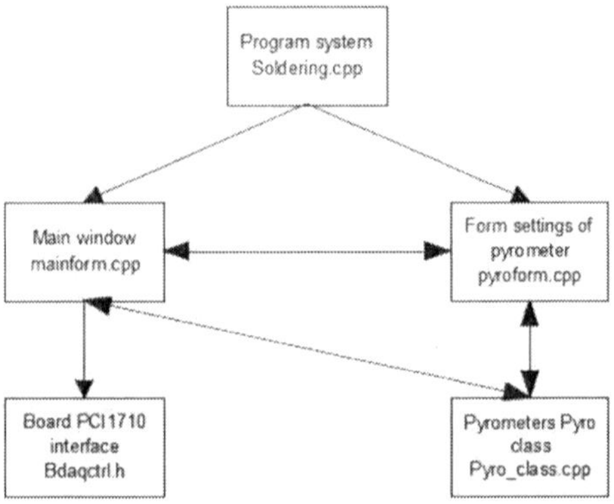

Figure 6. Software structural scheme

The software of control system allows defining the heating law of soldered elements of waveguide tracts to ensure required temperature distribution on the heated objects (Fig. 7), high quality of the soldering joints, eliminating the possibility of the overheating and destruction of the heated objects.

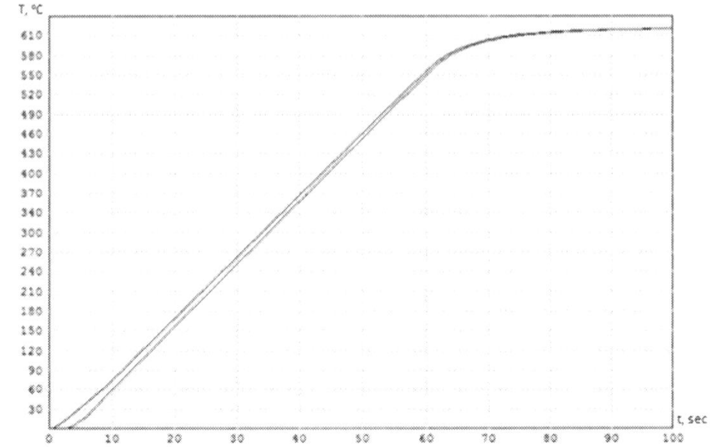

Figure 7. Law of temperature distribution in two characteristic points of the heated object with stabilization of temperature in the region of 600 ± 5 °C

3. Automated control system

A new double-circuit functional scheme of the automated control system of the induction soldering of the spacecraft (SC) antenna-feeder devices (AFD) elements (Fig. 8) process is developed, allowing to perform the heating of the waveguide tracts elements according to a given law by changing the power supplied to the inductor and the distance from the inductor to the heated product, providing the desired temperature distribution on heated objects (Fig. 7).

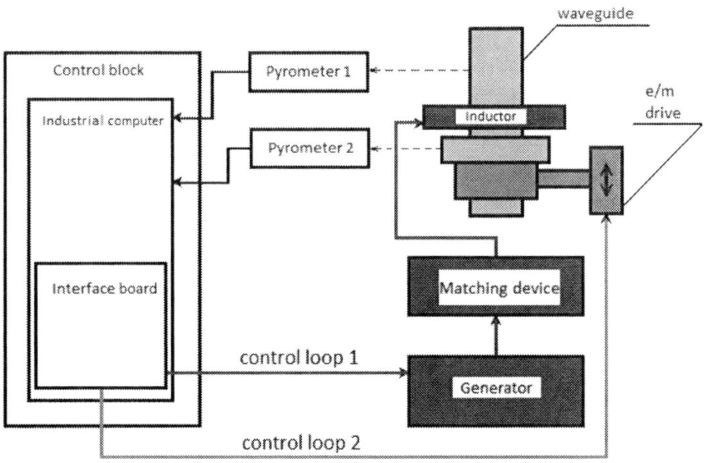

Figure 8. Structural scheme of the automated equipment complex.

The circuit 1 provides control of power supplied to the inductor. The contour 2 – control of the electromechanical manipulator drive – allows to change the position of the soldered waveguides elements relative to inductor. Control of the process technological parameters is performed by computer by using error signals obtained as a result of the comparison of the program temperature with pyrometers controlling temperature of waveguide tube and flange registrations.

4. Complex equipment

General view of working area of the automated equipment complex is shown in the Figure 9 [10], and general view of the entire complex is shown in the Fig.10.

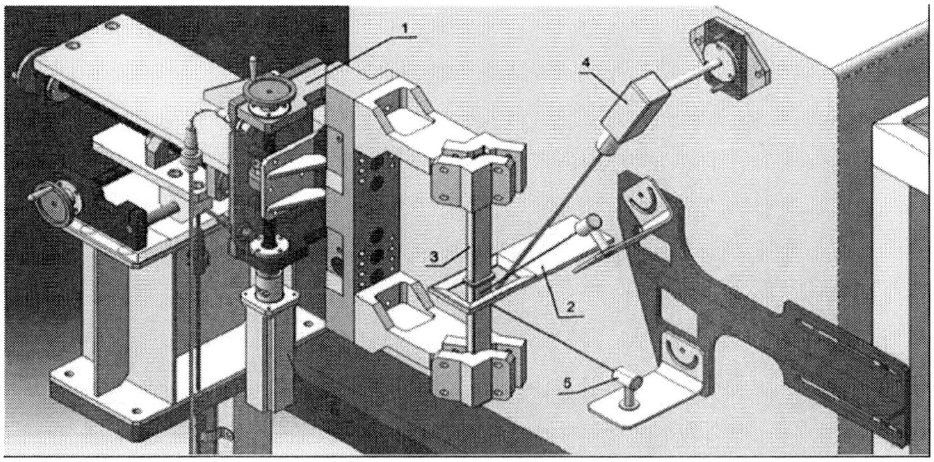

Figure 9. General view of the working area of induction heating facility: 1 – manipulator; 2 – inductor; 3 – waveguide; 4 – video camera; 5 – pyrometric sensor; 6 – electromechanical drive

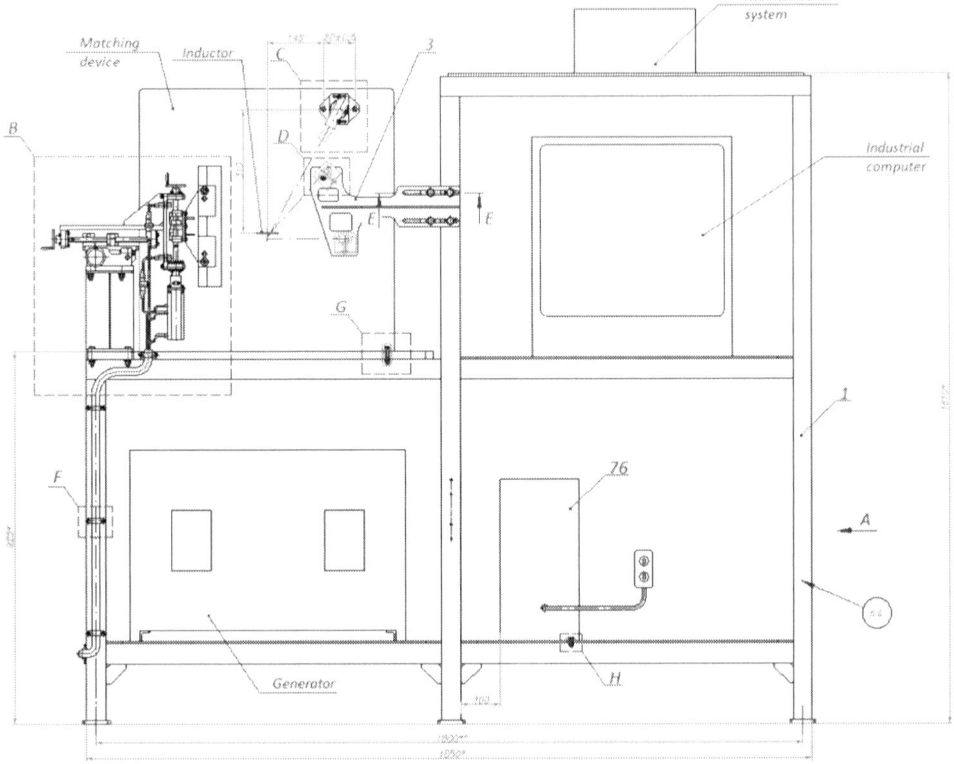

Figure 10. General view of the automated equipment complex for the waveguide tracts soldering

A scheme of equipment arrangement for the waveguide tracts induction soldering, which allows controlling the workpiece position relative to inductor freely, measuring the temperature in the heating area using two pyrometers and conducting video surveillance of the technological process, is shown on the above presented figures. Video surveillance allows staff to remotely control the technological process, thus completely eliminate influence of harmful factors on human health [11].

Conclusion

As a result, a set of automated assembly equipment for induction soldering AFD elements with pyrometric temperature control, which allows to manage the process of induction heating, to prevent overheating and destruction of the soldering products, to increase stability of the induction soldering process, to improve the quality of manufactured products, thereby to reduce the time and the material costs for the production, is developed. Developed automated equipment complex by remote control and monitoring of the soldering process, will reduce the impact of harmful production factors on personnel involved in soldering AFD elements.

References

[1] Zlobin S.K. *Features of soldering waveguide distribution paths of aluminum alloys using induction heating source* / S.K. Zlobin, M.M. Mihnev, V. D. Laptenok, R. V. Zaitsev // *Reshetnev reading : proceedings of the XVI Intern. scientific. conf.* in 2 p. Krasnoyarsk : SibSAU. - 2012. – pp 192-193.

[2] Silchenko P.N. *Ensuring the strength and accuracy of large-sized waveguide-distributive systems of space communication devices* / P.N. Silchenko, M.M. Mikhnev, A.V. Ankudinov, I.V. Kudryavtsev // Problemy mashinostroeniya i nadezhnosti mashyn. – 2012. - №1. – p. 112-117.

[3] Vologdin V. V. *Induction soldering* / V. V. Vologdin, E. V. Kushch, V. V. Assam. – L. : «Mechanical engineering», 1989. –72 p.

[4] Kudryavtsev I. V. *Mathematical model of heating of the waveguide during transmission of high signal power* / I. V. Kudryavtsev, E. S. Barykin, O. B., Gotselyuk // *Young scientist*. – 2013. - №9. – pp 52-57.

[5] Laptenok V. D. *Features of the production of the waveguide-junction of paths of antenna-feeder devices of spacecraft* / V. D. Laptenok, S. K. Zlobin, M. M. mihnev, A. N. Bocharov, B. G. // *Vestnik SibGAU*. – 2013. – №6(52). – pp 196-201.

[6] Laptenok V.D. *The technique of debugging of technological process of soldering the waveguide paths* / V.D. Laptenok, B.B. Dolgopolov, A.N. Bocharov // Krasnoyarsk : SibGAU. - 2009. – 112 p.

[7] Slugocki A. E. *Inductors for induction heating* / A. E. Slugocki, S. E. Ryskin. – Leningrad : "Energy", 1974. – 418 p.

[8] Slugocki A. E. *Inductors* / A. E. Slugocki – L.: *"Engineering"*, 1989. – 462 p.

[9] Tynchenko V.S. *The software of technological process of spacecraft's waveguide paths soldering* / V.S. Tynchenko, A.N. Bocharov, V.D. Laptenok, Yu.N. Seregin, S.K. Zlobin // Programmnye Produkty i Sistemy. – 2016. - №2(114). p. 128-134.

[10] Zlobin S.K. *Automated equipment and technology for soldering waveguide paths of spacecraft*/ S.K. Zlobin, M.M. Mikhnev, V.D. Laptenok, Yu.N. Seregin, A.N. Bocharov, V.S. Tynchenko, Yu.P. Dubets, B.B. Dolgopolov // Vestnik Sibirskogo gosudarstvennogo aerokosmicheskogo universiteta. – 2014. - №4(56). p. 219-229.

[11] Laptenok, V.D. *System of video observation for electron beam welding process* / V.D. Laptenok, Y.N. Seregin, A.N. Bocharov, A.V. Murygin, V.S. Tynchenko // IOP Conference Series: Materials Science and Engineering, Vol. 122, 2016. – 4 p.

IOP Conf. Series: Materials Science and Engineering **173** (2017) 012024 doi:10.1088/1757-899X/173/1/012024

Intelligent control system of autonomous objects

E. A. Engel[1], I. V. Kovalev[2], N. E. Engel[3], V.V. Brezitskaya[4], G.A. Prohorovich[5]

[1] Associate Professor, First Russian doctoral degree in computer sciences, Katanov State University of Khakassia, Abakan, RF,
[2] Professor, Second Russian doctoral degree in computer sciences, Siberian State Aerospace University, Krasnoyarsk, RF,
[3] Student, Katanov State University of Khakassia, Abakan, RF
[4] Scientific worker, Siberian State Aerospace University, Krasnoyarsk, RF
[5] Scientific worker, Siberian State Aerospace University, Krasnoyarsk, RF

E-mail: angel@khsu.ru

Abstract. This paper presents an intelligent control system of autonomous objects as framework. The intelligent control framework includes two different layers: a reflexive layer and a reactive layer. The proposed multiagent adaptive fuzzy neuronet combines low-level reaction with high-level reasoning in an intelligent control framework. The formed as the multiagent adaptive fuzzy neuronet the intelligent control system on the base of autonomous object's state, creates the effective control signal under random perturbations.

1. Introduction

Autonomous objects elaborate many practical tasks which are too dangerous or difficult for human to reach (e. g., military assignments, interstellar exploration, searching for victims after earthquake, and nuclear/biological accidents). Such autonomous objects are commonly suffering from restrictions due to the uncertainty of the environment and therefore unstable dynamics when designing control systems to provide stability, disturbance attenuation, and reference signal tracking. Within the research literature, a whole array of differing control strategies is proposed to deal with the control of the linearized system problem [1-6]. One of the most common control strategies is Proportional-Integral-Derivative (PID) controller due to its simplicity and applicability [7]. But the PID controllers for autonomous objects are often designed by hand, requiring extensive analysis of the system and dynamics. This process is generally difficult because it is hard to anticipate all operating conditions. The control algorithm must coordinate the autonomous objects properly, generating robust behavior to negotiate different terrains effectively while maintaining stability. Moreover, the autonomous objects should be robust to different environmental conditions, wear and tear, and even failure to reliably complete its mission. Therefore, automatic design algorithms base on intelligent techniques such as fuzzy neuronet and multi-agent system are a promising alternative [5-7]. In modern world, intelligent control has emerged as one of the most active and fruitful areas of research and development.

Monitoring and fault detection play an urgent role for autonomous objects due to the increasing demands on fault tolerant real-time tasks. This fault tolerant behavior of autonomous objects relates to the opportunity to automatically identify faults early before they result in disastrous failures as well as to the capability to continue operating after a fault has happened by switching to a safe state. However, fault detection for autonomous object is a complex problem because of the large space of possible faults (e.g. autonomous object's sensors, actuators, the uncertainty of the environment). Therefore autonomous objects are best modeled as intelligent systems because their behaviors result from the interaction between continuous and discrete dynamics. Several algorithms have been developed to effectively control of autonomous objects. The general approaches integrate the basic model, which are differential equations, with the intelligent models, which are fuzzy neuronet and multi-agent system. We are using intelligent agent as a generic modeling or abstraction mechanism. This research formalizes problem of autonomous object's control solving by multi-agent system. This paper presents

IOP Conf. Series: Materials Science and Engineering **173** (2017) 012024 doi:10.1088/1757-899X/173/1/012024

an intelligent control system of autonomous objects as framework.

2. The proposed architecture of the intelligent control system

The proposed architecture of the intelligent control system of autonomous objects includes two different layers: a reflexive layer and a reactive layer divided according to their complexity, speed and range of effect.

The reactive layer is faster and has a closer range of effects; it is still hierarchical in structure and requires some time for generating a control signal of the autonomous object. The reflexive layer has the closest range, the fastest effect and, most importantly, the highest priority. In cases where the higher level does not directly control the lower level it may train them to perform certain control actions. For example, a deliberative layer may cause a reactive layer to develop new condition-action behavior sequences, which may afterwards run without supervision. Usually, intelligent control system architecture involves several kinds of learning mechanisms in different parts of the system, e.g., neural nets, extendable knowledge bases, etc. The name intelligent is derived from this composite structure, in addition to the capabilities inherited from both the Hierarchical and the Layered Architectures. The components of the two above mentioned layers are described below.

3. Roadmap towards intelligent control system of autonomous objects

In this paper we are using intelligent agent as a generic modeling or abstraction mechanism, independently of whether autonomous object is implemented as object (using object-oriented programming techniques). This research formalizes problem of autonomous object's control solving by multi-agent system (MAS) with neuroevolution and student-teacher of-line learning. Unlike existing single-agent and team-search problems in proposed algorithm the agents collaborate through knowledge of other agents during the dynamic changes in the autonomous objects. The some main issues of the implementation intelligent multi-agent systems are shown in Figure 1.

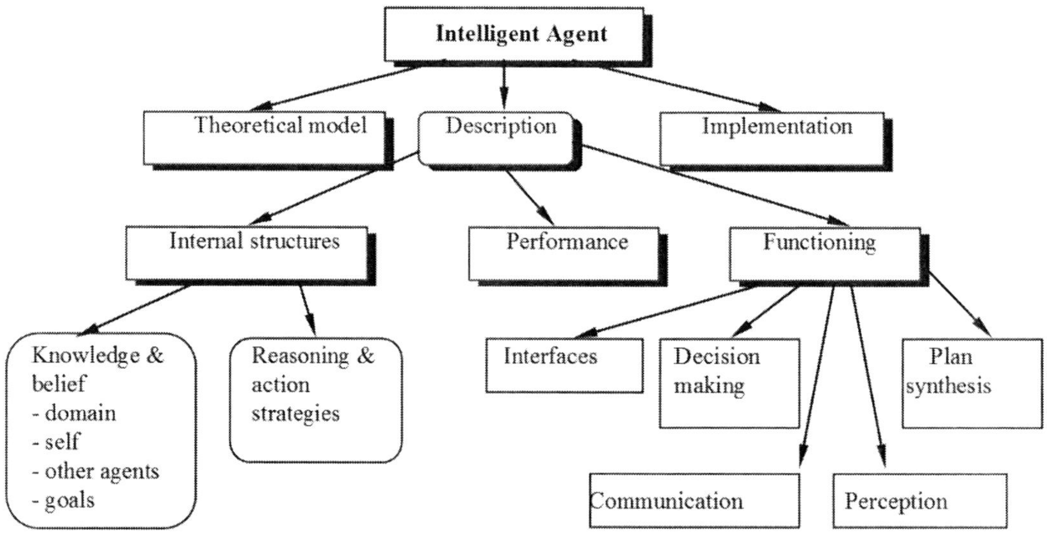

Fig. 1. Levels of specification and design of agents in intelligent control system of autonomous objects

Figure 2 represents a scheme of the basic aspects that correspond to intelligent interactions in cognitive MAS.

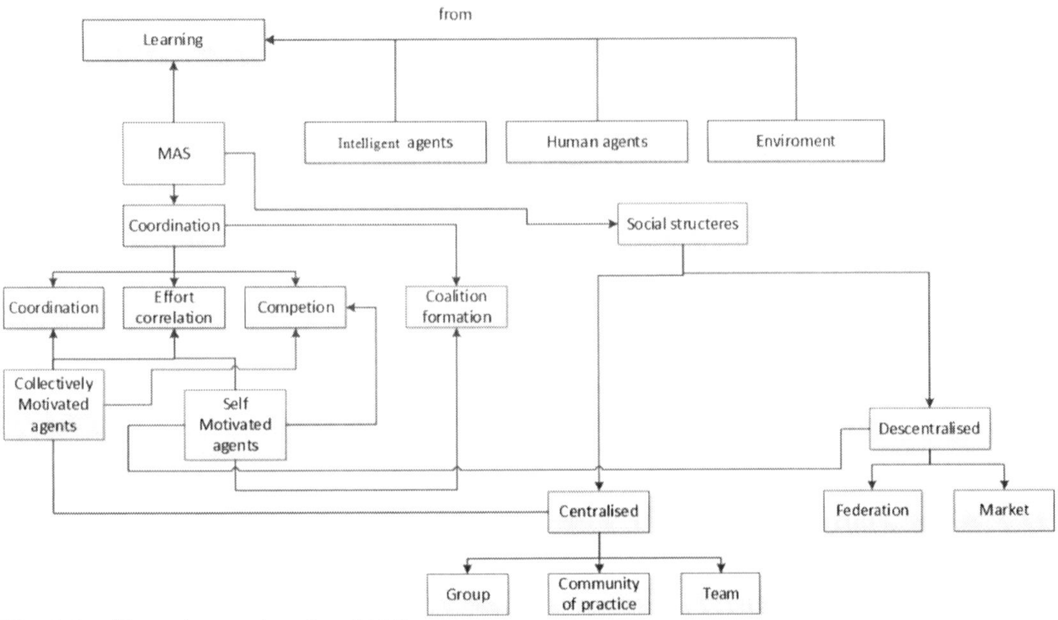

Fig. 2. Intelligent interactions in a MAS

The first two bundles of intelligent control system as MAS are Agent Interoperability and Agent Communication. First one covers agent identity, agent transport, agent registration and discovery. Second one covers agent communication language, ontology, and content language. The first bundle is in several ways orthogonal to the second. Also, within these bundles, the services should be separately specified as they are separable but they must work closely together so they are included in the same bundle.

In this paper a mobile robot prototype with four DC motor actuators was considered as an example of autonomous object. The drive and steering subsystem is included two independent wheels and motor [5].

The behavior of the motor speed for a given voltage is derived from physics law described in Laplace domain by the open-loop transfer function

$$\frac{\dot{\theta}(s)}{V(s)} = \frac{\frac{K_t}{JL}}{s^2 + \frac{(JR + bL)}{JL}s + \frac{bR + k_e K_t}{JL}} \tag{1}$$

where voltage $V(s)$ is input and shaft speed $\omega = \dot{\theta}(s)$ is output [5]. The relation between the reference speed $\dot{\theta}_{ref}$ and the output speed $\dot{\theta}_{out}$ is given by the closed loop transfer function

$$\frac{\dot{\theta}_{out}}{\dot{\theta}_{ref}} = \frac{K.\frac{K_t}{JL}}{S^2 + \frac{(JR + bL)}{JL}S + \frac{K.K_t + bR + k_e K_t}{JL}}$$

where K is a constant gain.

The state space representation of the robot's dynamic as follows

$$\begin{bmatrix} \frac{di}{dt} \\ \ddot{\theta} \end{bmatrix} = \begin{bmatrix} -\frac{R}{L} & -\frac{K_e}{L} \\ \frac{K_t}{J} & -\frac{b}{J} \end{bmatrix} \begin{bmatrix} i \\ \dot{\theta} \end{bmatrix} + \begin{bmatrix} \frac{1}{L} \\ 0 \end{bmatrix} V(t) \text{and } \dot{\theta} = [0\ 1] \begin{bmatrix} i \\ \dot{\theta} \end{bmatrix} \tag{2}$$

was used to confirm that the system is stable, controllable and observable.

4. PID controller design

The general transfer function for a PID controller in Laplace domain can be written as follows

$$PID(s) = K_p + \frac{K_I}{s} + K_D\, s = \frac{K_D\, s^2 + K_p\, s + K_I}{s} \tag{3}$$

where K_p is the proportional gain, K_D is the derivative gain and K_I is the integrator gain. The PID gains is choose by hand and are then tune by simulation.

This PID controller is designed by hand. Therefore, automatic design methods utilizing intelligent techniques such as fuzzy neuronet and multi-agent system are a promising alternative.

5. Intelligent control

An intelligent approach is proposed to address the weaknesses of the PID approach. The intelligent controller is carefully designed to correctly tackle the control of autonomous object task under uncertainty of the environment.

5.1. *Fuzzy logic control system design.*

A typical structure of a fuzzy logic control system is shown in Fig. 3.

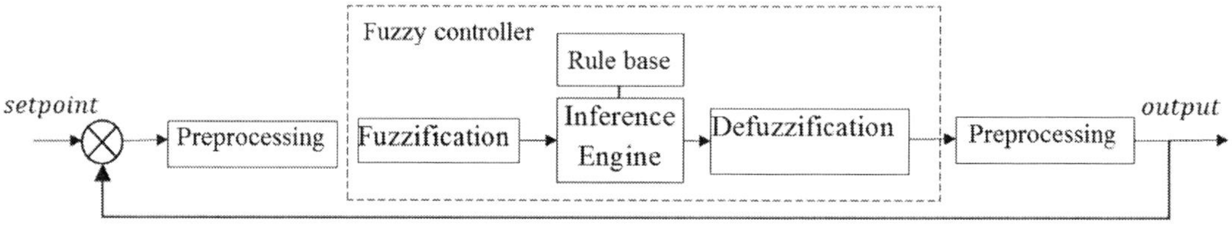

Fig. 3. Fuzzy logic control system

In our work, knowledge was borrowed from proportional integral control error and change of error to define fuzzy membership functions. The inputs that were in the form of crisp values generated from feedback error –e and change of error – de – were conditioned in terms of multiplying by constant gains before entering into the main control block. The fuzzification block converts input signals into appropriate way to membership functions, invokes appropriate rules, generates a result for each rule, and combines the results of those rules. The Mamdani-type inference engine combined the output state and assigned a membership value at the truth level of the premise. The truth values are then defuzzified using a discrete centroid computation. The post processing block then transforms these crisp values into control signals. The fuzzy logic controller's rules had triangular membership function and were implemented in the simulation.

These rules provide control signals based on several if-then statements about (e) and (de), i.e., if the error is equal Negative Big (NB) and change of error is equal to negative medium (NM), then the change in control (c) is positive big (PB). The numbers of fuzzy logic controller's rules is determining based on experiment and tuning of the system.

With onboard resources often limited, this paper considers the development of an effective control algorithm that remains easy to implement.

6. The intelligent control system on the base of the multi-agent adaptive fuzzy neuronet

A large body of literature on neuro-control exists on nonlinear systems of the form $\dot{x}=f(x)+g(x)u$,

IOP Conf. Series: Materials Science and Engineering **173** (2017) 012024 doi:10.1088/1757-899X/173/1/012024

where the neural nets are used to approximate *f(x)* and *g(x)* [5]. Linearization methods have been shown to be very effective for the control of a broad class of systems [5]. In contrast, in this paper, the function approximation capabilities of multi-agent adaptive fuzzy neuronet are exploited to approximate a nonlinear control law. The intelligent control system is capable of handling uncertainties in both the system parameters and the environment. The multi-agent adaptive fuzzy neuronet trained based on data

$$Z^i=(s_1^i, s_2^i, ..., s_r^i, u_1^i, u_2^i, ... u_b^i, \Omega_1^i, \Omega_2^i, ..., \Omega_q^i, Y), \quad (4)$$

where $i \in \{1,.., m\}$ – time, Y – response of the autonomous objects; s – input signal of the autonomous objects; u – control signal; Ω – influences of environment.

The multiagentny adaptive fuzzy neuronet implemented of-line learning. The algorithms of the multi-agent search and interaction employing the neuroevolution (Fig. 4-5).

```
Input: generations; maxTimesteps
population ← InitializePopulation()
g ← 0
t ← 0
loop
        while g < generations do
        subcultures ← FormSubcultures(population)
                while t < maxTimesteps do
                        for each subculture in subcultures do
                                for each agent in subculture do
                                        x ← ReceiveAgentInputs()
                                        f(x) ← GetResponse(agent; x)
                                        r ← TakeAction(f(x))
                                        agent.memory ← UpdateMemory({x; f(x); r})
                                end for
                                for each agent in subculture do
                                        if Acceptable(agent.memory) then
                                                for all observer in subculture do
                                                        Train(observer, agent.memory)
                                                end for
                                        end if
                                end for
                        end for
                        t ← t + 1
                end while
                population ← SelectAndReproduce(population)
                g ← g + 1
        end while
end loop
```

Fig. 4. Algorithm of the multi-agent search

IOP Conf. Series: Materials Science and Engineering **173** (2017) 012024 doi:10.1088/1757-899X/173/1/012024

```
for each agent in subculture S_k do
        x ← ReceiveAgentInputs()
        f(x) ← GetResponse(agent; x)
        r ← TakeAction(f(x))
        agent.memory ← UpdateMemory({x; f(x); r})
end for
for each agent in subculture S_k do
        if Acceptable(agent.memory) then
                for all observer in subculture do
                        Train(observer, agent.memory)
                end for
```

Fig. 5. Algorithm of the multi-agent interaction

Fulfilling of the multiagentny adaptive fuzzy neuronet as intelligent control system of the autonomous objects briefly can be described as follows (Fig 6).

I unit: Fulfilling of the multiagentny adaptive fuzzy neuronet

Step 1. All samples (*4*) were classified into groups by GNG. This classification generates vector with elements A_j.
Step2. Two-layer recurrent networks: $A_j = F_j(X)$ were trained based on of the data

$$X_j = (s_1^j, s_2^j, \ldots, s_r^j, \Omega_1^j, \Omega_2^j, \ldots, \Omega_q^j), \qquad (5)$$

$j=1..2$. Fuzzy sets A_j with membership function $\mu_j(X)$ are formed base on aforementioned two-layer recurrent networks $A_j = F_j(X)$.

↓

We train two-layered recurrent neural networks: $C_j = f_j(x_j)$ based on the data

$$x_j = (s_1^j, s_2^j, \ldots, s_r^j, u_1^j, u_2^j, \ldots u_b^j, \Omega_1^j, \Omega_2^j, \ldots, \Omega_q^j); Y.$$

These neural networks were evolved by multi-agent search. This step provides recurrent neural networks which create the control signal according identified system's state under random perturbations $C_j = f_j(x_j)$. Agent's subcultures S_j are formed base on aforementioned two-layer recurrent networks.

If-then rules are defined as:

$$\Pi_j\text{: IF X is } A_j \text{ THEN } C_j = f_j(x_j). \qquad (6)$$

↓

II unit: Simulation of the multiagentny adaptive fuzzy neuronet $\forall t \in \{0.T\}$

Aggregation antecedents of the rules (6) maps input data X into their membership functions and matches data with conditions of rules. These mappings are then activates the k rule, which indicates the k autonomous object's state and k agent's subcultures – S_k.

↓

According the k autonomous object's state the multiagentny adaptive fuzzy neuronet creates the effective control signal as a result of multi-agent interaction (Fig.5) of subculture S_k.

Fig. 6 The multi-agent adaptive fuzzy neuronet.

The formed as the adaptive fuzzy neuronet the intelligent control system creates the effective control signal according identified system's state under random perturbations (Fig. 6).

7. Conclusions

This paper presents an intelligent control system of autonomous objects as framework. The intelligent control framework includes two different layers: a reflexive layer and a reactive layer. The proposed multi-agent adaptive fuzzy neuronet combines low-level reaction with high-level reasoning in an intelligent control framework. The formed as the multi-agent adaptive fuzzy neuronet the intelligent control system on the base of autonomous object's state, creates the effective control signal under random perturbations. Unlike popular approaches to adaptive control, the multi-agent adaptive fuzzy neuronet is used to approximate the control law, and not the system nonlinearities, which makes it suitable to handle a wide range of nonlinearities. This shows the potential of the multi-agent adaptive fuzzy neuronet to control autonomous objects.

References

[1] Kovalev, I., Zelenkov, P., Ognerubov, S., Bahmareva, K., Denisova, E. 2015 The efficiency analysis of the automated plants // *IOP Conference Series: Materials Science and Engineering* 70 (1)

[2] Kovalev, I.V., Kovalev, D.I., Zelenkov, P.V., Voroshilova, A.A. 2015 The pseudo-Boolean optimization approach to form the N-version software structure // IOP Conference Series: Materials Science and Engineering 94 (1)

[3] Kovalev, I., Zelenkov, P., Ognerubov, S. 2014 The Efficiency Analysis of Automated Lines of Companies Based on DEA Method // Lecture Notes in Economics and Mathematical Systems 675 , pp. 107

[4] Kovalev, I.V., Zelenkov, P.V., Ognerubov, S The practical approach to the reliability analysis of the software architecture of a complex company control system// IOP Conference Series: Materials Science and Engineering 94 (1)

[5] Engel, E. , Kovalev, I.V. , Kobezhicov, V. 2015 Intelligent control of non-linear dynamical system based on the adaptive neurocontroller // IOP Conference Series: Materials Science and Engineering 94 (1)

[6] Engel, E. , Kovalev, I.V., D Karandeev 2015 Energy-saving technology of vector controlled induction motor based on the adaptive neuro-controller // IOP Conference Series: Materials Science and Engineering 94 (1)

[7] Engel E.A. 2012 Solving control problems, decision making and information processing by fuzzy selective neural network// Information Technology, 2

"The reported study was funded by Russian Foundation for Basic Research, Government of Krasnoyarsk Territory, Krasnoyarsk Region Science and Technology Support Fund to the research project № 16-47-242143"

IOP Conf. Series: Materials Science and Engineering **173** (2017) 012025 doi:10.1088/1757-899X/173/1/012025

Multiversion environment creation for control algorithm execution by autonomous unmanned objects

Kovalev I V[1], Zelenkov P V [2], Losev V V [3], Kovalev D I[4], Ivleva N V[5], Saramud M V[6]

[1]Dr. Sc., Professor of System analysis department, Rector, Reshetnev Siberian State Aerospace University, 31 "Krasnoyarsky Rabochy" Av., 660037 Krasnoyarsk, Russia.
[2]Cand. Sc., Vice-Rector for Informatization and Commecialization of Scientific Research Projects, Reshetnev Siberian State Aerospace University, 31 "Krasnoyarsky Rabochy" Av., 660037 Krasnoyarsk, Russia.
[3]Cand. Sc., Deputy Vice-Rector for Informatization and Commecialization of Scientific Research Projects, Reshetnev Siberian State Aerospace University, 31 "Krasnoyarsky Rabochy" Av., 660037 Krasnoyarsk, Russia.
[4]Postgraduate student, Reshetnev Siberian State Aerospace University, 31 "Krasnoyarsky Rabochy" Av., 660037 Krasnoyarsk, Russia.
[5]Senior English Instructor, postgraduate student, Reshetnev Siberian State Aerospace University, 31 "Krasnoyarsky Rabochy" Av., 660037 Krasnoyarsk, Russia.
[6]Engineer, Reshetnev Siberian State Aerospace University, 31 "Krasnoyarsky Rabochy" Av., 660037 Krasnoyarsk, Russia.

E-mail: basilos@mail.com, natalie_ivleva@mail.ru

Abstract. The article describes one of the approaches to increase resilience of software control systems for autonomous unmanned objects based on the multiversons methodology.

The task of piloting autonomous unmanned objects (space vehicles, including small spacecraft, the CubeSat, nano-, picosatellites, drones, quadrocopters, unmanned ground vehicles) is associated with a number of functions: providing telemetry; development of control actions; maintaining autonomy, including energetical, resource and functional one, which, in turn, imposes a number of higher requirements to the object to increase fault tolerance [1], as well as the ability to perform the functions assigned to it for the maximum operational period avaiable. The issue of increasing resiliency of objects of this class goes beyond the traditional "fault-tolerant computing" concept based on the introduction of the hardware redundancy in the system, for example, because of weight restrictions in the case of orbital objects. Thus, the focus of research is shifting towards software and software failures, namely, in the application of the multiversional programming aiming to develop design procedures failover software system of autonomous unmanned objects. The definition of theoretical and practical tools of choosing the multiversion model that meets the requirements, and the method of multiversion voting enabling at the stage of design and management systems modeling, as part of on-board equipment of autonomous unmanned facilities, taking into account its purpose and functioning, will improve the resilience of software systems of autonomous unmanned objects.

The main objective of the study is the use of the design methodology failover software to improve the quality of the algorithms in autonomous unmanned objects [2].

Given the technological complexity of the production cycle of final products as well as resource cost, the including temporary one, the task of ensuring the reliability and resiliency of autonomous unmanned aerial objects is of utmost importance. Since tasks, assigned to the spacecraft, become more complex and the number is growing in the algebraic progression, while the share of self-reliance features preserved and

IOP Conf. Series: Materials Science and Engineering **173** (2017) 012025 doi:10.1088/1757-899X/173/1/012025

functional gains achieved through payload (telecommunications, scientific and other equipment), the issue of failover control software systems of autonomous unmanned vehicles, as a result of the security and viability of objects and infrastructure formed by them [3].

Methodological basis of multiversion environment for autonomous unmanned objects algorithm performance uses multiversion technology to improve reliability of the software system of autonomous unmanned objects management system that allows you to analyze multiversion model of selective algorithm and their attribute differences to achieve the specified reliability characteristics of the software module designed [4] .

This formed the basis of the methodology designed to evaluate the effectiveness of multiversion models. This method implements the step by step mechanism for determining the value of the relationship between the correctness of Sn and the number of the multiverse n.

This method has an implementation model for recovering the blocks (DHS), and for multiversion program (IMP).

The main steps implementing this technique are:

a. classification of models, proposed for use in multiversion runtime;

b. situational probability estimation errors or return the correct result of software modules and a decision or an error verification module to DHS;

c. Review and assessment of exceptions to the DHS;

d. probabilistic assessment return errors or correct result of executable modules and a decision or an error of the algorithm for the MVP voting.

The following is an example of a technique for EBM and IMP.

Model recovering unit receives a "correct" result, when at least one of the multiverse returns "valid" and the result of the verification module receives the correct decision.

Suppose, we have n versions and their reliability is r1,r2, ..., rn respectively. Reliability verification module is b. Let's define the relation between the correctness of the system output (Sn) and the number of the multiverse (n). The following expression for Sn:

$$S_n = S_{n-1} + [r_1 (1-b) + (1-r_{1)} b] \cdot \ldots \cdot [r\,\mathrm{n}-_1 (1-b) + {}_{(1-r\,n-1)} b] \cdot r\, n\, b \ (1)$$

If all modules have a certain average reliability:

$$S_n = S_{n-1} + [(1-b) + (1-) b\,]\, n-^1 \cdot b \ (2)$$

Let us consider some special cases.

1) The reliability of the inspection module is equal to $(b = 1)$, then

$$S_n = 1 - (1-\bar{r})^n$$

Note that Sn→1, when n → ∞. In this case, we have formed a system to ensure guaranteed reliability which depends proportionally on their amount.

2) If average reliability software module is equal to one (= 1), then

$$S_n = 1 - (1-b)^n.$$

As in the previous case, Sn→1 for n → ∞. In this case the reliability of the verification module does not matter, as long as the amount of the multiversions was large enough.

By using the methodology multiversionnyh programming system reliability is mostly determined by the algorithm used by the multiverse matching results. The voting algorithm used in the theoretical study of the limit reliability and able to evaluate all the possible conditions that affect the output of the system is the algorithm of the vote by the absolute majority.

The vote by the absolute majority received the "correct" result, if more than half of the multiverse return "correct" the result of the vote and the algorithm selects one of them. Assuming that all program modules have a certain mean reliability, the reliability decision module - b, and the correctness of the output system - Sn, where n - the number in the multiverse system.

Assuming returned the "correct" result of m modules, and (n-m) - returned an error, then the probability of such a situation can be written as:

$$P_n(m) = C_n^m \cdot \bar{r}^m \cdot (1-\bar{r})^{n-m}.$$

Then Sn can be represented as the sum of the probabilities that m = $\left\lceil \frac{n}{2}+1 \right\rceil$ modules returned "right" result and the voting algorithm made the right decision:

$$S_n = b\sum_{j=m}^{n} C_n^j \bar{r}^j (1-\bar{r})^{n-j}.$$

It is also possible to estimate the ultimate reliability of the system through the improvement of all kinds used by the voting algorithm.

Our procedure allows to evaluate the effectiveness of the main multiversion models on different n - multiverses. The term "correctness" is an important property in the context of the development of fault-tolerant software. Higherthe value is, more reliable the system is. This property has a higher priority than the availability and reliability, as if we know the value of the "correctness", we can say with certainty how many results (multiverse outputs) contains no errors and only use them, thus increasing the accuracy of the system. And higher the accuracy of the system is, higher the availability and reliability are.

For modified versions of multiversion voting algorithms aimed at improving the stability of algorithms of multiversion errors of software modules (compared to the unweighted analogues) the key element of multiversion system is a decision block. This block divides outputs of multiple software versions on the "correct" and "incorrect". There are several methods of separating of outputs, while the most common of them are based on the classification of outputs, so the most promising of these techniques are voted by an absolute majority, a vote agreed by a majority, and fuzzy vote with majority agreement. As part of this research, a number of improvements are proposed to these techniques, namely, the weighted vote majority consistent and fuzzy weighted majority voting , based on the statistics of the program modules. The basic idea is that each module has a certain weight (a number from 0 to 1) during decision making. As a software weight the probability of return "correct" answer is used.

Functional expansion of the methodological framework of multiversion programming will deliver the reliability characteristics of software modules, and hence the predictability of their performance in multiversion environment [5].

Model-algorithmic component of multiversion environment autonomous unmanned facility management algorithms execution is formed from:

1. Algorithm of selecting multiversion vote methods to measure the effectiveness of algorithms. The study will be presented with an option to choose the method by voting multiversion by simulating the characteristics of the real system in a special simulation environment. The purpose of this

environment is to quantify the effectiveness of the method of voting multiversions used in the decision-making unit multiversion system, in particular a system with specified characteristics (number of modules, the average unit reliability, the probability of multiversional errors).

2. Software tools implementation (multiversion runtime), allowing to unify application multiversion approach to the various software complexes, which, unlike currently available allow executables are not only using multiversion programming methodology, but also with the other common multiversion models: recovering blocks in recovering units, *t / (n-1)* - versioned programming and multiversion programming with self-examination, etc. The principle of duality, laid in multiversionnyh execution environment autonomous unmanned facility management algorithms, is designed to provide not only the fault tolerance software management system kernel, but also to improve the accuracy of processed data results - algorithms provide telemetry [6].

Thus, the formed methodological basis is intended to not only provide a theoretical estimate of the limit of reliability of software modules, but also to perform model-algorithmic component in the design and implementation of software multiversion runtime.

References

[1] Kuznetsov PA, Kovalev IV, Losev VV, Kalinin AO, Murygin AV 2016 To the question about the states of workability for automatic control systems with complicated structure IOP Conf.Series: Materials Science and Engineering vol.45

[2] Kovalev IV, Zelenkov PV, Karaseva MV, Kovalev DI Ant algorithm modification for multi-version software building. *Lectute notes in computer science. Springer-Verlag, GmbH.* vol. 1940, 2015, 222-228 p.

[3] Engel EA, Kovalev IV 2011 Information processing using intelligent algorithms by solving WCCI 2010 *tasks* Vestnik SibGAU vol. 3, no. 36, pp.45

[4] Kovalev IV, Zelenkov PV, Ognerubov S. The minimization of inter-module the interface for the achievement of reliability of multi-version software IOP Conference Series: Materials Science and Engineering.(XVII International Scientific Conference «Reshetnev readings »). vol.45

[5] IV of Kovalev, Kovalev the DI, the PV Zelenkov, Voroshilova the AA of The Mathematical statement The for the solving of the problem of of N-version software system design *IOP Conference Series: Materials Science and Engineering Ser."International Scientific and Research Conference on Topical* Issues in Aeronautics and Astronautics (Dedicated to the 55th Anniversary from the Foundation of SibSAU)" in 2015. S

[6] Raspopin NA, Karaseva MV, Zelenkov PV, Kayukov EV, Kovalev IV 2012 Models and methods of optimization of information acquisition and processing Vestnik SibGAU vol. 2, no. 42 pp.45

"The reported study was funded by Russian Foundation for Basic Research, Government of Krasnoyarsk Territory, Krasnoyarsk Region Science and Technology Support Fund to the research project № 16-47-242143"

IOP Conf. Series: Materials Science and Engineering **173** (2017) 012026 doi:10.1088/1757-899X/173/1/012026

Modeling of thermal processes in waveguide tracts induction soldering [1]

A V Murygin, V S Tynchenko, V D Laptenok, O A Emilova, Yu N Seregin

Reshetnev Siberian State Aerospace University 31, Krasnoyarsky Rabochy Av., Krasnoyarsk, 660037, Russian Federation

E-mail: vadimond@mail.ru

Abstract. The problem solving of the induction heating models development, which describe the heating of the separate structural assembly components of the waveguide path and product generally, is presented in this paper. Proposed mathematical models are based on the thermodynamics equation and on the heat balance law. The system of the heating process mathematical models, such as surge tube and flange heating, and the mathematical model of the energy distribution are presented. During the modeling process with Matlab system by using mathematical models graphs of the tube, flange and coupling heating were obtained. These design charts are confirmed by the results of the experimental study. During the experimental studies pyrometers for temperature control and a video camera for visual control of the process parameters were used. On the basis of obtained models the induction soldering process features analysis is carried out and the need of its automation by the using of the information control systems for thermal management between the connection elements is revealed.

1. Introduction

Currently, restrictions on weight and size characteristics in the industry are tightened. This also applies to the waveguide channel antenna-feeder devices in radio engineering complexes. In this case the reliability of the elements must be high enough to ensure their operation during the period of active existence which is 12-15 years for the modern communication systems.

Waveguide can propagate different types of electromagnetic waves. The preferential propagation of waves of a particular type depends on the geometric dimensions of the waveguide, electromagnetic waves excited frequency and their excitation method [1].

Constituent elements of waveguides are connected to the assembly by the induction soldering, which has a number of advantages with respect to the other heating types, such as the open flame heating or heating in a furnace:

1) rapidity;
2) high performance;
3) stability;
4) unique accountability;
5) good working conditions;
6) efficient use of the space;
7) non-contact process.

However, using of this high-tech method is constrained by the number of limitations which are related to the high speed of the process that complicates manual control as well as the complex relationship between energy released by the inductor and the heat distribution on the constituent elements of the waveguide [2, 3]. Formation of the automated control system of induction soldering

[1] The reported study was funded by Russian Foundation for Basic Research, Government of Krasnoyarsk Territory, Krasnoyarsk Region Science and Technology Support Fund to the research project № 16-48-242029

process requires development of the models for describing the heating of assembly waveguide channel individual structural elements (tube and flange/coupling), and the product as a whole in figure 1.

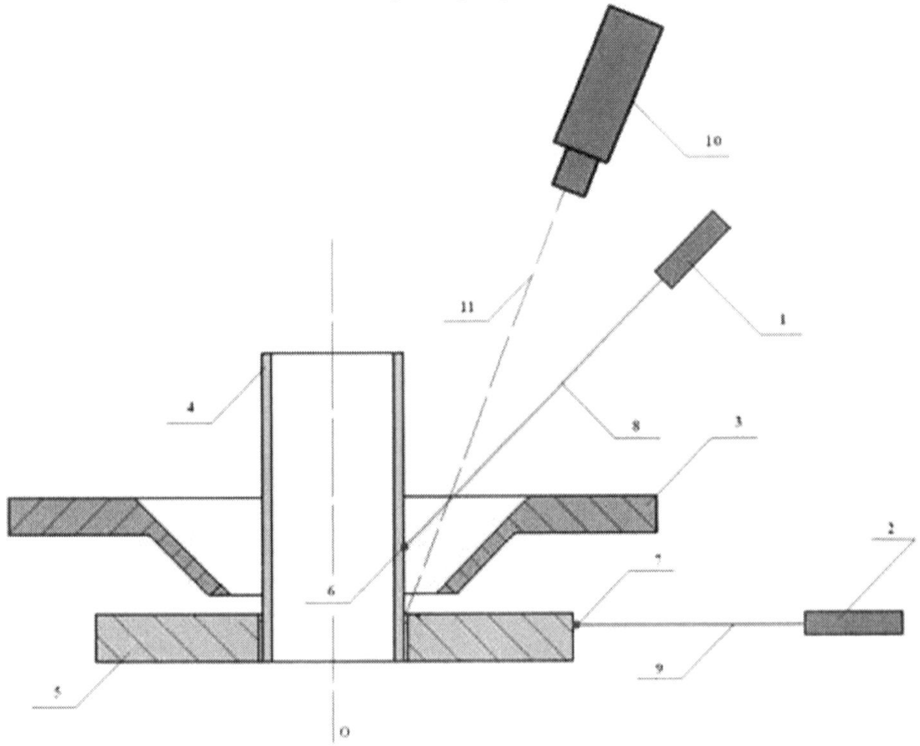

Figure 1. Map of the pyrometric sensors and video cameras
1 - pyrometer for monitoring the waveguide tube heating; 2 - pyrometer for monitoring the flange (coupling) heating; 3 - the inductor; 4 - waveguide tube; 5 - a flange (coupling); 6 - the control point of the waveguide tube heating; 7 - control point of the flange (coupling) heating; 8 - pyrometer beam 1; 9 - pyrometer beam 2; 10 - video camera; 11 – axis of the video capture

Developed mathematical models are based on the thermodynamics equation as well as on the thermal balance law; real system's parts, which have little influence on the heating process generally, are not considered, that allows to simplify the calculations, as well as to maintain the adequacy of the obtained models [4].

2. Mathematical modeling

2.1. Mathematical model of waveguide tube heating

Solution of the heating rod (linear heat flux) differential equation will be the mathematical model of the tube heating.

$$\frac{\partial \mathrm{T}}{\partial t} = \alpha \frac{\partial^2 T}{\partial x^2} - b(\mathrm{T} - \mathrm{T_c}) \tag{1}$$

$$b = \frac{\alpha p}{VhcF} \tag{2}$$

Infinite heat conducting rod with a section of F has an initial temperature $Tint$. At the initial time t = 0 in the volume element, which is a flat layer with the base, coinciding with the plane yOz in figure 2 and the infinitesimal height dx, has amount of heat equal to Q with uniform surface intensity Q/F.

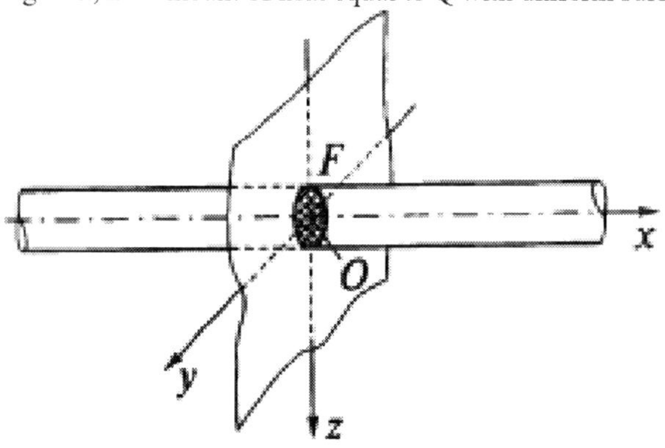

Figure 2. Scheme of the flat heat source in the rod

In this case the non-stationary temperature field of the flat heat source instantaneous action can be described by the expression:

$$T(x,t) = Tint + \frac{Q}{F*Vhc*\sqrt{4\pi at}} \exp\left(-\frac{x^2}{4at} - bt\right) \quad (3)$$

where $Tint$ – temperature in initial time [C^0];

Q – energy [J];

F – the tube cross section [m^2];

Vhc – volumetric heat capacity [J/m^3];

a – coefficient of thermal conductivity;

x – the distance from the heat source [m];

t – time [sec];

b – heat convection coefficient to the outside environment from the surface of the rod (formula 2);

x – distance from the arbitrary rod cross section to the flat heat source located in the plane yOz.

Temperature field is symmetrical with respect to the plane yOz and one-dimensional, i.e. at any moment temperature in the arbitrary point of the infinite rod can be determined only by its x-coordinate [5].

Based on the consideration that the waveguide tube is somewhat the same rod, namely:

- sufficiently long body of a homogeneous material;
- it has a relatively uniform cross section over the entire length;
- it has a similar mechanism of heat transfer and thermal conductivity.

It can be concluded that the mathematical model of a flat heat source in a rod is also valid for the flat heat source in a rectangular tube [6]. Therefore, in our system this model is adopted for the calculations with only one simplification, namely the material heat transfer of the temperature to the environment will not be considered on the assumption that it is negligible and thus can be neglected without risking losing the model adequacy. This allows simplifying the calculations associated with calculating unsteady factor «bt», which will have a variable value with respect to time. Thus the model takes the following form:

$$T(x,t) = Tint + \frac{Q}{Fc\rho\sqrt{4\pi at}} \exp\left(-\frac{x^2}{4at}\right) \quad (4)$$

2.2. Flange heating mathematical model

It was established due to the experiments that the flange (coupling) heating is close to the linear on condition that the maximum heating zone is chosen optimally, which can be achieved by the using of inductors with specific form with an oblique window and constrained heating rate.

Research shows that the temperature gradient in the flange (coupling) is very high due to the small product proportions and the high thermal conductivity of the product material. Based on these data model of the instantaneous heat source action on the flange (coupler) can be considered as the effect of the heat flow on the concentrated mass, which in turn makes model very simple for calculations. Accordingly, the heat transfer surface is quite small that allows ignoring this aspect of the real system, without accuracy loss.

$$T = \frac{Q}{c * p * m} \tag{5}$$

where Q - energy [J];

c – specific heat per unit mass [J\kg];

p – density [kg/m³];

m – weight [kg].

It is assumed that the temperature is spread over the body evenly.

2.3. Mathematical model of the energy distribution

Mathematical model of the energy distribution is presented below:

$$q_{gen} = q_{tu} + q_{fl} \tag{6}$$

$$q_{fl} = q_{tu} * K_h \tag{7}$$

$$K_h = \alpha * h \tag{8}$$

where q_{gen} - power supplied by the generator;

q_{tu} - power obtained by the tube;

q_{fl} - power obtained by the flange (coupler);

K_h - dependency coefficient of the energy transferred from distance;

h - the distance from the inductor to the flange (coupler);

α - coefficient, which reflects the dependence of the distance between the product and the inductor and the power transferred to him.

3. Results of the experimental study

During the modeling process with Matlab system by using mathematical models, described in the Section 2, graphs of the tube and flange/coupling heating were obtained, they are shown in figure 3. When modeling on the heating stage capacity was set as a constant, and the power setting was carried off when the temperature reached the 580^{0}C.

Figure 3 shows a comparison of the dynamics of heating tube and flange/coupling, from which it follows that at the heating stage when the same power is applied, the lag flange/coupling temperature on the tube temperature is occurred, but at the cooling stage due to the smaller volume of the flange the tube cooling rate exceeds that of the flange/coupling [6].

Comparison of the graphs shows that tube is more susceptible to the cooling process due to the small cross section at the large area. The flange or coupling has more smooth temperature curve because of the substantial material mass concentration within the small volume.

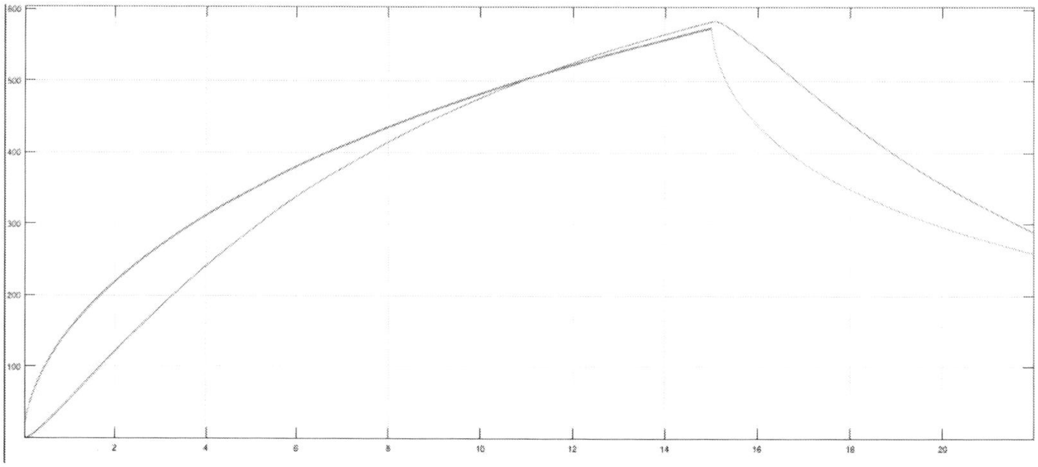

Figure 3. Simulation of the tube and the flange/coupling heating process with applying the constant power
Red - flange/coupling, blue - tube

Presented charts in the figures 3 are confirmed by the results of experimental studies, which prove the adequacy of the proposed mathematical models to the real object. During the experimental studies pyrometers for temperature control and a video camera for visual control of the process parameters were used.

Conclusion

Discussed mathematical models of the waveguide elements heating show that in the induction soldering process discrepancy of the waveguide tube and flange temperatures happens, which can lead to defects in the solder joints [7]. Therefore, measures to control the heating process, which allows conducting the heating process with the same temperature conditions for connection elements, are needed.

References

[1] Zlobin S.K. Features of soldering waveguide distribution paths of aluminum alloys using induction heating source / S.K. Zlobin, M.M. Mihnev, V. D. Laptenok, R. V. Zaitsev // *Reshetnev reading : proceedings of the XVI Intern. scientific. conf.* in 2 p. Krasnoyarsk : SibSAU. - 2012. – pp 192-193.

[2] Vologdin V. V. *Induction soldering* / V. V. Vologdin, E. V. Kushch, V. V. Assam. – L. : «Mechanical engineering», 1989. –72 p.

[3] Slugocki A. E. *Inductors for induction heating* / A. E. Slugocki, S. E. Ryskin. – Leningrad : "Energy", 1974. – 418 p.

[4] Kudryavtsev I. V. Mathematical model of heating of the waveguide during transmission of high signal power / I. V. Kudryavtsev, E. S. Barykin, O. B., Gotselyuk // *Young scientist*. – 2013. - №9. – pp 52-57.

[5] Rykalin N.N. Calculations of thermal processes in welding / N.N. Rykalin. – M. : Mashgiz, 1951. – 296 p.

[6] Laptenok V. D. Features of the production of the waveguide-junction of paths of antenna-feeder devices of spacecraft, V. D. Laptenok, S. K. Zlobin, M. M. mihnev, A. N. Bocharov, B. G. // *Vestnik SibGAU*. – 2013. – №6(52). – pp 196-201.

[7] Slugocki A. E. Inductors / A. E. Slugocki – L.: *"Engineering"*, 1989. – 462 p.

Institute of Physics
Dirac House, Temple Back
Bristol BS1 6BE UK

ISSN: 1757-8981
ISBN 978-1-5108-3676-1

CURRAN ASSOCIATES INC.
proceedings.com

10th Aceh International Workshop and Expo on Sustainable Tsunami Disaster Recovery (AIWEST-DR 2016)

IOP Conference Series: Earth and Environmental Science
Volume 56

Bada Aceh, Indonesia
22 – 24 November 2016

IOP
Institute of Physics